Carbon Nanostructures

More information about this series at http://www.springer.com/series/8633

Sumanta Sahoo · Santosh Kumar Tiwari ·
Ganesh Chandra Nayak
Editors

Surface Engineering of Graphene

 Springer

Editors
Sumanta Sahoo 🆔
Department of Chemistry
Indian Institute of Technology (ISM)
Dhanbad, Jharkhand, India

Santosh Kumar Tiwari
Hanyang University
Seoul, Korea (Republic of)

Ganesh Chandra Nayak
Department of Chemistry
Indian Institute of Technology (ISM)
Dhanbad, Jharkhand, India

ISSN 2191-3005   ISSN 2191-3013   (electronic)
Carbon Nanostructures
ISBN 978-3-030-30206-1   ISBN 978-3-030-30207-8   (eBook)
https://doi.org/10.1007/978-3-030-30207-8

This Springer imprint is published by the registered company Springer Nature Switzerland AG
The registered company address is: Gewerbestrasse 11, 6330 Cham, Switzerland

Preface

Science is becoming more and more dominant in our civilization and incessantly making the life of human beings more expedient. Along with the several vital discoveries and innovations, the twenty-first century will be reminded as technological boom for the centuries. With extraordinary and distinctive properties, graphene has been considered as the miracle material of the twenty-first century and a door-opening agent for the other 2D materials and many exceptional innovations. The enthrallment of graphene is rapidly increasing, and the surface engineering of graphene has become one of the fast-growing topic in materials chemistry. The properties and application of graphene are expanding day by day and achieving new milestones. Since the discovery of graphene, various types of graphene materials with improved properties have been synthesized, and therefore, graphene family has become wider now. So, it is important to have the knowledge of current state of the art of graphene research. But, due to the rapid progress in graphene research, it is impossible to cover all the prospects of graphene in a single book. In this book, we have presented an overview of current graphene research of a certain arena, and many things will covered in our new books of this series. This book comprises ten chapters based on recent experimental and theoretical works. Chapter "Present Status and Prospect of Graphene Research" describes the modern graphene technology. This chapter mainly focused on the properties and applications of modern graphene materials. Chapter "Graphene-Based Advanced Materials: Properties and Their Key Applications" deals with the properties and applications of graphene-based blends and nanocomposites. Mainly, mechanical properties of graphene are the main focus of this chapter. Chapter "Graphene and Its Derivatives for Secondary Battery Application" accounts for a brief introduction to the graphene material followed by a brief discussion on the recent advances in the field of its derivatives. This chapter describes the application of graphene and graphene-derived materials in the field of energy storage, specifically batteries in various forms like lithium-ion, sodium-ion, lithium–air, and lithium–sulfur batteries. Chapter "Recent Progress in Graphene Research for the Solar Cell Application" is dedicated to the solar cell application of graphene. This chapter provides a comprehensive overview of the applications of graphene and its derivatives, namely graphene oxide and reduced graphene oxide in the field of organic, perovskite, and dye-sensitized solar cells. Chapter "Graphene and Its Modifications for Supercapacitor Applications" is concerned detailing the variety of chemical modifications routes of graphene reported so far, and their effect on the electrochemical properties of graphene and the applicability of the developed material as a supercapacitor electrode material. Chapter "Functionalization of Graphene—A Critical Overview of its Improved Physical, Chemical and Electrochemical Properties" deals with the functionalization of graphene for the improvement of physical, chemical, and electrochemical properties. Chapter "Synthesis and Properties of Graphene and Graphene Oxide-Based Polymer Composites" reviews and explores the progress of fabrication of graphene and graphene oxide-based polymer composites. Chapter "Application of Reduced Graphene Oxide (rGO) for Stability of Perovskite Solar Cells" deals with the application of reduced graphene oxide for perovskite solar cell. Chapter "Graphene and Graphene Oxide as Nanofiller for Polymer Blends" describes the electrical, mechanical, thermal, and barrier properties of graphene-based polymer blends, and lastly, chapter "Facile Room Temperature Synthesis of Reduced Graphene Oxide as Efficient Metal-Free Electrocatalyst for Oxygen Reduction Reaction" demonstrates the use of reduced graphene oxide as efficient metal-free electrocatalyst for oxygen reduction reaction. Overall, the book deals with the different aspects of graphene, which will provide a concise outline of modern graphene science and technology. We hope, after reading the first book of this series, the reader will get adequate information to move forward independently in the ocean of current graphene research and to apply the knowledge within. Moreover, each chapter of this book has abundant references associated with the cutting-edge research around the globe, which enables further reading and exploration.

Dhanbad, India   Sumanta Sahoo, Ph.D.
Seoul, Korea (Republic of)   Santosh Kumar Tiwari, Ph.D.
Dhanbad, India   Ganesh Chandra Nayak, Ph.D.

Contents

Present Status and Prospect of Graphene Research

Sumanta Sahoo and Ganesh Chandra Nayak

Abstract Among various carbon allotropies, graphene has been considered the most attractive one, till date. After its discovery, graphene research has been gone through different phases. With tremendous progress in graphene research, various types of advanced graphene materials have been developed depending on the specific application. Different synthetic approaches have been employed to synthesize high-quality graphene materials. Recently, graphene has been successfully combined with other promising 2D materials to form multifunctional 2D hybrids. In this chapter, the recent progress on graphene research has been emphasized. The current trends of the synthesis, properties, application, and commercialization of graphene materials have been briefly discussed.

Keywords Graphene derivatives · Graphene synthesis · Graphene functionalization · Graphene doping · Heterostructures of graphene

1 Introduction

Carbon materials are one of the most significant abundant materials of this century. With their remarkable properties, these materials exhibited huge potentiality in different areas of applications. Among different types of carbon allotropies, graphene is the most promising one. This 2D sp^2-hybridized carbon allotropy has grown momentous attention due to its extraordinary properties including ultrahigh surface area, enhanced thermal and electrical conductivity, high ionic mobility, superior mechanical strength, etc. [1–5]. After its discovery in the year of 2004, the so-called graphene gold rush is continuing till date. Owing to its attractive properties, intensive researches are being carried out worldwide in recent times. However, the current research trends are mainly focused on the modification of graphene materials to improve its properties

S. Sahoo (✉) · G. C. Nayak
Department of Chemistry, Indian Institute of Technology (Indian School of Mines) Dhanbad, Dhanbad 826004, India
e-mail: sumanta95@gmail.com

G. C. Nayak
e-mail: gcnayak@iitism.ac.in

© Springer Nature Switzerland AG 2019
S. Sahoo et al. (eds.), *Surface Engineering of Graphene*, Carbon Nanostructures,
https://doi.org/10.1007/978-3-030-30207-8_1

and enhance its application potential in different areas of nanoscience and nanotechnology. As a result, various graphene-based materials like graphene nanoribbon, 3D graphene, graphene aerogel, graphene hydrogel, and nanoporous graphene have been synthesized and their structure properties have been extensively studied. The surface functionalization of graphene was also performed to expand the application potential. Apart from that, heteroatom doping of graphene was also introduced. Most recently, various graphene derivatives are also synthesized for multiple application purposes. At the initial stage of graphene research, the synthetic processes were also time-consuming, hazardous, and explosive.

But, in recent times, the researchers can able to synthesize graphene at room temperature. Even, various low-cost, eco-friendly approaches have been introduced to prepare high-quality graphene. The prime objectives of these studies are the improvement of the properties of graphene. In this chapter, we mainly focused on the recent developments in graphene research. The main object of this chapter is to provide a brief overview of the recent trends in graphene synthesis, properties, and application.

2 Graphene Synthesis

With rapid progress in graphene research, various synthetic approaches have been introduced to synthesize high-quality graphene. Those synthetic processes can be classified mainly into two types: top-down and bottom-up. While the top-down approaches mainly focused on the separation of stacked graphite layers, the bottom-up approaches are mainly concentrated on the alternating carbon sources to synthesize graphene sheets [6–8]. In terms of cost-affectivity, the top-down approaches are much more favourable than the bottom ones. But these processes generally suffer from few factors including the re-agglomeration of the sheets, low yield, multi-step reactions, etc. Bottom-up approaches produce high-quality graphene, but require high temperature. Top-down methods include the micromechanical cleavage, electrochemical exfoliation, mechanical exfoliation of graphite oxide and graphite intercalation compounds (GICs), arc discharge process, unzipping of carbon nanotubes, etc. On the other, bottom-up approaches mainly include the epitaxial growth of graphene on SiC and chemical vapour deposition (CVD) method (Fig. 1).

Besides these two approaches, miscellaneous approaches have been introduced in recent times to synthesize graphene and related materials. Further, the problems involved in the previous synthetic approaches have also been solved. For example, graphene nanoplatelets were synthesized with ~100% yield from graphite through top-down approach within few hours [9]. Recently, graphene with micrometric lateral sizes was prepared at room temperature from benzene through a simple electrophilic aromatic substitution reaction [10]. As demonstrated in Fig. 2, the chemical reaction was carried out in liquid–liquid interface through the polymerization of benzene through the proposed mechanism. Even though the produced graphene contained large amount of impurities, the synthetic strategy opened a new door for future graphene research.

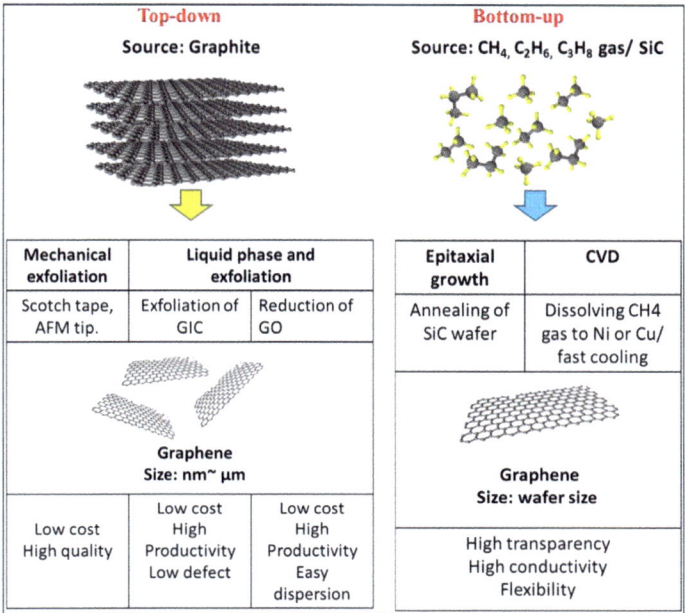

Fig. 1 Schematic diagram for the top-down and bottom-up approaches for the synthesis of graphene. Reproduced from Ref. [6]

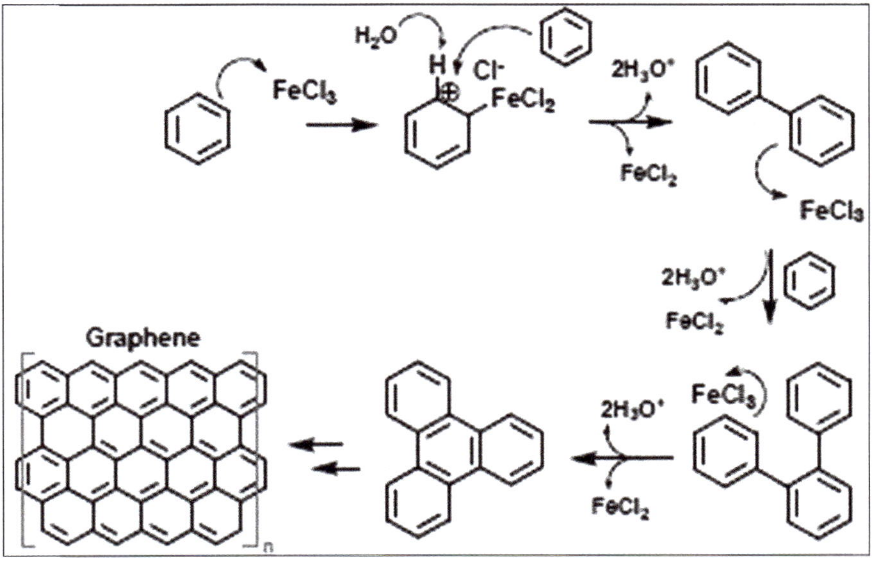

Fig. 2 Mechanism for the conversion of graphene from benzene. Reproduced from Ref. [10]

Graphene has also been synthesized from natural carbon-containing materials such as food (cookies, chocolate, etc.), insects (roaches), and waste (grass, plastics, dog faeces, etc.) [11]. The growth of graphene was conducted on Cu foil from the carbon-containing natural resources through CVD process. Recently, researchers have paid huge attention on the synthesis of graphene from biomass and biowaste. The growth of graphene on Cu substrate from mango peels through plasma treatment has been demonstrated in a recent article. According to the report, at high temperature, pyrolysis of the pectin-rich biomass (mango peels) has occurred in the presence of plasma and the corresponding single and double bonds were broken. Further, under the plasma treatment, the carbon ions/atoms/radicals were deposited in the empty spaces of the substrate surface unit cells and the graphene layers were grown [12]. Graphene materials were also synthesized from a verity of carbon-containing natural sources such as rice husk, rice straw, plant leaves, waste peanut shell, human hair, and waste wheat straw [13–16]. As shown in Fig. 3, the few-layer graphene was synthesized from waste peanut shell through high-temperature pyrolysis process, followed by activation and mechanical exfoliation process, which exhibited superior electrochemical performance as the electrode material for solid-state supercapacitor [16].

Ball milling is another process for graphene synthesis. The process involves the exfoliation of graphite layers by wakening the van der Waals interaction using various precursors such as triazine derivatives, melamine, ammonia borane, KOH, N,N-dimethylformamide, dry ice etc. [17–19]. Compared to other synthesis processes, this process is inexpensive and produces high-quality graphene with fewer defects.

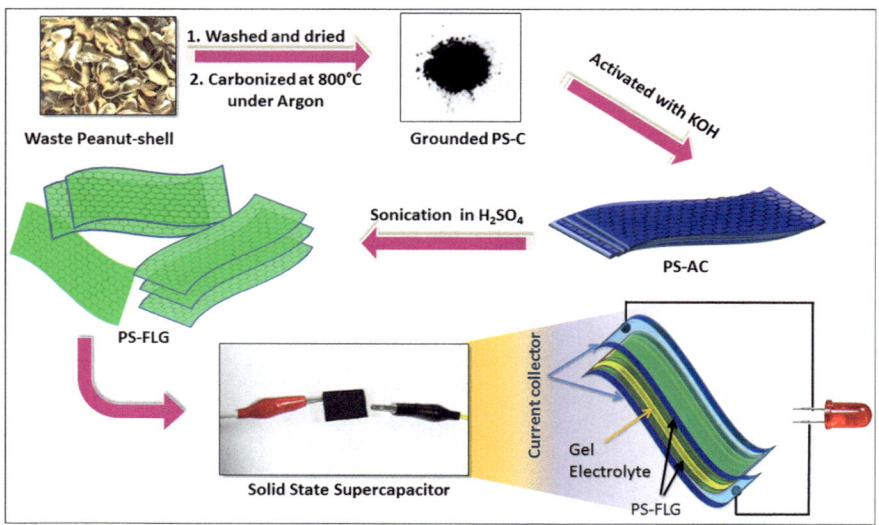

Fig. 3 Schematic diagram for the synthesis of waste peanut shell-derived graphene and its configuration for solid-state supercapacitor device. Reproduced from Ref. [16]

Graphene has also been synthesized through microwave route. In a typical process, metal phthalocyanine was synthesized and carbonized by microwave heating. Finally, the obtained graphite was exfoliated to form graphene by rapid cooling [20]. The synthesis process is useful for large-scale production of graphene with controlled structure. In another report, the reduction of graphene oxide, N-doping of graphene, and inclusion of metal atoms into graphene lattices were performed through single-step microwave heating [21]. Microwave-assisted solvothermal exfoliation of graphene using nontoxic solvents was also reported to prepare chemically modified graphene at low temperature within a short duration of time [22]. The synthetic process has advantages as compared to the conventional solvothermal process in terms of toxicity, reaction time, and reaction temperature. In general, for research purpose graphene is synthesized through Hummers'/modified Hummers' method. In these processes, the chemically reduced graphene is formed by the reduction of graphene oxide. Among the common methods of graphene synthesis, CVD process of hydrocarbons is the well-established one in industries, but the mass production of graphene through this method is not favourable due to high cost. Similarly, mechanical exfoliation of graphene and the synthesis of graphene on SiC processes are also expensive. However, the mechanical exfoliation and chemical reduction methods are the most efficient processes for mass production of graphene. A comparison of all the methods in terms of cost-affectivity, purity, scalability, and yield is shown in Fig. 4 [23]. Although a lot of research has already been done for the graphene synthesis, the large-scale production of high-quality graphene is still remained as one of the greatest challenges.

3 Graphene Materials

For the betterment of the properties of pristine graphene, lots of research has been carried out and a series of graphene materials have been produced. To be suitable for various applications, numerous graphene modification and functionalization were also performed. Moreover, in recent times, different graphene derivatives have been synthesized, which exhibited excellent potentiality for various applications. In this section, the recent progress on graphene materials is emphasized.

3.1 Common Graphene Materials

Graphene can be defined as "*a single-atom-thick sheet of hexagonally arranged, sp^2-bonded carbon atoms*" [24]. However, monolayer/single-layer graphene (SLG) and multi-layered graphene (MLG) are the most common graphene materials, which are generally derived from the exfoliation of graphite. But, it has been observed that the properties of SLG, bilayer graphene (BLG), and MLG are different from each other in terms of the number of well-defined stacked graphene

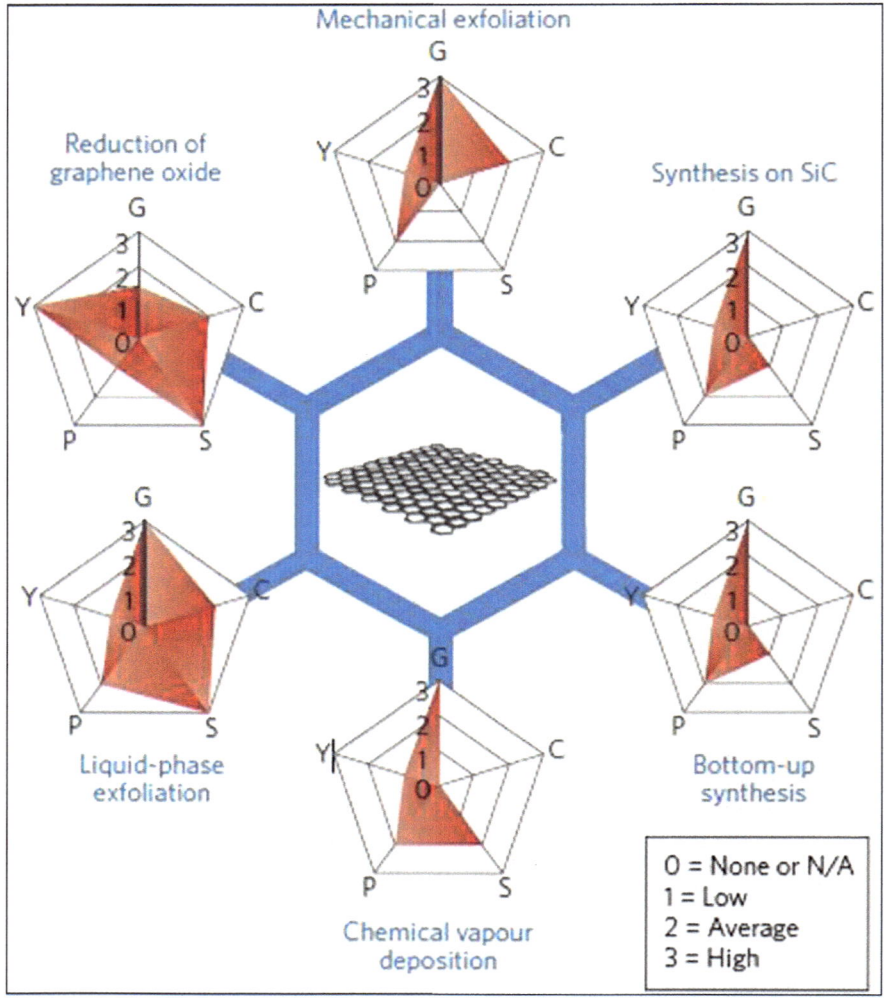

Fig. 4 Schematic representation of the comparison of common graphene synthesis methods in terms of graphene quality (G), cost aspect (C: a low value corresponds to high cost of production), scalability (S), purity (P), and yield (Y). Reproduced from Ref. [23]

layers of extended lateral dimension. Graphene layers of extended lateral dimension. Few-layer graphene (FLG) can be identified as the graphene sheets with 2–5 layers. By varying the layers, one can change the electronic properties. For example, the quantum capacitance of SLG and BLG is found to be higher than MLG [25]. Even, the number of layers affects the dielectric properties and mechanical properties. From the theoretical calculation also, it has been observed that work function varies with the graphene layers due to the shifting of Fermi energy with respect to the Dirac point [26]. Another common type of graphene material is graphene oxide (GO), which are generally utilized as the precursor for graphene synthesis through chemical or thermal reduction process. In general, among various carbon materials, GO is characterized by the C/O ratio, which is less than 3.0. The structure of GO contains various oxygen-containing functional groups including the hydroxyl and epoxy groups on the basal plane and a little quantity of carboxy, carbonyl, phenol groups at the edges. The presence of these oxygen-containing functional groups differentiates GO from pristine graphene and influences the electrical, electrochemical, and mechanical properties [27]. It has been reported that GO can be acted as semiconducting, semimetallic, or insulating material through proper tuning of oxidation or reduction parameters. Even, the presence of larger number of defects enables GO to exhibit higher chemical activity than pristine graphene. Various types of GO such as GO thin film, free-standing GO paper, GO membrane have also been explored, which exhibited outstanding mechanical strength. One of the important characteristics of GO is that it can easily form colloidal solution due to hydrophilic nature. The reduction of GO produces reduced graphene oxide (rGO), which is one of the fascinating members of graphene family owing to its large-scale production capability. In general, the reduction of GO is performed through *"chemical, thermal, microwave, photo-chemical, photo-thermal, or microbial/bacterial methods to reduce its oxygen content"* to provide high-quality rGO [24]. It has also been designated as chemically converted graphene, chemically derived graphene, functionalized graphene, or reduced graphene in literature. The rGO sheets are generally composed of graphene domains with oxygen functionalities. However, as compared to GO, rGO is less hydrophilic because of the elimination of oxygen moieties during reduction process. It is important to note that, although GO is insulating in nature, the rGO shows prominent electrical conductivity because of the removal of oxygen groups, which enabled them to serve as suitable fillers for composites. Besides this, other fascinating properties of rGO have also been reported for various applications.

3.2 *Wrinkled, Rippled, and Crumpled Graphene*

As a prominent member of 2D nanomaterials' family, graphene has the intrinsic properties of exhibiting surface corrugations. However, through different processing techniques, crumpled graphene was also formed. The formation of wrinkles, ripples, crumples on graphene alters the electrical, mechanical, optical, and chemical properties. Figure 5 represents the formation mechanism, properties, and application of different corrugations on graphene [28]. Various mechanisms have been proposed for the formation of corrugations on graphene. While thermal fluctuations and edge instability are responsible for ripples formation, the negative thermal expansion coefficient and dislocations and strain-induced formation mechanism are responsible for wrinkles formation on graphene. Crumbled graphene was also formed through strain-induced formation mechanism. If we look into the applications of these corrugation-induced graphene structures, the random nature and small size of the ripples restrict

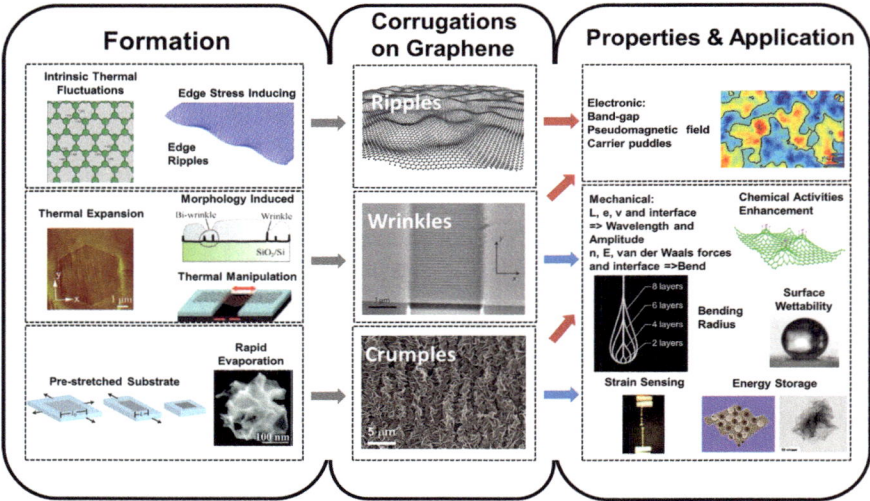

Fig. 5 A summary illustration of corrugations on graphene formation, properties, and application. Grey arrows stand for formation of corrugations on graphene, red arrows indicate electronic properties, and blue arrows for other properties. Reproduced from Ref. [28]

the rippled graphene to be utilized in various application fields except the electronic ones. The previous research on rippled graphene was mainly focused on its electronic properties. But, the wrinkled and crumpled graphene has modified several properties of graphene including the band gap opening, induction of pseudomagnetic field in bilayers, creation of electron–hole puddles, suppression of weak localization. Therefore, these wrinkled and crumpled graphenes showed wide range of applications including energy storage/conversion, nanoscale devices, strain sensing, tunable wettability surfaces, etc. However, for proper tuning of the properties of graphene, it is necessary to control the physical attribute of these corrugations.

3.3 3D Graphene

Even though 2D planar structure of graphene is suitable for conventional electronic devices, it can be utilized for semiconductor devices since graphene is a zero band gap semi-metal. Therefore, it is necessary to open up the band gap of graphene to utilize it in different field of applications including energy, environment, and biological applications. Here comes the idea of developing 3D graphene [29–32]. In general, the 3D graphene can be defined as the assembly of 2D graphene sheets into 3D architectures. In combination of the high surface area and 3D porous network, this type of newly invented graphene materials exhibited exceptional mechanical strength, superior electrical conductivity, and rapid mass transfer kinetics. Different kinds of 3D graphene networks have been synthesized in recent years. However, these can be classified mainly into three types—(1) graphene foams (GFs), (2) graphene sponges (GSs), and (3) graphene aerogels (GAs). The GFs were synthesized through template-assisted process using Ni foam, carbon foam, etc. Similarly, GSs were prepared by using spongy-type carbon materials as templates. However, the GAs were produced using sol–gel chemistry including the gelation process of GO, supercritical drying, freeze-drying, etc. GFs are composed of organized and uninterrupted 3D graphene network. But, GSs exhibited anisotropic lamellar structure. CVD process is the most commonly used method for the synthesis of 3D graphene networks. Various metal substrates like Ni foam, MgO, ZnO, anodic aluminium oxide, metal nanostructures have been utilized as the template for growing 3D graphene by CVD technique. Besides this, the self-assembly of graphene was also executed by chemical, electrochemical, and hydrothermal routes. Cross-linking agents like DNA, metals, organic compounds, polymers have been used to produce 3D graphene through gelation process of GO. In comparison with pristine graphene, 3D graphene showed improved properties owing to its high surface area and 3D porous network. GAs showed high electrical conductivity up to 1 S/cm, which is much higher than graphene/rGO. Moreover recently reported 3D-printed GF exhibited a superior conductivity of 8.7 S/cm [31]. It has also been proposed that the 3D graphene assembly is 10 times strong as mild steel [33]. Due to the exceptional properties, 3D graphene and its composites have been reported for different applications including supercapacitor, Li-ion batteries, catalyst, thermal management devices, solar cell, etc. For example, a 3D honeycomb-like structured graphene exhibited high energy conversion efficiency of 7.8% as the counter electrode in dye-sensitized solar cell (DSSC), which is comparable with the efficiency of costly Pt electrode [34].

3.4 Graphene Nanoribbons (GNRs)

Another important member of graphene materials family is graphene nanoribbons (GNRs), which has been defined as *"a single-atom-thick strip of hexagonally arranged, sp²-bonded carbon atoms that is not an integral part of a carbon material, but is freely suspended or adhered on a foreign substrate"* [24]. In short, the graphene strips with width less than 50 nm can be considered as the GNRs. Based on the edge geometries, GNR can be classified into two types—(1) zigzag and (2) armchair. On the basis of boundary conditions, these two types of GNR show different electronic properties. Like graphene, GNR can also be synthesized by top-down approaches such as lithographic patterning method, chemical methods, graphene cutting with catalytic nanoparticles and bottom-up approaches including molecular precursor-based growth, CVD, epitaxial graphene on SiC sidewall growth, etc. [35–37]. To become very effective for different energy- and environment-related applications, GNRs should have low level of disorders. Concisely, GNR have the following advantages—(1) GNRs have nearly ideal 1D character; (2) adaptable synthesis to create the network of 1D quasi-channels; and (3) through proper tuning of edge and width, the electronic properties can be easily altered. The electronic properties have been further tuned by the heteroatom doping of GNRs. Even, for further enhancement of electronic properties, the GNRs have also been combined with other 2D materials including BN.

3.5   Graphene Quantum Dots (GQDs)

Small pieces of graphene with 2D lateral size of less than 100 nm have been classified as the graphene quantum dots (GQDs). Generally, GQDs described as an alternative term of graphene nanosheets, which have the lateral dimensions of less than 10 nm at the lower end of the for the graphene nanosheets [38–40]. This special member of graphene family is popular for its remarkable fluorescence properties. However, besides the fluorescence properties, GQDs also possess low toxicity, fine surface grafting through the conjugated π-π network, high surface area, unique spin property, high photostability, etc. The GQDs have been synthesized by top-down approaches, like oxidative cleavage, hydrothermal/solvothermal method, microwave-assisted process, ultrasonic-assisted process, electrochemical oxidation, CVD, electron-beam irradiation, pulsed laser ablation process, and bottom-up approaches, like controlled synthesis in organic solvents, molecular carbonization organic molecules, and polymers. The basic idea of synthesizing GQDs is the breakdown of carbon-based starting materials through frequent oxidation and reduction process. However, a scalable synthesis process for the production of high-quality GQDs is still a major challenge for the scientific community. The GQDs have highly crystalline structure. Moreover, the band gap energy of the GQDs tuned from 0 to 6 V by altering their 2D size or by modifying their surface properties. These are the newly synthesized fluorescent materials, which have effectively used in sensors, bio-imaging, drug delivery, solar cell, supercapacitor, secondary batteries, etc. Even for further improvement of the properties including catalytic activities, electrical, electrochemical properties, these GQDs have also been doped with heteroatoms like N, B, S.

3.6   Holey Graphene (HG)

Holey graphene (HG) is a newly introduced graphene derivative which has in-plane porosity. In some literature, it is also classified as graphene mesh/nanomesh. Generally, HG has been produced by removing a large number of atoms from the graphitic plane to create holes all along the graphene sheets. Due to the presence of holes, the HG has several advantages like quantum confinement in the graphitic structure, large number of active sites, easy transport of ions and electron. High-quality HG has been synthesized by various high-precision methods like electron or ion beam bombardment, nanolithographic methods, template-assisted growth process including CVD, and scalable process like liquid-phase oxidation, gaseous-phase etching, chemical activation, guided etching with catalytic or reactive nanoparticles. Due to its porous nature, HG has been considered as a superior candidate for energy-related applications. Besides this, recently heteroatom like N-, P-, B-, S-doped HG have been synthesized, which exhibited better catalytic activities and considered as a suitable replacement for expensive noble metal catalytic system. Being a defective derivative of graphene, HG also displayed better mechanical properties than pristine graphene. Moreover, the hole-based structure of HG allowed better polymer–filler interaction in HG-based nanocomposites, resulting in inferior properties. However, depending on the types of starting graphene materials and the synthesis process, the properties of HG such as hole size, doping extent, hole distribution, surface area, porosity defect density changes [41]. Most recently, hybrid film based on modified MXenes (family of 2D transition metal carbides, carbonitrides, and nitrides) and HG has been introduced for supercapacitor application, which exhibited excellent electrochemical properties like ultrahigh volumetric capacitance, good rate capability, and excellent volumetric energy density [42].

3.7 Graphene Functionalization

'Graphene functionalization' is the recently introduced hot topic in graphene research. In terms of properties and applications, functionalization approach of graphene played an important role. The driving force for the graphene functionalization is the band gap opening. Since graphene possesses zero band gap, it shows minimal application potentiality in semiconductor and sensor fields. All the functionalization techniques have been categorized into two main types—covalent and non-covalent functionalization [43–45]. The examples of both types of functionalization are shown in Fig. 6a, b. In addition, the functionalization possibilities on graphene basal plane are schematically represented in Fig. 7.

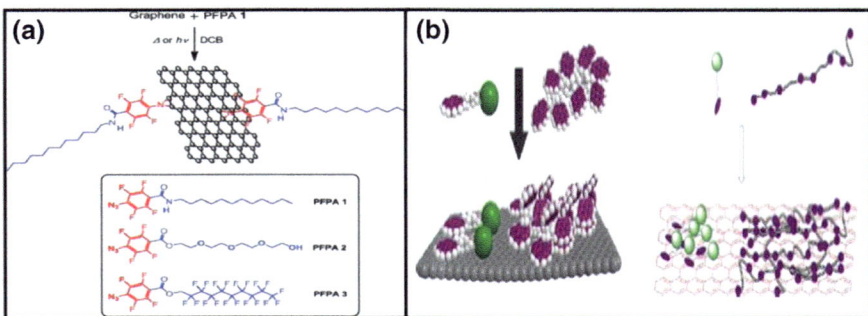

Fig. 6 Functionalization of graphene: **a** covalent functionalization of graphene with PFPA. Reproduced from Ref. [43]; **b** schematic for the non-covalent functionalization of graphene with polymers and small molecules. Reproduced from Ref. [44]

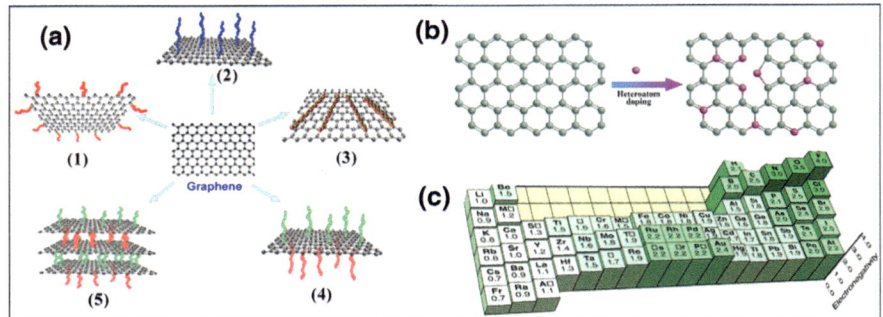

Fig. 7 a Functionalization possibilities for graphene: (1) edge functionalization, (2) basal-plane functionalization, (3) non-covalent adsorption on the basal plane, (4) asymmetrical functionalization of the basal plane, and (5) self-assembling of functionalized graphene sheets; **b** heteroatom doping of graphene, **c** electronegativity of elements increases along the *Y* axis. Reproduced from Ref. [45]

3.7.1 Covalent Functionalization of Graphene

The covalent functionalization is generally performed by forming new covalent bonds between graphene and guest functional groups. On the other hand, non-covalent functionalization is based on the physical interaction between graphene and guest molecules. The covalent functionalization enhances the solubility and processability of graphene. Additionally, these types of graphene materials exhibited better thermal and chemical stabilities than pristine graphene. Various approaches have been performed for covalent functionalization of graphene. However, the organic covalent functionalization has been performed mainly by two synthetic routes—(1) the formation of covalent bonds between free radicals and C=C bonds of graphene and (2) the covalent bond formation between the organic functional groups and the oxygen groups of GO. It is important to note that the dispersion of graphene in the organic solvents is one of the critical steps for the synthesis of graphene-based nanocomposites. Through the free radical path, graphene has been decorated by organic functional groups like nitrophenyls, dihydroxyl phenyl, tetra-phenylporphyrin (TPP), palladium-TPP, para-substituted perfluorophenylazides (PFPA). Additionally, graphene was also functionalized with the alkylazides, where the alkyl chains include the hexyl, dodecyl, hydroxylundecanyl, carboxy-undecanyl groups. Through covalent attachment path, generally GO was grafted with amine-terminated oligothiophenes, amine-terminated polyethylene glycol, the polymeric chains having reactive species like hydroxyls, amines, i.e. polyethylene glycol, polylysine, polyallylamine, and polyvinyl alcohol.

3.7.2 Non-covalent Functionalization of Graphene

Non-covalent functionalization is necessary to make graphene (hydrophobic in nature) soluble in common organic solvents. In addition, the functionalization through π interaction is also an attractive approach. The non-covalent intermolecular interactions including various types of π interactions like gas–π interaction, H–π interaction, π–π interaction, cation–π interaction, and anion–π interaction have been extensively studied in the past decades. These π systems play a vital role in stabilizing the proteins; the formation of enzyme–drug complexes, DNA–protein complexes; synthesis of organic supramolecules and functional nanomaterials.

Graphene has also been functionalized with metal nanoparticles, metal oxide nanoparticles, quantum dots, etc. In fact, pristine graphene has been categorized as the ideal substrate for the dispersion of the nanoparticles because of its huge active surface area. Novel metals like Pt, Au, Ag, Rh, Pd, and their alloys have been dispersed in graphene to form graphene/nanoparticle nanocomposites. Nanocomposites based on graphene and the nanoparticles of metal oxides like SnO_2, Mn_3O_4, Co_3O_4, TiO_2 exhibited superior capacitive performance for Li-ion batteries.

3.7.3 Heteroatom Doping of Graphene

Heteroatom doping of graphene is classified as another type of graphene functionalization route. Basically, the process is referred to the replacement of the carbon atoms of the graphitic structure of graphene by heteroatoms. The change in the electronegativity of carbon and heteroatoms is responsible for the alteration of work function. The introduction of heteroatoms like N [3.04], B [2.04], O [3.44], S [2.58], P [2.19] produces strong polarization in the carbon network due to the difference in electronegativity [C-2.55]. As a result, the electronic, optical, magnetic, electrochemical properties of doped graphene change from the pristine graphene [45–48]. In general, the doping (both n-type and p-type) of graphene has been conducted by two synthetic routes—(1) replacement of carbon atoms with heteroatoms and (2) integration of dopants by physical or chemical adsorption on the surface of graphene.

For doping of graphene, various synthetic approaches have been introduced like CVD, hydrothermal/solvothermal, microwave process, electrical annealing, arc discharge technique, plasma-assisted synthesis process, layer-by-layer doping process, adsorption, UV irradiation. Among various heteroatoms, N-doped graphene exhibited superior properties than the others because of the similarity in atomic size with carbon, hole acceptor, and electron donor properties. Even though B is also similar in atomic size with carbon, the B-doping approach is not widely utilized because of the difficulty in substitution of B into graphene. The doping of graphene can be confirmed by the Raman spectra and XPS spectra through the shifting/appearing of characteristic peaks. The theoretical studies confirmed that a minute doping level of heteroatom is enough for altering the sophisticated properties of graphene. Herein, it is noteworthy to mention that the N-doping of graphene generally creates different types of N atoms like pyrrolic N, pyridinic N, and quaternary N, which are dissimilar

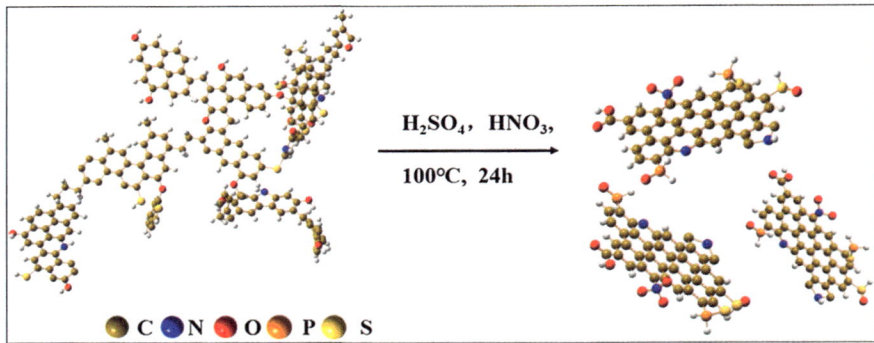

Fig. 8 Schematic diagram of the synthesis process of N, P, S co-doped GQD. Reproduced from Ref. [48]

from each other in terms of bonding configuration. Recently, dual-doped/co-doped like B-N-, N-P-, B-S-, N-S-, B-P-doped graphenes have also been synthesized, which exhibited better properties than the single-doped and pristine graphene. Moreover, multi-heteroatom-doped graphene also reported. For example, N, P, and S co-doped GQD have been synthesized from coal, which exhibited promising fluorescence properties (Fig. 8).

3.8 Graphene Derivatives

After the successful discovery of graphene, in recent times researchers have focused on the development of graphene derivatives. In this area, few new graphene-related materials have been proposed and simultaneously synthesized too. The 2D carbon allotropies of graphene have been designated as graphyne and graphdiyne. On the other hand, two hydrogenated derivatives have been proposed, which are named as graphone and graphane. Fluorinated graphene (also known as fluorographene) is another important structural derivative of graphene. Even, the oxidized form of graphene, i.e. GO, can also be considered as a graphene derivative. In this section, the structure, properties, and applications of few significant graphene derivatives have been discussed [48–53].

3.8.1 Graphyne

Graphyne can be defined as the one-atom-thick planar sheet of sp- and sp^2-hybridized carbon atoms arranged in a honeycomb fashion in a crystal lattice. It has one acetylenic linkage in each unit cell of its structure. With the variation of the arrangement of double and triple bonds, different forms of graphyne are predicted like α-graphyne, β-graphyne, and γ-graphyne. Among these, γ-form is the most common one. The electronic structure of graphyne is quite different from the common carbon materials like graphite and diamond. The graphyne ribbon shows a band gap in semiconductor range. However, it possesses promising mechanical and remarkable optical properties. The graphyne has been synthesized by replacing one-third of the carbon–carbon bonds of graphene by acetylenic linkages. Because of the possibility of the in-plane and out-plane diffusion of Li-ions in graphyne, it has been considered as a promising material for Li-ion batteries. Additionally, the improved mechanical properties of graphyne can be used as nanofiller for polymer-based composites. Moreover, the semiconducting nature of graphyne enabled it to be used in transistors. Further, these can be useful for other applications like sensors, anisotropic conductors.

3.8.2 Graphdiyne (GD)

The name "graphdiyne" has been proposed due to the presence of two acetylenic (di-acetylenic) linkages in each unit cell of its structure. However, because of the presence of these two linkages, the chain length has become doubled with respect to the graphyne. As a result, GD exhibits poorer mechanical properties than graphyne. In fact, it is much softer than graphene too. Even though it shows deprived mechanical properties, the electronic properties of GD are exceptional. It has low effective electron mass, which resulted from the Dirac cone structure of the electron bands, and exhibits band gap in semiconductor range. Additionally, GD exhibits high stability, superior electrical conductivity, uniform pore distribution, and high surface area. However, the stable form of GD has not been synthesized till date. But, the fabrications of GD-based composites and films have been reported for different applications. For example, graphydine film was fabricated on Cu surface using hexaethynylbenzene by cross-coupling reaction. This high surface area—film exhibited high conductivity of 2.516×10^{-4} S/m. Moreover, GD nanosheets showed enhanced Li storage capacity. The N-doped GD exhibited superior hydrogen purification. Nanocomposites based on metal oxide/mixed metal oxide and GD have been reported in recent times. ZnO/GD hybrid displayed promising photo-catalytic activities on the degradation of methylene blue and rhodamine B dyes. MnO_2-supported GD oxides showed superior electrochemical performance for supercapacitors. Moreover, GD-supported $NiCo_2S_4$ nanowires exhibited exceptional catalytic activity and stability towards OER and HER.

3.8.3 Fluorographene (FG)

Fluorographene (also known as fluorinated graphene) is the monolayer of graphite fluoride. As it is the perfluorinated hydrocarbon, FG was designated as the thinnest 2D insulator. It has structural similarity with graphane. Different conformations of

FG have been proposed like chair-type, zigzag-type, and boat-type. It is important to mention that the FG demonstrates lower friction than graphene. Moreover, FG displays prominent band gap opening with a little amount of fluorine. FG also exhibits transparency in the range of visible light in absorption spectrum. Its chemical properties are similar to that of poly-tetrafluoroethylene (Teflon). FG has been synthesized mainly by two approaches—(1) the exfoliation of fluorine-containing graphitic materials like graphite fluoride and (2) fluorination of graphene using fluorinating agents. The exfoliation process includes mechanical and thermal exfoliation. The fluorination process comprises CVD and plasma fluorination techniques. The theoretical studies revealed that the FG is the most thermodynamically stable derivative among the hypothetical derivatives like graphane, graphene bromide, chloride, fluoride, and iodide. FG modified electrode exhibited better catalytic performance than its graphene counterpart in DSSC. Moreover, it showed substantial potentiality towards the electrochemical sensing of ascorbic acid and uric acid. Additionally, FG also showed significant electrocatalytic properties and can be applied as competent metal-free catalyst for ORR.

3.8.4 Graphane

The fully hydrogenated derivative of graphene, which is only consisted of sp^3 carbon–carbon bonds, is defined as graphane. It has two conformations—(1) chair-type and (2) boat-type. The calculated carbon–carbon chain length in the chair-type conformer of graphane is similar to that of diamond and much higher than graphene because of the sp^3 bonding characteristics. On the other hand, the binding energy of boat-type graphane is calculated to be higher than other hydrocarbons like benzene, acetylene. The formation of graphane can be confirmed from the Raman spectroscopy. In comparison with the Raman spectrum of graphene, the G and 2D bands broaden and additional E band appears for graphane. The previous studies revealed that the band gap steadily increases with increasing the degree of hydrogenation. Graphane is much softer than graphene and exhibits lesser Poisson ratio. The fully hydrogenated graphane generally shows semiconducting nature and magnetic behaviour, while the normal graphane exhibits metallic and non-magnetic characteristics. Additionally, the half-hydrogenated form exhibits ferromagnetic behaviour. Heteroatom-doped graphane is reported to exhibit semiconducting nature. Li-doped graphane exhibited high hydrogen storage capacity. Graphane shows promising application potentiality in the field of hydrogen storage, biosensing, and spintronics.

3.8.5 Graphone

The semi-hydrogenated derivative of graphene is named as graphone. It can be considered as the graphene sheets with stoichiometry C_2H. Therefore, unlike graphane, it has a mixture of sp^2 and sp^3 carbon atoms. In general, graphone is stable in room temperature. Replacing hydrogen in the structure of graphone by fluorine alters the

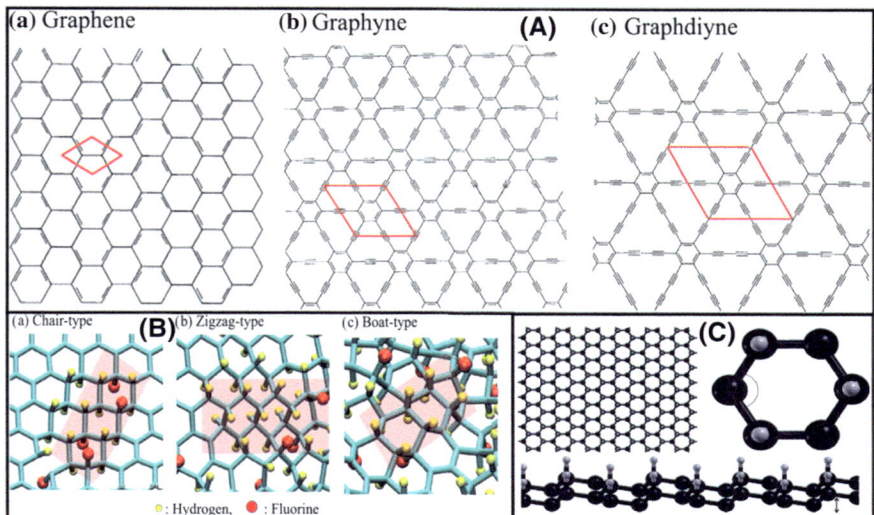

Fig. 9 Structure of few graphene derivatives: **A** structure of (a) graphene, (b) graphyne, (c) graphdiyne (the parallelogram with red line indicates a unit cell); **B** proposed hybrid structure of fluorographene; (a) chair-type, (b) zigzag-type, and (c) boat-type [50]; **C** geometry of a sheet of graphone. Reproduced from Ref. [51]

electronic structure of graphone. This fluorinated graphone showed antiferromagnetic properties and exhibited opening in the energy gap. Facile synthesis process of graphone has not been reported yet. However, the theoretical studies revealed the possible applicability of graphone in the field of field-effect transistors (FETs), organic ferroelectrics, molecular packing, etc. The possible structure of different kinds of graphene derivatives is shown in Fig. 9. Despite these materials, graphene has few other common derivatives, which are emphasized in Fig. 10.

3.9 Rebar Graphene and Rivet Graphene

Rebar graphene is identified as the CNT-toughened graphene, where CNTs acted as the reinforcing bar (rebar). Rebar graphene was introduced by famous graphene scientist James M. Tour and his research group in the year of 2014 [54]. This is nothing but the planar graphene/CNT hybrid, in which graphene is toughened by the macroscopic bar (CNT) through π−π stacking interaction and covalent bonding. This new type of graphene material was synthesized by annealing the functionalized CNT on Cu foil. Later on, the same research group also synthesized rebar graphene from functionalized BN nanotubes [55]. The presence of covalent BCN hybridization in BN-based rebar graphene is confirmed from electron energy loss spectroscopy (EELS) elemental mapping and Raman spectroscopy. The planar

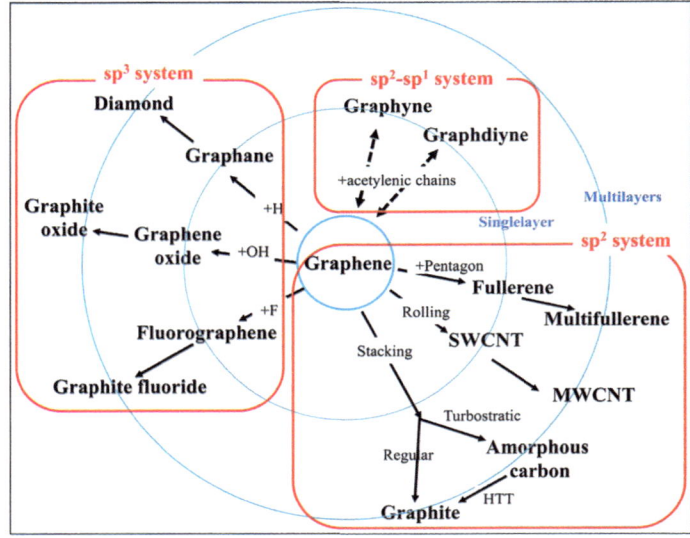

Fig. 10 Graphene derivatives family. Reproduced from Ref. [50]

CNT-introduced rebar graphene exhibited higher electronic conductivity than the CVD-grown graphene because of its network-like structure. Therefore, it can be applied to construct flexible transparent conductive electrode. On the other hand, the BN-introduced rebar graphene demonstrated superior mechanical strength than graphene. As a whole, rebar graphene is one type of graphene-based 2D heterostructures. As described by Tour's group, rebar graphene materials can be described as the *"in-plane marriage of 1D nanotubes and 2D-layered materials might herald an electrical and mechanical union that extends beyond carbon chemistry"* [54] (Fig. 11). A detailed discussion of graphene-based other 2D heterostructures is given in next section.

Later on, Tour's group also fabricated the carbon nano-onion-encapsulated Fe nanoparticles on rebar graphene, which was termed as rivet graphene [56]. The term "rivet" has been referred due to the structural similarity of this type graphene with between the rivet joints in metals. Like rebar graphene, this hybrid material has the unique ability to float on water without breaking. This exceptional hybrid material exhibited elevated optical transparency, excellent electronic conductivity, and superior electron mobility under tensile and compressive strains. Because of these properties, rivet graphene has been considered as a promising candidate in flexible and transparent electronics.

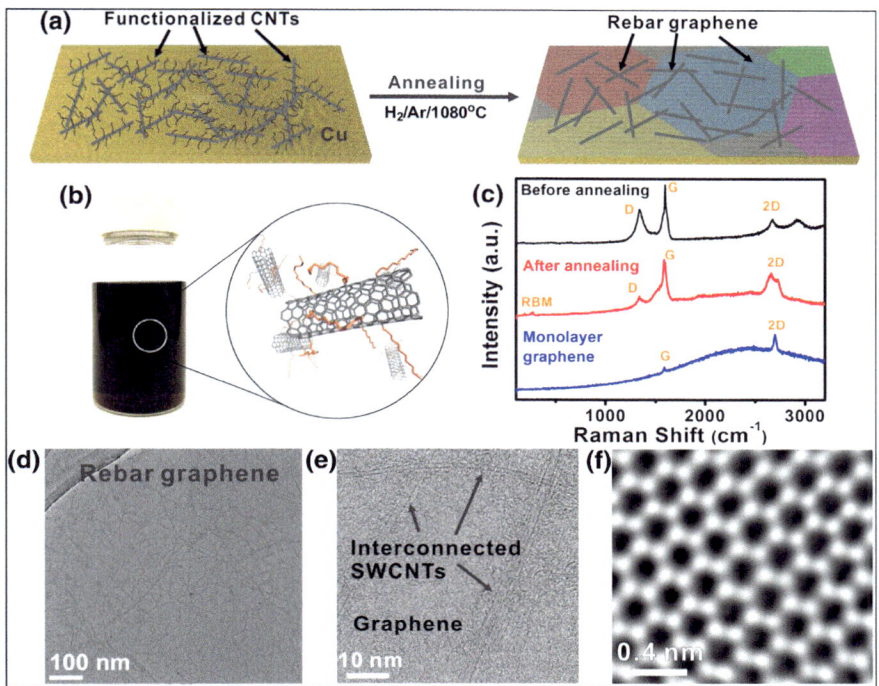

Fig. 11 Synthesis and characterization of RG: **a** synthesis process, **b** optical image of the DF-SWCNT chloroform solution and the related structural models, **c** Raman spectra, **d, e** TEM and BF-STEM images indicating the formation of interconnected SWCNT networks in rebar graphene sheets, **f** an atomic-resolution ADF-STEM image showing the defect-free hexagonal lattice of monolayer graphene. Reproduced from Ref. [54]

4 Graphene-Based 2D Heterostructures

In recent years, 2D materials have gained a great deal of research interest. With the discovery of new 2D materials, the 2D family has been expanding day by day. Verities of synthetic technologies have been developed to combine graphene with these 2D materials to form heterostructures. Both the top-down and bottom-up approaches have been employed to produce these graphene-based 2D heterostructures. The 2D family includes insulator like h-BN, wide band gap semiconductor like MoS_2, metal like NbS_2, newly developed Mxenes, superconductors like $NbSe_2$, FeSe, and few transition metal dichalcogenides. One of the major drawbacks of graphene is its zero band gap property, which restricts its application potential in electronic and optoelectronic field. This limitation of graphene can be overcome by combining it with 2D materials of wide band gap. However, through appropriate combination, the heterostructures not only conquer the limitation of graphene, but also enhance the properties of graphene. Figure 12 demonstrates a schematic representation of various

Fig. 12 A schematic representation of different types of graphene heterostructures and the synthetic approaches for the synthesis of these heterostructures. Reproduced from Ref. [57]

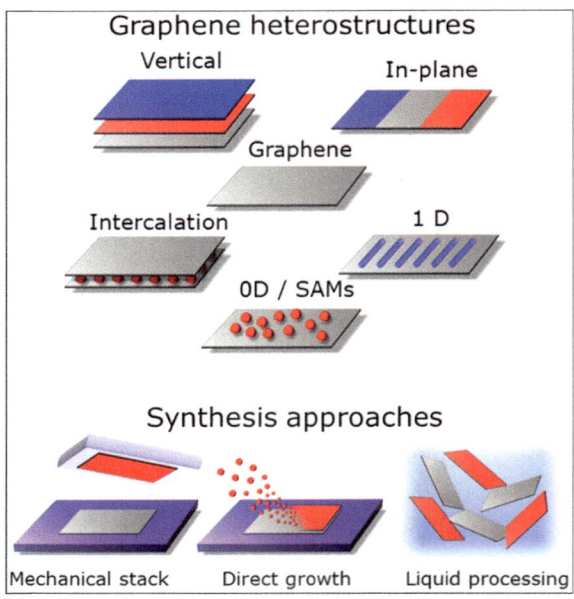

graphene-based 2D heterostructures and synthetic approaches to fabricate those.

In general, three types of synthetic approaches have been obtained—(1) mechanical stack (mainly referred to the mechanical exfoliation process), (2) direct growth (majorly CVD process), and (3) liquid processing (liquid exfoliation cum self-assembly process). A comparative study of these synthetic processes is summarized in Table 1 [58]. While mechanical stack process results in low yield of heterostructures, the other two processes produce high yield. However, in terms of cost-affectivity, mechanical stack process is superior. On the other hand, CVD process generally produces both the vertical and in-plane types of heterostructures. But, it is difficult to synthesize in-plane heterostructures through liquid exfoliation. Overall, all the processes are effectively employed for different fields of applications including energy storage/conversion, optoelectronics, spintronics. In this section, we will briefly discuss few widely applicable graphene-based 2D heterostructures.

4.1 2D Heterostructures Based on h-BN and Graphene

h-BN has the unique structural similarity with graphene (having hexagonal lattice similar to graphene), which enabled them to form promising heterostructures in combination with graphene. However, because of the ionic bonding between B and N, BN is insulator in nature with wide band gap (~5.97%), which is different from graphene having zero band gap. The minimal lattice mismatch (~1.8%) between these two 2D materials is responsible for the significant changes in the electronic

Table 1 A comparison of the synthetic methods for the preparation of 2D heterostructures. Reproduced from Ref. [58]

Method	Quality defects/impurities	Number of layers	Heterostructure yield	Typical applications	Cost	Type of heterostructures	Flake sizes/structure
Mechanical exfoliation and aligned transfer	High-quality, mechanical defects, polymer residues	Virtually any low number of layers	Low	Fundamental research of all types	Low	Vertical	Several to tens of micrometer flat or wrinkled flakes
CVD	High-quality chemical impurities, polymer residues	Two few layers	Moderate to high	Electronics, optoelectronic, spintronics	High	Vertical and in-plane	Continuous films of wafer size flat, porous, complex
Liquid exfoliation and self-assembly	Unintentional functionalization composites instead of heterostructure may be formed	Few layers and high more (LbL allows lower number of layers)	High	Energy storage, energy conversion	Low to moderate	Vertical (in-plane very difficult)	Hundreds of nanometre to several micrometer flat, wrinkled, porous, complex

structure of graphene in the h-BN/graphene heterostructures. The most frequently employed process for synthesizing this type of heterostructures is the CVD process. The two-step CVD technique is not suitable for the synthesis of the heterostructures on Cu foil as graphene has been damaged during the second CVD process. The growth mechanism of graphene on h-BN is not clearly interpreted yet, but it has been observed that some catalytic surface is required for the growth of graphene. In this circumstances, it has also been witnessed that nickelocene has enhanced the growth rate of graphene by catalytic activity. In the h-BN/graphene heterostructures, the encapsulation of h-BN layers prevents the degradation of graphene against harsh environmental condition, and therefore, the chemical stability has been increased.

4.2   2D Heterostructures Based on MoS$_2$ and Graphene

2D layered structures with the formula MX$_2$ (M denotes transition metal elements and X denotes chalcogenide element) are one of the next-generation smart 2D materials because of their enhanced physicochemical properties, high surface area, improve chemical stability, easy synthetic process, and cost-affectivity [59]. Among different 2D transition metal chalcogenides, MoS$_2$ is the most widely studied one. In recent times, the fabrication of hetrostructured hybrid of graphene and MoS$_2$ as the multifunctional material has been considered as one of the most effective approaches in both the academic and industry sectors for performance enhancement. The crystals of MX$_2$ are available in three different phases—trigonal (1T), hexagonal (2H), and rhombohedral (3R). However, MoS$_2$ is available in first two phases mostly. The hybrid heterostructures have been synthesized by different synthetic approaches including hydrothermal, solvothermal process, thermal as well as chemical reduction, microwave synthetic route, CVD process, dry/wet transfer method, etc. The hybrid structure showed much higher mechanical properties than MoS$_2$. For example, the layer-by-layer assembled hybrid of the single layers of graphene and MoS$_2$ exhibited Young's modulus of ~511.8 GPa, which is higher than the value of MoS$_2$ (128.7 GPa). Because of the lattice mismatch of these two, the hybrid also exhibited spontaneous strain energy at the interface. The thermal stability of the hybrid has also been improved due to the presence of graphene, which could easily carry the heat energy. Moreover, the synergistic interactions between these two 2D materials enhanced the electroactive sites and therefore improved the electrochemical properties. Even, the 3D architectures of graphene and MoS$_2$ also enhanced the charge transfer process during redox reaction, which is benefitted from the superior structural compatibility and excellent interaction between the individual components.

4.3 2D Heterostructures Based on MXenes and Graphene

Another important 2D material is MXene, which has grown a great deal of research interest after its discovery in the year of 2011 [60, 61]. Basically, MXenes denote the transition metal carbides, nitrides, and carbonitrides, which have been synthesized from a bulk crystal called MAX. The most important property of MXene is its high conductivity, which is unusual for 2D ceramics. These 2D materials are generally synthesized by extorting the A element from MAX phases, where M indicates early transition metal, A denotes the group IIIA or IVA element, and X designates the C or N element. Because of its high conductive nature, it has been considered as a promising candidate for the fabrication of van der Waals heterostructures with graphene. This smart combination not only enhanced the surface area of the hybrid, but also improved the electronic and electrochemical properties. Within a short duration of time (2011–2018), the researchers have successfully synthesized a series of MXenes like Ti_3AlC_2, $(Mo_2Ti)AlC_2$, $Ti_3(Al_{0.5}Si_{0.5})C_2$, $Ti_2Al(C_{0.5}N_{0.5})$. In a recent study, it has been observed that the incorporation of 0.25 wt% of graphene doubled the conductivity and tripled the Hall mobility of $Ti_3C_2T_x$. A sandwich-type all-solid-state supercapacitor has also been fabricated with electrochemically exfoliated graphene and HF-etched $Ti_3C_2T_x$, which exhibited a high volumetric capacitance of 184 F/cm^3 at the current density of 0.2 A/cm^3 and showed a high capacitance retention of 82.5% after 2500 cycles. Even, the insertion of graphene into MXene layers restricts the restacking in hybrid and enhanced the conductivity, Li adsorption strength, and Li mobility, which are beneficial for Li-ion battery application.

5 Application of Graphene-Based Materials

In this decade, graphene has been considered as the most wonderful multifunctional 2D material, which has a wide range of applications (Fig. 13) [62–64]. Owing to its unique structure and exceptional properties, graphene showed its applicability in almost every sector of our day-to-day life. Because of the formation of ohmic contact on the metallic surface and the presence of sharp edges, graphene has been applied in field emission applications. On the other hand, in addition to high surface area, graphene exhibits high electrical and thermal conductivities, which enables it to be considered as a promising candidate for biosensor and gas sensor applications. In particular, graphene-based electrodes exhibited better biosensing performance than CNTs towards dopamine detection. Moreover, graphene has also been tested for photovoltaic applications. More specifically, because of high surface area, enhanced carrier ability, and superior chemical stability, graphene has been considered as a promising candidate for the fabrication of advanced transparent and flexible electrodes. However, energy storage is another sector, where graphene showed a promising applicability. Graphene in combination with mixed metal oxides, metal nanoparticles, or other 2D materials demonstrated superior electrochemical performance for

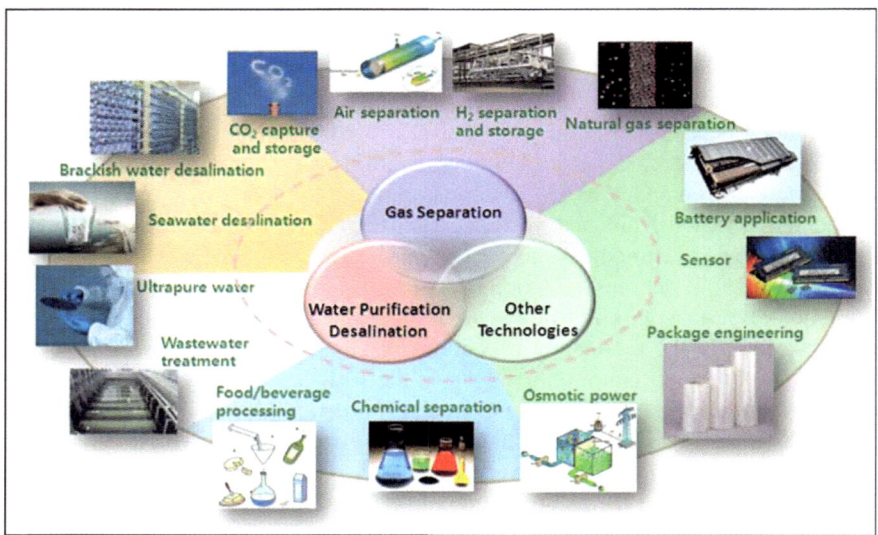

Fig. 13 Various applications of graphene. Reproduced from Ref. [64]

various secondary batteries like Li-ion batteries, Na-ion batteries, Na–air batteries. The introduction of graphene enhances the storage capacity, cycling stability, and columbic efficiency. Graphene-based materials have been comprehensively investigated for the fabrication of both anode and cathode. One of the promising energy storage devices of this century is supercapacitor. Due to its porous nature, high surface area, and enhanced electrochemical properties, graphene has been widely investigated for the fabrication of solid-state symmetric and asymmetric supercapacitor, micro-supercapacitor, binder-free supercapacitor, etc. Graphene-based transparent electrodes have been fabricated for DSSC application. Functionalized graphene and graphyne have been utilized for water desalination application. Recently, graphene LED bulbs have been developed, which demonstrated a great prospect in energy consumption. Moreover, graphene chips have been considered as a promising replacement of Si chip. Graphene also served as an excellent drug carrier because of its high biocompatibility. Nevertheless, it has also been considered as a suitable hydrogen storage material due to its superior gas adsorption properties. Herein, it is important to mention that recently a research group from China developed a unique graphene hair dye which exhibited better antistatic performance and efficient heat as well as electrostatic charge dissipation than the commercial black dyes [65]. Overall, graphene displayed a wide range of application potential, for which the current era has been considered as the graphene era.

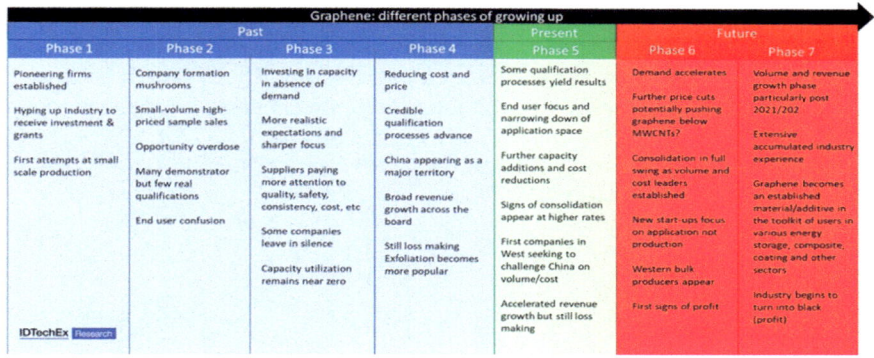

Fig. 14 Different phases of the evaluation of commercial graphene. Reproduced from Ref. [66]

6 Commercialization of Graphene-Based Materials

In the field of nanoscience and nanotechnology, graphene is the most attractive material of today's world. After its discovery in 2004, within the short time duration, it has become very popular. Many companies and institutes have invested billion dollars on graphene research. A huge number of graphene-based patents have been filed worldwide within the last few years. However, most of these patents are based on the application of graphene. In the current situation, it has been realized that the fabrication techniques of graphene are one of the fields, which require more attention, not the application part. The graphene market is growing day by day, but it should also be noticed that not all graphene materials are same. Therefore, the production of high-quality graphene is one of the major targets of current graphene research. The market research data indicates that there has been a drastic growth on the income of graphene-based companies since 2013 and it will be in the peak point until 2020/2021. For reference, a market research company called IDTechEx analysed different phases of graphene commercialization and summarized in a table, which is shown in Fig. 14 [66]. It is expected that, in terms of commercialization, graphene can overtake MWCNT in future, but it has not been proven till date.

7 Conclusion

It is highly acceptable that graphene has opened a new door in the field of nanoscience and nanotechnology because of its multifunctional properties. However, it also brought a bag full of challenges. Current graphene research is mainly focused on the quality. Even though, it will take few more years for the commercialization. Graphene-based 2D hybrids are the newly introduced materials which exhibited a wide range of application. Recently invented MXenes are also considered as multifunctional 2D material, but research on this material is still in preliminary stage. Overall, the prospect of graphene research is increasing day by day. It is expected that, within a couple of years, we will have the commercialized graphene product.

Acknowledgements S. Sahoo acknowledges DST-SERB, India, for the national postdoctoral fellowship (NPDF File No.: PDF/2017/000328).

References

1. Geim, A.K., Novoselov, K.S.: The rise of graphene. Nat. Mater. **6**, 183–191 (2007)
2. Meyer, J.C., Geim, A.K., Katsnelson, M.I., Novoselov, K.S., Booth, T.J., Roth, S.: The structure of suspended graphene sheets. Nature **446**, 60–63 (2007)
3. Rao, C.N.R., Sood, A.K., Subrahmanyam, K.S., Govindaraj, A.: Graphene: the new two-dimensional nanomaterial. Angew. Chem. **48**, 7752–7777 (2009)
4. Castro Neto, A.H., Guinea, F., Peres, N.M.R., Novoselov, K.S., Geim, A.K.: The electronic properties of graphene. Rev. Mod. Phys. **81**, 109–162 (2009)
5. Ghosh, S., Bao, W., Nika, D.L., Subrina, S., Pokatilov, E.P., Lau, C.N., Balandin, A.A.: Dimensional crossover of thermal transport in few-layer graphene. Nat. Mater. **9**, 555–558 (2010)
6. Mahmoudi, T., Wang, Y., Hahn, Yoon-Bong: Graphene and its derivatives for solar cells application. Nano Energy **47**, 51–65 (2018)
7. Edwards, R.S., Coleman, K.S.: Graphene synthesis: relationship to applications. Nanoscale **5**, 38–51 (2013)
8. Allen, M.J., Tung, V.C., Kaner, R.B.: Honeycomb carbon: a review of graphene. Chem. Rev. **110**, 132–145 (2010)
9. Dimiev, A.M., Ceriotti, G., Metzger, A., Kim, N.D., Tour, J.M.: Chemical mass production of graphene nanoplatelets in~ 100% yield. ACS Nano **10**, 274–279 (2015)
10. Lopes, L.C., da Silva, L.C., Vaz, B.G., Oliveira, A.R., Oliveira, M.M., Rocco, M.L., Orth, E.S., Zarbin, A.J.: Facile room temperature synthesis of large graphene sheets from simple molecules. Chem. Sci. **9**, 7297–7303 (2018)
11. Ruan, G., Sun, Z., Peng, Z., Tour, J.M.: Growth of graphene from food, insects, and waste. ACS Nano **5**, 7601–7607 (2011)
12. Shah, J., Lopez-Mercado, J., Carreon, M.G., Lopez-Miranda, A., Carreon, M.L.: Plasma Synthesis of Graphene from Mango Peel. ACS Omega **3**, 455–463 (2018)
13. Chen, F., Yang, J., Bai, T., Long, B., Zhou, X.: Facile synthesis of few-layer graphene from biomass waste and its application in lithium ion batteries. J. Electroanal. Chem. **768**, 18–26 (2016)
14. Ding, Z., Li, F., Wen, J., Wang, X., Sun, R.: Gram-scale synthesis of single-crystalline graphene quantum dots derived from lignin biomass. Green Chem. **20**, 1383–1390 (2018)
15. Wang, Z., Yu, J., Zhang, X., Li, N., Liu, B., Li, Y., Dissanayake, S., Appl, A.C.S.: Large-scale and controllable synthesis of graphene quantum dots from rice husk biomass: a comprehensive utilization strategy. Mater. Interfaces **8**, 1434–1439 (2016)
16. Purkait, T., Singh, G., Singh, M., Kumar, D., Dey, R.S.: Large area few-layer graphene with scalable preparation from waste biomass for high-performance supercapacitor. Sci. Rep. **7**, 15239 (2017)
17. León, V., Rodriguez, A.M., Prieto, P., Prato, M., Vázquez, E.: Exfoliation of Graphite with Triazine Derivatives under Ball-Milling Conditions: Preparation of Few-Layer Graphene via Selective Noncovalent Interactions. ACS Nano **8**, 563–571 (2014)

18. Jeon, I.Y., Shin, Y.R., Sohn, G.J., Choi, H.J., Bae, S.Y., Mahmood, J., Dai, L.: Edge-carboxylated graphene nanosheets via ball milling. Proc. Natl. Acad. Sci. **109**, 5588–5593 (2012)
19. Yan, L., Lin, M., Zeng, C., Chen, Z., Zhang, S., Zhao, X., Guo, M.: Electroactive and bio-compatible hydroxyl-functionalized graphene by ball milling. J. Mater. Chem **22**, 8367–8371 (2012)
20. Xu, Z., Li, H., Li, W., Cao, G., Zhang, Q., Li, K., Wang, J.: Large-scale production of graphene by microwave synthesis and rapid cooling. Chem. Comm. **47**, 1166–1168 (2011)
21. Fei, H., Dong, J., Wan, C., Zhao, Z., Xu, X., Lin, Z., Zhao, S.: Microwave-assisted rapid synthesis of graphene-supported single atomic metals. Adv. Mater. **30**, 1802146 (2018)
22. Murugan, A.V., Muraliganth, T., Manthiram, A.: Rapid, facile microwave-solvothermal synthesis of graphene nanosheets and their polyaniline nanocomposites for energy strorage. Chem. Mater. **21**, 5004–5006 (2009)
23. Raccichini, R., Varzi, A., Passerini, S., Scrosati, B.: The role of graphene for electrochemical energy storage. Nat. Mater **14**, 271 (2015)
24. Bianco, A., Cheng, H.M., Enoki, T., Gogotsi, Y., Hurt, R.H., Koratkar, N., Zhang, J.: All in the graphene family – A recommended nomenclature for two-dimensional carbon materials. Carbon **65**, 1–6 (2013)
25. Zhao, S., Lv, Y., Yang, X.: Layer-dependent nanoscale electrical properties of graphene studied by conductive scanning probe microscopy. Nanoscale Res. Lett. **6**, 498 (2011)
26. Ziegler, D., Gava, P., Güttinger, J., Molitor, F., Wirtz, L., Lazzeri, M., Stampfer, C.: Variations in the work function of doped single-and few-layer graphene assessed by Kelvin probe force microscopy and density functional theory. Phys. Rev. B **83**, 235434 (2011)
27. Chen, D., Feng, H., Li, J.: Graphene oxide: preparation, functionalization, and electrochemical applications. Chem. Rev. **112**, 6027–6053 (2012)
28. Deng, S., Berry, V.: Wrinkled, rippled and crumpled graphene: an overview of formation mechanism, electronic properties, and applications. Mater. Today **19**, 197–212 (2016)
29. Li, C., Shi, G.: Three-dimensional graphene architectures. Nanoscale **4**, 5549–5563 (2012)
30. Ma, Y., Chen, Y.: Three-dimensional graphene networks: synthesis, properties and applications. Natl. Sci. Rev. **2**, 40–53 (2015)
31. Sha, J., Li, Y., Villegas Salvatierra, R., Wang, T., Dong, P., Ji, Y., Ajayan, P.M.: Three-dimensional printed graphene foams. ACS Nano **11**, 6860–6867 (2017)
32. Chen, K., Shi, L., Zhang, Y., Liu, Z.: Scalable chemical-vapour-deposition growth of three-dimensional graphene materials towards energy-related applications. Chem. Soc. Rev. **47**(9), 3018–3036 (2018)
33. Qin, Z., Jung, G.S., Kang, M.J., Buehler, M.J.: The mechanics and design of a lightweight three-dimensional graphene assembly. Sci. Adv. **3**, 1601536 (2017)
34. Wang, H., Sun, K., Tao, F., Stacchiola, D.J., Hu, Y.H.: 3D honeycomb-like structured graphene and its high efficiency as a counter-electrode catalyst for dye-sensitized solar cells. Angew. Chem. **52**(35), 9210–9214 (2013)
35. Celis, A., Nair, M.N., Taleb-Ibrahimi, A., Conrad, E.H., Berger, C., de Heer, W.A., Tejeda, A.: Graphene nanoribbons: fabrication, properties and devices. J. Phys. D Appl. Phys. **49**, 143001 (2016)
36. Dutta, S., Pati, S.K.: Novel properties of graphene nanoribbons: a review. J. Mater. Chem. **20**, 8207–8223 (2010)
37. James, D.K., Tour, J.M.: Macromol. The chemical synthesis of graphene nanoribbons-a tutorial review. Chem. Phys. **213**, 1033–1050 (2012)
38. Chen, W., Lv, G., Hu, W., Li, D., Chen, S., Dai, Z.: Synthesis and applications of graphene quantum dots: A review. Nanotechnol. Rev. **7**, 157–185 (2018)
39. Kaur, M., Kaur, M., Sharma, V.K.: Nitrogen-doped graphene and graphene quantum dots: A review on synthesis and applications in energy, sensors and environment. Adv. Colloid Interface Sci. **259**, 44–64 (2018)
40. Bak, S., Kim, D., Lee, H.: Graphene quantum dots and their possible energy applications: A review. Curr. Appl. Phys. **16**, 1192–1201 (2016)

41. Lin, Y., Liao, Y., Chen, Z., Connell, J.W.: Holey graphene: a unique structural derivative of graphene. Mater. Res. Lett. **5**, 209–234 (2017)
42. Fan, Z., Wang, Y., Xie, Z., Wang, D., Yuan, Y., Kang, H., Liu, Y.: Modified MXene/holey graphene films for advanced supercapacitor electrodes with superior energy storage. Adv. Sci. **5**, 1800750 (2018)
43. Georgakilas, V., Otyepka, M., Bourlinos, A.B., Chandra, V., Kim, N., Kemp, K.C., Kim, K.S.: Functionalization of graphene: covalent and non-covalent approaches, derivatives and applications. Chem. Rev. **112**, 6156–6214 (2012)
44. Wang, X., Shi, G.: An introduction to the chemistry of graphene. Phys. Chem. Chem. Phys. **17**, 28484–28504 (2015)
45. Hu, C., Liu, D., Xiao, Y., Dai, L.: Functionalization of graphene materials by heteroatom-doping for energy conversion and storage. Prog. Nat Sci-Mater. **28**, 121–132 (2018)
46. Lee, H., Paeng, K., Kim, I.S.: A review of doping modulation in graphene. Synth. Met. **244**, 36–47 (2018)
47. Wang, H., Maiyalagan, T., Wang, X.: Review on recent progress in nitrogen-doped graphene: synthesis, characterization, and its potential applications. ACS Catal. **2**, 781–794 (2012)
48. Xu, Y., Wang, S., Hou, X., Sun, Z., Jiang, Y., Dong, Z., Cao, Y.: Coal-derived nitrogen, phosphorus and sulfur co-doped graphene quantum dots: A promising ion fluorescent probe. Appl Surf Sci. **445**, 519–526 (2018)
49. Li, G., Li, Y., Liu, H., Guo, Y., Li, Y., Zhu, D.: Architecture of graphdiyne nanoscale films. Chem. Commun. **46**, 3256–3258 (2010)
50. Inagaki, M., Kang, F.: Graphene derivatives: graphane, fluorographene, graphene oxide, graphyne and graphdiyne. J. Mater. Chem. A **2**, 13193–13206 (2014)
51. Peng, Q., Dearden, A.K., Crean, J., Han, L., Liu, S., Wen, X., De, S.: New materials graphyne, graphdiyne, graphone, and graphane: review of properties, synthesis, and application in nanotechnology. Nanotechnol. Sci. Appl. **7**, 129 (2014)
52. Chronopoulos, D.D., Bakandritsos, A., Pykal, M., Zbořil, R., Otyepka, M.: Chemistry, properties, and applications of fluorographene. Appl. Mater. Today **9**, 60–70 (2017)
53. Paupitz, R., Autreto, P.A.S., Legoas, S.B., Srinivasan, S.G., van Duin, A.C.T., Galvao, D.S.: Graphene to fluorographene and fluorographane: a theoretical study. Nanotechnol. **24**, 035706 (2012)
54. Yan, Z., Peng, Z., Casillas, G., Lin, J., Xiang, C., Zhou, H., Yang, Y., Ruan, G., Raji, A.-R.O., Samuel, E.L.G., Hauge, R.H., Yacaman, M.J., Tour, J.M.: Rebar graphene. ACS Nano **8**, 5061–5068 (2014)
55. Li, Y., Peng, Z., Larios, E., Wang, G., Lin, J., Yan, Z., Tour, J.M.: Rebar graphene from functionalized boron nitride nanotubes. ACS Nano **9**, 532–538 (2014)
56. Li, X., Sha, J., Lee, S.-K., Li, Y., Ji, Y., Zhao, Y., Tour, J.M.: Rivet Graphene. ACS Nano **10**, 7307–7313 (2016)
57. Solís-Fernández, P., Bissett, M., Ago, H.: Synthesis, structure and applications of graphene-based 2D heterostructures. Chem. Soc. Rev. **46**, 4572–4613 (2017)
58. Das, P., Fu, Q., Bao, X., Wu, Z.S.: Recent advances in the preparation, characterization, and applications of two-dimensional heterostructures for energy storage and conversion. J. Mater. Chem. A **6**, 21747–21784 (2018)
59. Thanh, T.D., Chuong, N.D., Van Hien, H., Kshetri, T., Kim, N.H., Lee, J.H.: Recent advances in two-dimensional transition metal dichalcogenides-graphene heterostructured materials for electrochemical applications. Prog. Mater Sci. **96**, 51–85 (2018)
60. Ng, V.M.H., Huang, H., Zhou, K., Lee, P.S., Que, W., Xu, J.Z., Kong, L.B.: Recent progress in layered transition metal carbides and/or nitrides (MXenes) and their composites: synthesis and applications. J. Mater. Chem. A **5**, 3039–3068 (2017)
61. Anasori, B., Lukatskaya, M.R., Gogotsi, Y.: 2D metal carbides and nitrides (MXenes) for energy storage. Nat. Rev. Mater. **2**, 16098 (2017)
62. www.pgmcapital.com/why-investing-in-graphene-can-be-lucrative/
63. Ghany, N.A.A., Elsherif, S.A., Handal, H.T.: Revolution of Graphene for different applications: State-of-the-art. J. Surf. Interfac. **9**, 93–106 (2017)

64. Ren, S., Rong, P., Yu, Q.: Preparations, properties and applications of graphene in functional devices: A concise review. Ceram. Int. **44**, 11940–11955 (2018)
65. Luo, C., Zhou, L., Chiou, K., Huang, J.: Multifunctional graphene hair dye. Chem **4**, 784–794 (2018)
66. www.idtechex.com/research/articles/graphene-commercialization-a-look-back-at-the-story-so-far-00015453.asp

Graphene-Based Advanced Materials: Properties and Their Key Applications

Santosh Kumar Tiwari, Nannan Wang and Sung Kyu Ha

Abstract Since the last 500 years, science is becoming more and more dominant in our civilization and continuously making the life of human beings more convenient. Along with the numerous fundamental discoveries and innovations, twenty-first century will be evoked as technological achievements for a long time. Among the many outstanding scientific achievements, the introduction of graphene can be considered as one of the most important breakthroughs for this century. This single-atom thin 2D carbon nanomaterial is the foundation of all graphitic structures. Owing to its amazing physical and chemical properties, graphene has found applications in many scientific and technological fields, from medical science to aerospace engineering. However, scientists of the various disciplines are working hard individually and in collaborations around the globe to utilize and explore application potentials of the graphene and its derivatives (graphene oxide, graphene quantum dot, graphene nanoribbon, functionalized graphene etc.). In this chapter, some novel discoveries and innovations closely related to the graphene-based advanced nanomaterials for the real-time applications have been reviewed in detail, especially in contest of high-performance polymer blends, nanocomposites for catalysis, water splitting and 3D printings. In addition, a brief outline for the fabrication of graphene-based polymer blends and nanocomposites has also been discussed with appropriate citations for the further reading.

Keywords Graphene · Composite materials · Polymer blends · Thermo-mechanical properties and 3D printing

S. K. Tiwari · N. Wang (✉)
Key Laboratory of New Processing Technology for Nonferrous Metals and Materials, Ministry of Education, School of Resources, Environment and Materials, Guangxi University, Nanning, China
e-mail: wangnannan@gxu.edu.cn

S. K. Tiwari (✉) · S. K. Ha
Department of Mechanical Engineering, Hanyang University, Seoul, South Korea
e-mail: ismgraphene@gmail.com

1   Introduction

Innovations are governing our civilization and continuously making world more and more comfortable [1]. One of the most important successes of scientific innovations is in the characterization techniques, which allow us to investigate materials up to the molecular and atomic levels [1]. Moreover, a clear knowledge on the structure of materials gives us the opportunity to modify these structures to meet our needs. That is why material scientists and nanochemists have oppressed the surface modification of different materials at the nanoscale in recent years [2]. Graphene, a single-atom thin 2D nanosystem of carbon atoms organized in a hexagonal atomic orientation, has engrossed great consideration in the past 20 years due to its ultrahigh specific surface area (~2630 $m^2\,g^{-1}$), extraordinary thermal and electrical conductivity (~3000–5000 $W\,m^{-1}K^{-1}$), high chemical inertness and outstanding mechanical properties (transparency of ~97.7% and Young's modulus of ~1 TPa) [3]. These exceptionalities have made graphene a promising candidate for a wide range of applications including electronics, catalysis, sensors, energy storage systems, energy conversion materials, carbon capture systems etc. [1, 4, 5].

Compared to other carbonaceous nanomaterials (such as carbon nanofibers, carbon nanotubes, fullerenes), graphene-based nanostructures have superior advantages such as low-weight, very high available surface area, high electrical, optical and thermal properties [1–4]. The surface modification of graphene nanosheets possesses great capability to reinforce different matrixes [6]. It is believed that the surface functionalization of graphene nanosheets is much easier than CNTs and nanofibers and has unlocked an array of bids across the wide range of research interest [7]. Due to the easy and controllable functionalization, many graphene-based functional materials are in the pipeline for the real practice and commercialization [8]. However, few of them are in the market for the commercial uses. For example, graphene-based flexible supercapacitor, solar cells, batteries, composites for aerospace applications, etc., are the major outcome of graphene research. Catalysis and water purifications research are also of the fast-growing application potentials of graphene-based nanostructures [8, 9].

The instance collaborative investigations on graphene by the physicists, chemists, engineers and even social scientists have blossomed the uniqueness of graphene and its derivatives over the last two decades [8, 9]. This is a main reason for the fast-growing interest for the graphene-based nanomaterials especially for the robotics, computing, genetic engineering, sensors, biomedical, thermal management, aviation, coatings, antennas, wave absorption systems, lubricants, metamaterials, water filtration, piezoelectric systems, etc. [10, 11]. By considering the electronic devices as an example, it can be seen that the pristine graphene and the few-layers graphene are very important for the electronic industries. On the other hand, graphene oxide and other functionalized graphene nanostructures have been mostly used for the innovation in the field of polymer engineering, biomedical and sensing [10, 11]. The recent studies on physicochemical properties of polymer nanocomposites with different loading contents of graphene oxide and reduced graphene oxide and different polymeric matrixes (PC, PP and Nylon 66) have suggested that graphene derivatives may be one of the best candidate for polymeric systems due to higher compatibility between the graphene oxide nanosheets and polymers, compared to pristine graphene nanosheets [12, 13]. The behavior of compatibility can be correlated to the nature of interactions between nanosheets and the matrix, which can be engineered by controlling the structure of nanosheets and the polymeric matrix for the desired applications [14].

In this chapter, recent progresses in the fabrication of graphene-based materials for the numerous prototypes and commercial applications have been reviewed in details with proper citations. These applications include thermo-mechanically stabilized polymer systems, catalysis, homogenous distribution of graphene and graphene oxide into various polymeric materials, water splitting and 3D printings.

2 Recently Developed Graphene-Based Materials and Their Application

As mentioned in the introduction section, the utilization of graphene and its derivatives growing day by day and numerous paths have been developed for the mass production of graphene and graphene oxide [1, 9]. The bulk production of pure graphene, graphene oxide and derivatives is one of the great achievement in the line of its real-time applications [8, 9]. Just 10–15 years ago, there were only 2–3 methods (chemical exfoliation, mechanical exfoliation, and thermal expansion) for the production of graphene, and selective surface modification was very limited and therefore restricted application of the same [1–9]. But, within the last 6–7 years, many facile and green approaches have been developed for the large-scale synthesis of graphene which includes, sonochemical, solvothermal, electrochemical and chemical vapour deposition (CVD) on a suitable thin metallic layers, which greatly triggered research on practical as well as commercial applications [15, 16]. It is surprising and notable that, the commercialization and real-world applications of graphene and their derivatives are still not satisfactory. In the case of polymer composites, the compatibility of graphene (or its derivatives) with the polymeric chins is one of the most critical issue to be resolved and it depends on the nature of the surface modification of graphene system and its uniform distribution into the polymer matrices [3, 9, 15]. To improve commercial aspect of graphene, the European Union announced a €1 billion grant in 2013 for research to explore potential of graphene applications [15, 16]. China, India, Japan, and South Korea also investing hugely for the graphene research especially for the composites, renewable energy and biomedical applications. The following newly fabricated graphene-based materials for the different applications are listed below with brief descriptions.

2.1 GO-Based Polyamide/Polyphenylene Oxide Polymer Blends

Carbon nanofiller reinforced polymeric materials offer the opportunity to obtain materials with desired properties, and several carbon-based nanofillers (like graphite, carbon black CNTs, graphene, graphene oxide etc.) were used by different research groups to reinforce properties of polymeric composites and blends [1, 7]. The first well-organized study concerning to the use of graphene oxide as nanofiller for the polymer blends was reported by the Cao and co-worker where they focused on the advantage of the unique amphiphilic nature of graphene oxide and their interaction with selected polymers [17]. They testified a very effective tactic to compatibilized immiscible (polyamide: PA/polyphenylene oxide: PPO) polymer blends using graphene oxide [17]. For this work, they used 9:1 of PA/PPO along with 0.5 wt% GO as nano compatibilizer [17]. In this work, a very drastic decrease in the droplet size and huge change in surface morphology of the dispersed minor phase (PPO) has been reported owing to the incorporation of GO into the blend components during the processing [17]. In the neat PA/PPO blend, PPO holes are prominent with a diameter of more than 10 μm and these holes can be easily seen even at the low magnification SEM images (Fig. 1a) [17]. Surprisingly, the incorporation of GO gives tremendous shrinkage in the size of the PPO domains and the big holes convert into minute pores that are not visible at a low magnification consequently, the ductility of fabricated blends notably increased.

They have explained such a great compatibility between PA/PPO after the incorporation of the minute amount of GO due to the strong interactions between blend

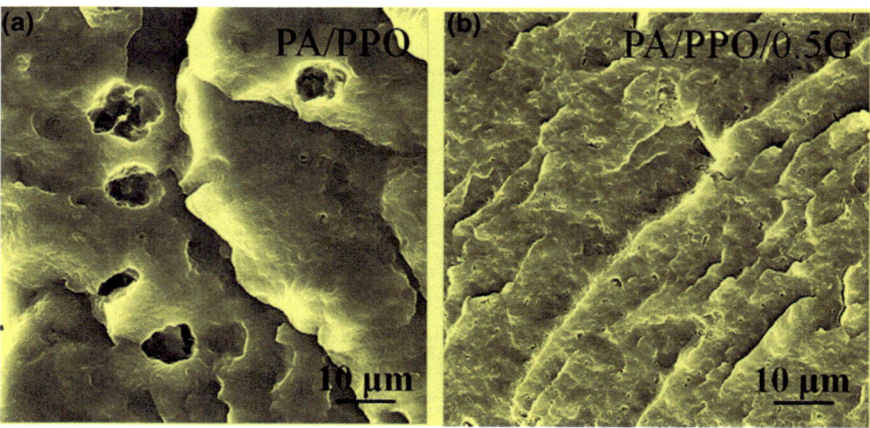

Fig. 1 SEM micrographs for the cryogenically fractured surfaces of neat blend and GO compatibilized PA/PPO blends. Reproduced with permission from Ref. [17]

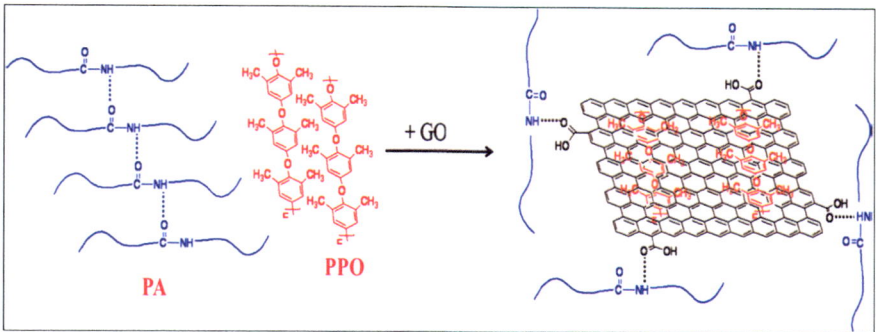

Fig. 2 Schematic diagram of for the interaction between PA, PPO, and GO. Reproduced with permission from Ref. [17]

components and therefore a great reduction in interfacial tension [7, 17]. Furthermore, herein GO is also acting as reinforcing fillers for polymer blends, thus extraordinarily improvement in their thermo-mechanical stability. The interfacial interaction behavior of blend and bridging effect of GO with PA/PPO is presented in the Fig. 2.

2.2 Graphene-Polyurethane Nanocomposites

Polyurethane is one of the most frequently used polymeric materials for domestic and engineering applications. It composed of organic units joined by carbamate covalent linkage [18, 19]. As we know that most polyurethanes are thermosetting type polymeric materials and do not melt when heated [19]. However, thermoplastic polyurethanes are also available for the specific applications when it blended with other polymeric units [18, 19]. In this line, Lee and co-worker synthesized waterborne polyurethane functionalized graphene nanocomposites by in situ process [19]. The electrical conductivity of the polymer nanocomposite was increased 105-fold compared to pure polymer owing to homogeneous dispersion of graphene nanosheets in polyurethanes matrices [18–20]. It has been concluded that presence of functionalized graphene nanosheets also increased the melting temperature and heat of fusion of the soft segment of polymeric chains in the nanocomposites [19]. Thus thermo-mechanical as well electrical properties of nanocomposites get enhanced [19, 20]. Furthermore, Liang and co-worker prepared three types of polymer nanocomposites by solution mixing procedure to verify the work of Lee and co-worker and they used isocyanate-modified graphene, sulfonated graphene and reduced graphene oxide as nanofiller and thermoplastic polyurethane as the matrix polymer [19, 20]. They reported amazing enhancement in thermo-mechanical properties of fabricated nanocomposites [19, 20]. From these two works, we can say graphene and their derivatives can be used as nanofiller to increase thermal and mechanical stability of polymer.

The degree of enhancement depends on processing method, distribution of graphene sheets into polymer matrices, type of functional groups on the surface of graphene and interaction of filler and matrices [12, 19]. The melt compounding approach for the production of graphene-based polymer blend nanocomposites possesses several advantages over the solution mixing especially in terms of bulk production, cost-effectiveness and viscosity controlled selective distribution of fillers in into the matrices [12].

2.3 Poly(3,4-Ethyldioxythiophene)/Graphene Nanocomposites

Xu and co-worker tested thermal stability by homogeneous dispersion of very low amount of graphene-based filler [21]. The nanocomposites using Poly (3, 4-ethyldioxythiophene) and sulfonated graphene were fabricated via situ polymerization then it was found that processing can be easily done in both aqueous and organic solvents. The physical properties of the fabricated nanomaterial were tested for various applications [18, 21]. The prepared polymer nanocomposite showed outstanding transparency, high electrical conductivity, and decent flexibility along with high thermal stability [21]. The conductivity of fabricated nanocomposites film deposited on PMMA is greater than that deposited on quartz owing to the proper templating of polymeric chains [21]. Moreover, the Poly (3,4-ethyldioxythiophene) and sulfonated graphene nanocomposites have high thermal stability and showed a very small weight loss up to 300 °C [18, 21]. Due to such a small weight loss even at high temperature this polymeric material can be used in production of underground oil supply pipes and also for other similar applications [18, 21]. In addition, the minute weight loss even at 300 °C makes this nanocomposite a promising candidate for aerospace and other engineering applications.

2.4 Epoxy/Graphene Nanocomposites

Polyepoxides, commonly called as epoxy resins, are a class of reactive prepolymers and polymers which comprise epoxide groups and extensively been used in the field of composite science and technology since last 50 years. To tune the properties of epoxy resins several strategies have been applied and a lot of literature for the same is available [22, 23]. For a specific application, graphene-based epoxy nanocomposites were prepared using in situ polymerization method and their electromagnetic interference shielding was examined in detail [22, 23]. It has been noted that the newly developed materials can be applicable for a wide frequency range. Nonetheless, the electromagnetic interference shielding effectiveness increased with increasing graphene loading and a critical optimization is a matter of further study

[23]. Thus, epoxy nanocomposites can be used as effective lightweight shielding materials for electromagnetic radiation in various devices and equipment [22, 23]. Recently, a graphene oxide incorporated epoxy composites were prepared and their thermal expansion was inspected using TGA/DSC instruments [24]. Interestingly, the used epoxy resin showed very weak thermal conductivity; however, the incorporation of graphene nanosheets showed a noteworthy development for the same [23, 24]. In an optimized result, it has been found that 5 wt% GO loaded epoxy resin showed nearly 4–5 times advanced thermal conductivity than that of the pure epoxy resin [25, 26]. Thus, graphene-based nanocomposites are promising thermal interface material for heat dissipation and they can be used for several applications [26].

2.5 Polyvinyl Alcohol/Graphene Nanocomposites

Polyvinyl alcohol and graphene oxide both are significantly soluble in water; therefore, Bao and co-workers produced graphene-based polyvinyl alcohol nanocomposites by incorporating graphene oxide into the polyvinyl alcohol matrix using water as solvent [18, 27]. They investigated thermo-mechanical properties of resulted materials and reported that mechanical performance of polymer nanocomposites much superior to that of pure PVA [18, 27]. This dramatic improvement in thermal and mechanical properties is owing to the homogeneous dispersion of graphene nanosheets into polyvinyl alcohol matrix at molecular level because of proper hydrogen bonding between graphene oxide and polyvinyl alcohol [25, 28]. The hydrogen bonding between graphene oxide and polyvinyl alcohol schematically is shown in Fig. 3 [29]. Similar work is performed by Zhao and co-workers to improve tensile strength of polyvinyl alcohol using exfoliated graphene nanosheets as nanofiller in facial aqueous solution [28]. In addition to this, Zhao and co-workers evaluated DSC thermogram of fabricated nanocomposites and noted very high degree of crystallization owing to the incorporation of graphene oxide nanosheets [30]. This work opens a new way for the fabrication of polymer nanocomposites using aqueous solutions instead of hazardous organic and ionic solvents. Moreover, this work very clearly explains the role of hydrogen bonding in the case of graphene-based polymer blends and impact of hydrogen bonding on the thermo-mechanical properties of polymer nanocomposites.

Bao and co-workers successfully incorporated graphene and graphene oxide into polyvinyl alcohol and made an extensive study on properties of nanocomposites and their mechanism for the property augmentations owing to the use of nanofiller [29, 31, 32]. It has been found that graphene oxide fashioned better dispersion and exfoliation while pure graphene sheet caused more property improvements including thermal stability, mechanical properties and electrical conductivity [29]. The mechanical strength of the graphene/graphene oxide nano-layers is attributed to be the fundamental cause for the enhancements in crystallinity and mechanical properties; the

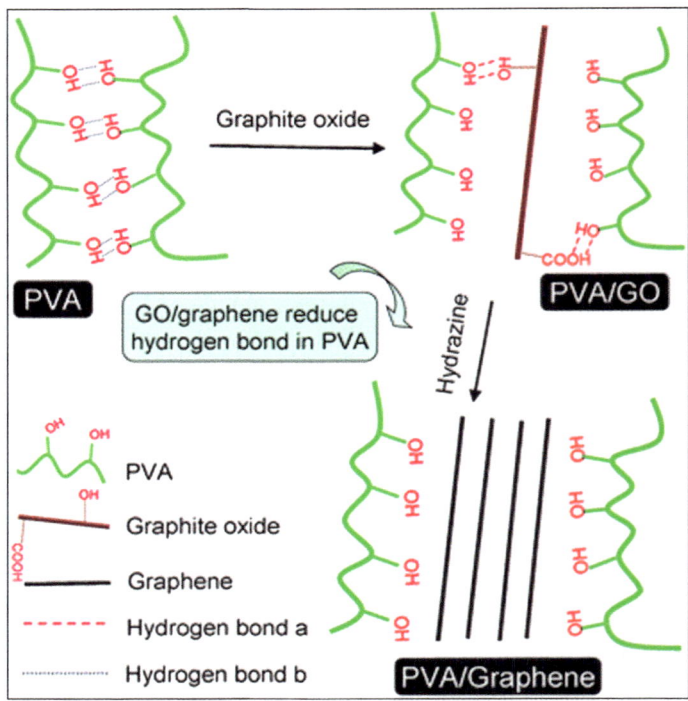

Fig. 3 PVA/GO nanocomposite showing hydrogen bonding between polymeric chains of PVA and graphene oxide. Reproduced with permission from Ref. [29]

hydrogen bond among the PVA molecules is the key factor to influence the glass transition temperatures [29]. Recently, Tiwari and co-workers reported a facile method for the bulk rGO production and to verify the applicability of the same they prepared rGO-epoxy nanocomposites [30]. They have reported a notable improvement in the thermo-mechanical and morphological properties of epoxy nanocomposites even at very small loading of rGO. In brief, the few-layered rGO-based Epoxy nanocomposites were obtained via solution blending method and subsequent hot pressing. They observed that rGO apparently filled in the interspaces of polymeric chains, which were helpful for the synchronously ~10–15% increment in thermo-mechanical properties of fabricated nanocomposites even with an eminently low loading of rGO [30]. The storage modulus and T_g at different temperatures of the studied nanocomposites is tabulated in Table 1. It seems quite difficult to incorporate every single example for the explanation purpose. A brief information about some notable works are presented in Table 2 for further reading, which also shows the effect of graphene and their derivatives on the properties of polymer blends.

Table 1 Storage modulus and glass transition temperature of prepared epoxy composites reinforced with rGO3	Sample code	30 °C (MPa)	At 60 °C	T_g (°C)
	Epoxy	2175	1190	95.3
	E_0.1rGO3	2450	2100	97.1
	E_0.2rGO3	2580	2330	98.8
	E_0.3rGO3	2732	2485	104.4

2.6 Highly Conductive Graphene Foam-Based Polymer Composite

As discussed in the above sections that the control of the nanostructure and the incorporation of nanoparticles to polymeric systems have led to morphological and functional property enhancements owing to the interfacial interaction between the polymer chains and nanoparticles. Such materials can cater the need of continuous requirements various industries and can our environmental issues [31]. The accessibility of newly studied nanoparticles such as graphene, graphene oxide, nanocellulose, metal oxide, carbon nanotubes, boron nitride and many more with astonishing properties have determined new and exciting possibilities for a continuous enlargement of polymers application in different fields [31]. However, the potentialities of these new materials are still strappingly reliant on the manufacturing processes and scaling-up for the common people and day-to-day applications [31]. Consequently, the purpose of this section of the chapter is to review the most recent progress happened in the field of fabrication of nanostructured polymers using functional nanomaterials to explore the state of art for polymer–particle interaction [31].

In this line, Jun and co-workers reported an interesting work-related incorporation of graphene into the poly (dimethylsiloxane). Herein, they explored influence of graphene sheet size on the electrical conductivity of interconnected graphene foam polymer composite along with their morphological and physical properties [31]. In this work, graphene oxide solution was produced from small flake graphite (SFG) of size 2–15 mm and large flake graphite (LFG) of less than 100 mm, respectively [31]. Each solution was used to produce 3D graphene oxide foam, which is subsequently heat-treated to produce reduced graphene oxide foam. The reduced graphene oxide foams are then infiltrated with poly (dimethylsiloxane) [PDMS] to produce graphene-PDMS (G-PDMS) nanocomposites. The in-plane electrical conductivity of the G-PDMS composite (0.4 wt%) from LFG reaches ~3.2 S/m, which is more than two orders of magnitude greater than that of G-PDMS (1.9 wt%, 1.4×10^{-2} S/m) from SFG [31]. This value is also four times greater than that of the G-PDMS composite prepared from mechanical mixing of 4 wt% reduced graphene oxide powder made from SFG with PDMS (4.2×10^{-5} S/m). The through-plane electrical conductivity of the investigated system followed the same trend for SFG and LFG [31]. This discloses that the interconnected graphene foam supplies more paths for the electron transfer inside the polymer than conventional graphene powder and the use of large-sized graphene sheets can significantly improve the electrical properties of G-PDMS.

Table 2 List of recently used GO/rGO/fGO-based polymer nanocomposite [32–59]

Matrices	Filler	Focused properties	References
Poly(vinyl chloride)	Graphene nanosheets	Thermo-mechanical properties	[32]
PANI/PMMA/POM/PPY	Reduced graphene oxide	Mechanical and conducting properties	[39]
Polystyrene/polypropylene	PP-g-rGO	Thermo-mechanical and surface properties	[36]
Silica/PP	Graphene oxide	Nanocompatibility and mechanical property	[37]
Polyolefin	PP-g-rGO	Thermo-mechanical properties	[38]
Nylon 6	Pristine graphene sheets	Electrical and mechanical properties	[40]
Polymethyl methacrylate	Expanded graphite	Electrical and mechanical properties	[41]
Poly(lactic acid)	Few-layers graphene	Thermo-mechanical and surface properties	[33]
Polyphenylene oxide/polyamide	Graphene oxide	Thermo-mechanical properties	[42]
Polypropylene	Graphene oxide	Effect of nanofiller on thermal properties and crystallization	[43]
Epoxy resin	Reduced graphene oxide	Mechanical properties	[44]
Polyurethane	Graphene oxide	Mechanical properties	[45]
Polymethyl methacrylate	Pure graphite powder	Conducting properties	[46]
Oligo(phenylene-ethynylene)/Polyethylene	Graphene nanosheets	Mechanical strength and conductivity	[47]
Polyethylene terephthalate	Graphene nanosheets	Mechanical properties	[48]
CS/Polyvinylpyrrolidone	Graphene oxide	Mechanical strength and flexibility	[49]
Polyvinyl alcohol	Reduced graphene oxide	Thermo-mechanical	
Polymethyl methacrylate	Reduced graphene oxide	Electrical conductivity	[50]
Polyimide	Phenyl functionalized graphene oxide	Electrical conductivity	[51]
Polystyrene	Graphene	Electrical conductivity	[1]
Chitosan	Cryomilled graphene	Mechanical properties	[53]
Polymethyl methacrylate	Graphene sheets	Thermal stability	[54]
Unsaturated polyester resin	Graphene nanosheets	Mechanical and dielectric properties	[55]

(continued)

Table 2 (continued)

Matrices	Filler	Focused properties	References
Isotactic polypropylene	Graphene nanosheets	Thermal and electrical conductivity	[34]
Polyethylene terephthalate	Graphene nanosheets	Electrical conductivity	[56]
Polystyrene	Graphene nanosheets	Thermal stability	[57]
Polyvinyl alcohol	Tryptophan functionalized graphene	Thermo-mechanical properties	[58]
Polystyrene	Graphene-polystyrene nanoparticles	Thermo-mechanical properties	[59]
Polymethyl methacrylate	Chemically modified graphene	Thermo-mechanical properties	[35]

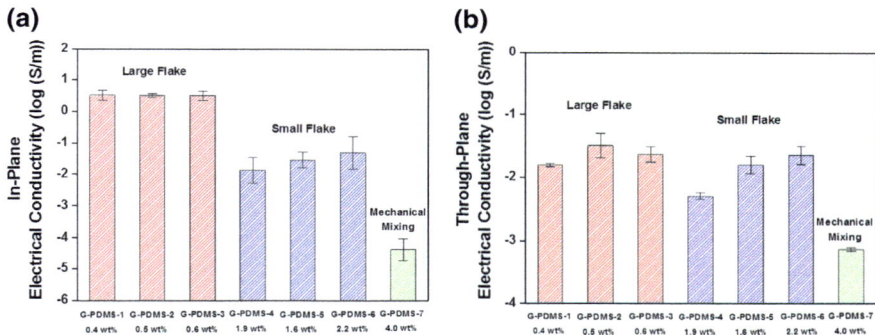

Fig. 4 **a** In-plane electrical conductivity (sI) and **b** through-plane electrical conductivity (sT) of G-PDMS composites produced from LFG and SFG, respectively. G-PDMS-7 was produced by mechanical mixing of reduced graphene oxide powder produced from SFG with PDMS. Reproduced with permission from Ref. [31]

This work further implies, that the use of two types (small and large size) of flake graphite provides specific path for the electronic conduction within the polymer matrices (Fig. 4).

2.7 Polyurethane/Polypropylene Composites with Selectively Distributed Graphene

Selective dispersion and controlled surface manipulation nanoparticles as filler in the polymeric systems (in neat polymer, blends and composites) greatly affects thermodynamic and kinetic properties of fabricated materials [60]. Though, homogeneous dispersion of nanofiller in the polymer matrices is one of the great obstacles in the path commercialization of several graphene-based polymer nanocomposites [60]. An expedient and operative strategy to produce polymeric materials with the co-continuous structure (commonly known as the double-percolation structure) is the blending of two immiscible polymer components using different methodologies [32]. To solve dispersion issues of nanofillers in polymer matrices, Lan and co-workers recently reported the electrically conductive nanocomposites comprising of different loading (0.05–1.5 wt%) of reduced graphene oxide in the thermoplastic polypropylene/polyurethane [33, 34] matrices with a fine co-continuous structure were invented by melt compounding and solution-flocculation methodologies [60]. For the same, Lan and co-workers evaluated both thermodynamic and kinetic properties of prepared nanocomposites and provided very strong support for property enhancement and dispersion of graphene nanosheets onto the interfaces of polymer components [60]. They also verified theoretical results using optical microscope and field emission scanning electron microscopy [60]. The uniform dispersion of reduced graphene oxide in the polymer matrices was confirmed by wide-angle X-ray diffraction patterns and transmission electron microscope analysis [60]. Moreover, a very low percolation threshold of 0.054 wt% was achieved owing to high conductivity of reduced graphene oxide and promising double-percolation consequence. The tensile strength and elongation at break of the composites with reduced graphene oxide loading of only 0.5 wt% were enhanced by 341.9 and 354.3%, respectively [60]. Thus, we can say the work performed by Lan and co-workers provides a recommendation for a well-organized, simplistic fabrication technique of reduced graphene oxide-based conductive polymer nanocomposites with high electrical conductivity and enhanced thermo-mechanical properties [60]. The authors believed that the reported method can also be useful to manufacture many other polymer nanocomposites reinforced with 2D nanomaterials [60, 61]. Therefore, this section of chapter discloses the employment of efficiently commercial techniques for formulating graphene oxide/reduced graphene oxide-based nanocomposites, which can bridge the gap between nanoscience and nanotechnology [60]. The reported strategy can also be used for the production of similar nanomaterials using 3D printing techniques (Fig. 5).

2.8 Graphene an Ideal Platform Catalytic Sites for Oxygen Reduction

Amplified production of green energy intended at satisfying the world's increasing energy demand while dropping greenhouse gaseous emissions is a foremost challenge [60, 35, 62]. Effective oxygen reduction reaction via catalytic assistance is principal for metal-air batteries and fuel cells, and these two efficient technologies having theoretical specific energies enough to power bikes, cars, and other vehicles in real-time use [62]. Although the polymer electrolyte membrane-based fuel cell technology is presently more advanced, its need Pt-based catalysts are a long-term

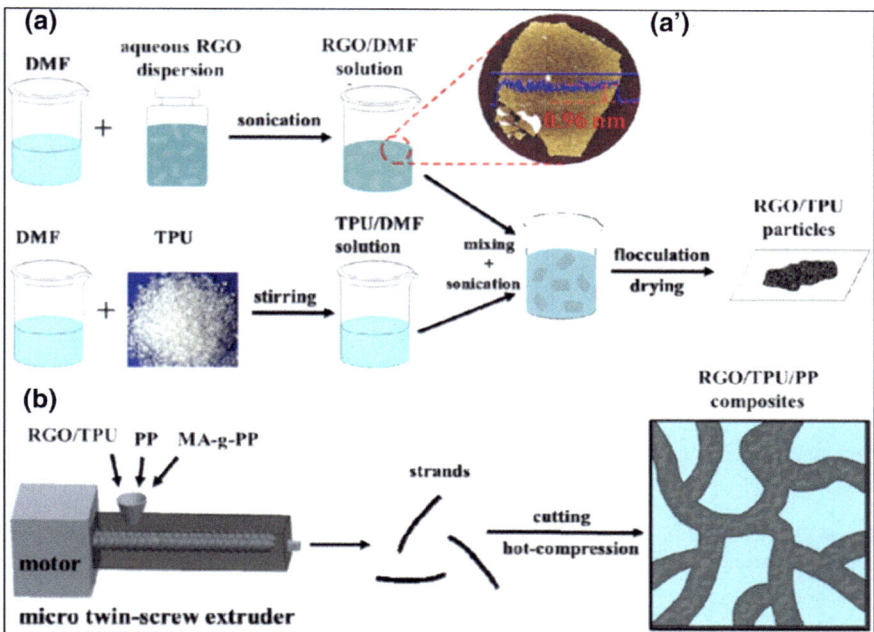

Fig. 5 Schematic illustration of the process for the preparation of reduced graphene oxide/polyurethane/polypropylene composites by **a** solution-flocculation and **b** melt-mixing, **a′** is the non-tapping-mode AFM image of reduced graphene oxide. Reproduced with permission from Ref. [32]

issue [62]. Moreover, it is well known that Pt is typically used at the cathode due to slow and effective oxygen reduction reaction kinetics. This is why a considerable inducement to search for Earth-abundant materials to catalyze such chemical reactions [63]. To improve oxygen reduction efficiency, several strategies have been employed, and a significant progress has been achieved [63]. To achieve the same, a groundbreaking work carried out by Zitolo and co-workers where they identified the catalytic sites for oxygen reduction in iron and nitrogen-doped graphene nanosheets [63]. In this particular work, researchers created unique Fe–N–C nanomaterials quasi-free of crystallographic iron structures after argon or ammonia pyrolysis [62]. These nanomaterials display approximately indistinguishable Mössbauer spectra and identical X-ray absorption near-edge spectroscopy (XANES) spectra, skimpy the same Fe-cantered moieties [63]. However, the excellent higher activity and basicity of ammonia-pyrolyzed Fe–N–C nanomaterials proves that the turnover frequency of Fe-centered moieties be contingent on the physicochemical possessions of the assistance. They carried out careful XANES analysis of the fabricated samples and reported that two FeN_4 porphyrinic architectures with different molecular oxygen adsorption modes were then recognized. These porphyritic

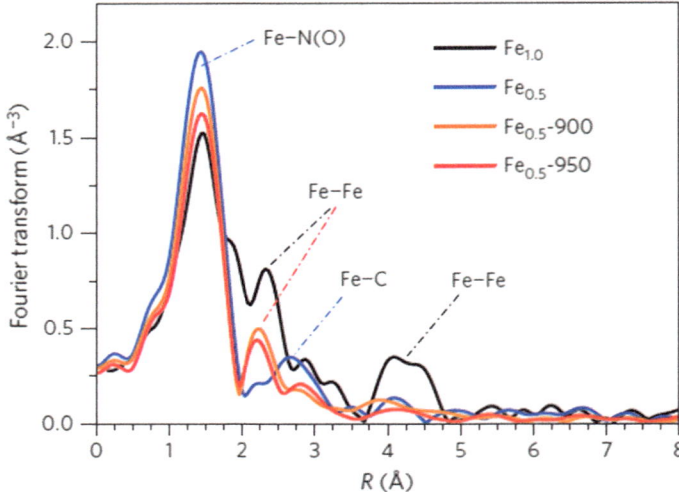

Fig. 6 Fourier transforms of the experimental EXAFS spectra of fabricated Fe–N–C nanomaterials showing generated catalytic active site. Reproduced with permission from Ref. [63]

systems are not effortlessly combined in graphene nanosheets, in contrast with Fe-centered moieties assumed hitherto for pyrolyzed Fe–N–C nanomaterials [63]. This discovery opens many possibilities for the application of nanosystem consisting Fe–N–C moiety along with graphene skeleton. Moreover, such new discoveries opened the path to bottom-up synthesis approaches and studies on the site-support interactions which not explored much in details especially for the graphene and similar structures [62, 63]. For the detailed story about this work, readers may see original work of Prof. Zitolo et al. [63]. After this groundbreaking discovery, several works for the various oxidation-reduction processes have been investigated for the similar applications with better efficacies [63]. The catalytic studies of graphene and their derivates are opening several unique pathways for electron and holes transfer and seem very valuable to produce renewable energies. For instance, a huge research work around the globe is going on for catalysis of oxygen reduction and water splitting and one of them discussed in the next section [63] (Fig. 6).

2.9 Catalysis for the Oxygen Reduction and Water Splitting

The imminent global energy catastrophe has provoked intense research work on energy storage systems and conversion as a source of renewable energy [62, 63]. The electrocatalytic oxygen reduction reaction (ORR), hydrogen evolution reaction (HER), and oxygen evolution reaction (OER) are playing important parts in numerous renewable energy technologies, including water splitting and fuel cells [62, 63].

Fig. 7 Selected FESEM images of newly prepared nanomaterials **b–c** ZIF-67, **d–e** N/Co-doped PCP, and **f–g** N/Co-doped PCP//NRGO (the corresponding EDX spectra). Reproduced with permission from Ref. [64]

Thus, further advancement in electrocatalysts research for ORR, OER, and HER is crucial for these technologies [63]. To increase efficacy and practical validity of such electrochemical reactions, several materials including metallic and non-metallic materials have been identified and tested [63, 64]. In a recent work, Chen and co-workers reported a new nanohybrid electrocatalyst containing of atomic N-doped graphene/cobalt embedded with highly porous carbon polyhedron (say N/Co-doped PCP//NRGO composite). They synthesized this engineered nanomaterial through a facile and one step pyrolysis of graphene oxide-supported cobalt-based zeolitic imidazolate frameworks [64]. The chemical and physical properties of the prepared samples were investigated using different spectroscopic and micrographic techniques. The characteristic FESEM images of same are presented in the Fig. 7.

The high surface area of the prepared nanomaterials can be easily seen in the Fig. 7 along with well-defined cubic shape of the nanoparticles [64]. The extraordinary topographies of the porous nanostructures, N/Co-doping effect, introduction of NRGO, and good contact between N/Co-doped PCP and NRGO seems to be mainly responsible for high catalytic affectivity [64, 36]. It has been investigated that as-synthesized nanohybrid shows outstanding electrocatalytic actions and kinetics for oxygen reduction reaction only in the basic media, which associates positively with those of the Pt/C catalyst, together with greater durability, a four-electron pathway, and exceptional methanol tolerance [64]. This nanohybrid, likewise, displays great performance for H_2 evolution response, offering a very low onset overpotential of 58 mV and stable current density of 10 mA cm^{-2} at 229 mV in acid conditions. Moreover, such a hybrid system can present good catalytic performance for oxygen evolution process (a small overpotential of 1.66 V for 10 mA cm^{-2} current density as established by the researchers). Interestingly, Chen and co-workers reported a dual-active-site reaction mechanism originating from synergic possessions between N/Co-doped PCP and NRGO which indeed responsible for the excellent performance of the invented nanohybrid [63]. Therefore, we can say that work reported by Chen et al. offers a very attractive catalyst material for large-scale fuel cells, water splitting technologies, and related device fabrication [63]. Thus, peculiar surface morphology, simple composition of nanohybrid and efficient efficiency in basic reaction conditions makes this investigation unique.

2.10   3D Printing-Based CNT-Graphene Hybrid Nanocomposite

As we all well aware about 3D (three-dimensional) printing revolution, the process to build a 3D object or body on the basis of computer-aided design, typically by continually adding material layer by layer, this is why it is also called 3D additive manufacturing. Nowadays, the 3D printing technologies are greatly helping to nanotechnology in creating specific internal morphologies for the desired applications. This allows for larger design, flexibility and enables improved performance, often at reduced weight. The integration of fused deposition modelling (FDM) tool with 3D printing technology has pervaded most into science, education, and various domestic applications. In FDM, the feedstock is a thermoplastic polymer filament, which is heated above its glass transition temperature (T_g) and extruded through a movable nozzle in the X–Y plane to form a 3D structure by layer-wise addition [64, 65]. While almost every base material, that is, metals, ceramics, and polymers, can be 3D printed into complex shapes, nowadays a strong demand exists for 3D printable materials having additional functionality [64–66]. Thus, we can say that improvement in FDM technologies will add additional wing into 3D printing for more diverse applications.

Fused deposition modelling (FDM) is limited by the availability of application-specific functional materials. Here, we illustrate printing of non-conventional polymer nanocomposites (CNT and graphene-based polybutylene terephthalate–PBT) on a commercially available desktop 3D printer leading toward printing of electrically conductive structures [37, 38]. The printability, electrical conductivity, and mechanical stability of the polymer nanocomposites before and after 3D printing were evaluated. The results show that 3D printed PBT/CNT objects have better conductive and mechanical properties and a better performance than 3D printed PBT/graphene structures [64, 65]. In addition to that, printing more than one material (multi-materials) and challenges in using abrasive conductive fillers (i.e., CNT and graphene) are also discussed. Overall, this study demonstrates that a commercially available desktop 3D printer can be used to fabricate low-cost functional objects [64–66] (Fig. 8).

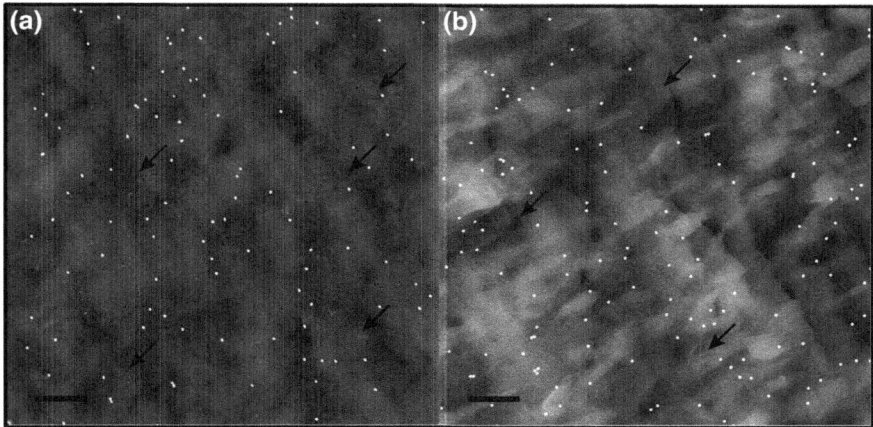

Fig. 8 STEM images of 3D printed PBT/CNT and PBT/G composites. White dots represent the gold markers used for tracking. Scale bar: 500 nm. Reproduced with permission from Ref. [65]

2.11 Summary and Perspectives

In summary, this book chapter has reviewed and summaries the processing behavior of graphene-based nanomaterials especially polymer nanocomposites and blends. These newly studied materials have expanded their acceptance within 2–5 years owing to their wide range applications. Unquestionably, nanocomposites based on polymeric systems gained momentous attention after one of the first industrial applications when Toyota testified the use of nanoclays in Nylon-6 matrix in 1993. From this chapter, it is very clear that the control of the nanostructure of polymers and the incorporation of nanoparticles has led to structural and functional property augmentations in several polymeric systems as a material answer to continuous requirements for the application in different industrial areas. The availability of new nanoparticles with extraordinary physical and chemical properties (graphene, graphene oxide, nanocellulose, metal oxide, carbon nanotubes, boron nitride including many inorganic nanomaterials) have determined new and exciting possibilities for a continuous enlargement of polymer markets owing to the growing demands. Graphene-based nanocomposites for water oxidation, as catalyst, and for thermo-mechanical applications have been briefed for the readers and several new citations have been added for the further readings. The use of graphene and derivatives as nanofiller can be an excellent alternative for the recycling various engineering polymers is highly expected in the coming future. Moreover, we have also discussed some real-times applications of graphene-based materials for high-performance polymer blends, nanocomposites, 3D printing, and water purification. The continuous marketing of polymer-based nanotechnologies will be certainly associated to combined research efforts on processing-structure-properties relationships and on safe procedures for manipulation and disposal of nanoparticles.

Acknowledgements Dr. Nannan is grateful to Key Laboratory of New Processing Technology for Nonferrous Metals and Materials, Ministry of Education, School of Resources, Environment and Materials, Guangxi University, Nanning, China for all kinds of support in the articulation of this chapter.

References

1. Tiwari, S.K., Kumar, V., Huczko, A., Oraon, R., Adhikari, A.D., Nayak, G.C.: Magical allotropes of carbon: prospects and applications. Crit. Rev. Solid State Mater. Sci. **41**, 317 (2016)
2. Wang, S., Morris, W., Liu, Y., McGuirk, C.M., Zhou, Y., Hupp, J.T., Farha, O.K., Mirkin, C.A.: Surface-specific functionalization of nanoscale metal–organic frameworks. Angew. Chem. Int. Ed. **54**, 14742 (2015)
3. Liu, G., Jin, W., Xu, N.: Graphene-based membranes. Chem. Soc. Rev. **44**, 5030 (2015)
4. Li, X., Zhi, L.: Graphene hybridization for energy storage applications. Chem. Soc. Rev. **47**, 3216 (2018)
5. Dimiev, A.M., Eigler, S.: Mechanism of formation and chemical structure of graphene oxide. In: Graphene Oxide: Fundamentals and Applications, pp. 36–84. 1st edn. New York (2016)
6. Novoselov, K.S., Fal, V.I., Colombo, L., Gellert, P.R., Schwab, M.G., Kim, K.: A roadmap for graphene. Nature **490**, 192 (2012)
7. Cheng, Y., Lu, S., Zhang, H., Varanasi, C.V., Liu, J.: Synergistic effects from graphene and carbon nanotubes enable flexible and robust electrodes for high-performance supercapacitors. Nano Lett. **12**, 4211 (2018)
8. Zurutuza, A., Marinelli, C.: Challenges and opportunities in graphene commercialization. Nat. Nanotechnol. **9**, 730 (2014)
9. Tiwari, S.K., Mishra, R.K., Ha, S.K., Huczko, A.: Evolution of graphene oxide and graphene: from imagination to industrialization. ChemNanoMat **4**, 620 (2018)
10. Higgins, D., Zamani, P., Yu, A., Chen, Z.: The application of graphene and its composites in oxygen reduction electrocatalysis: a perspective and review of recent progress. Energy Environ. Sci. **9**, 390 (2016)
11. Fan, Z., Pereira, L.F.C., Hirvonen, P., Ervasti, M.M., Elder, K.R., Donadio, D., Ala-Nissila, T., Harju, A.: Thermal conductivity decomposition in two-dimensional materials: application to graphene. Phys. Rev. B **95**, 144309 (2017)
12. Tiwari, S.K., Verma, K., Saren, P., Oraon, R., De Adhikari, A., Nayak, G.C., Kumar, V.: Manipulating selective dispersion of reduced graphene oxide in polycarbonate/nylon 66 based blend nanocomposites for improved thermo-mechanical properties. RSC Adv. **7**, 32731 (2017)
13. Tiwari, S.K., Hatui, G., Oraon, R., De Adhikari, A., Nayak, G.C.: Mixing sequence driven controlled dispersion of graphene oxide in PC/PMMA blend nanocomposite and its effect on thermo-mechanical properties. Curr. Appl. Phys. **17** (2017)
14. Bai, H., Li, C., Wang, X., Shi, G.: On the gelation of graphene oxide. J. Phys. Chem. C **115**(13), 5545–5551 (2011)
15. Zhang, Y.I., Zhang, L., Zhou, C.: Review of chemical vapor deposition of graphene and related applications. Acc. Chem. Res. **46**, 2339 (2013)
16. Liu, P., Cottrill, A.L., Kozawa, D., Koman, V.B., Parviz, D., Liu, A.T., Strano, M.S.: Emerging trends in 2D nanotechnology that are redefining our understanding of "Nanocomposites". Nano Today **21**, 40 (2018)
17. Cao, Y., Zhang, J., Feng, J., Wu, P.: Compatibilization of immiscible polymer blends using graphene oxide sheets. ACS Nano **5**, 5927 (2011)
18. Das, T.K., Prusty, S.: Graphene-based polymer composites and their applications. Polym.-Plast. Technol. Eng. **52**, 331 (2013)

19. Lee, Y.R., Raghu, A.V., Jeong, H.M., Kim, B.K.: Properties of waterborne polyurethane/functionalized graphene sheet nanocomposites prepared by an in situ method. Macromol. Chem. Phys. **210**, 1254 (2009)
20. Liang, J., Xu, Y., Huang, Y., Zhang, L., Wang, Y., Li, Y., Ma, F., Guo, T., Chen, Y.J.: Infrared-triggered actuators from graphene-based nanocomposites. Phys. Chem. C, **113**, 9927 (2009)
21. Xu, Y., Wang, Y., Liang, J., Huang, Y., Ma, Y., Wan, X., Chen, Y.: A hybrid material of graphene and poly (3, 4-ethyldioxythiophene) with high conductivity, flexibility, and transparency. Nano Res. **2**, 348 (2009)
22. Wang, S., Tambraparni, M., Qiu, J., Tipton, J., Dean, D.: Thermal expansion of graphene composites. Macromolecules **42**, 5255 (2009)
23. Yu, J., Lu, K., Sourty, E., Grossiord, N., Koning, C.E., Loos, J.: Characterization of conductive multiwall carbon nanotube/polystyrene composites prepared by latex technology. Carbon **45**, 2903 (2007)
24. Kuila, T., Srivastava, S.K., Bhowmick, A.K.: Rubber/LDH nanocomposites by solution blending. J. Appl. Polym. Sci. **111**, 641 (2009)
25. Yu, A., Ramesh, P., Itkis, M.E., Bekyarova, E., Haddon, R.C.: Graphite nanoplatelet−epoxy composite thermal interface materials. J. Phys. Chem. C **111**, 7569 (2007)
26. Yu, A., Ramesh, P., Sun, X., Bekyarova, E., Itkis, M.E., Haddon, R.C.: Enhanced thermal conductivity in a hybrid graphite nanoplatelet–carbon nanotube filler for epoxy composites. Adv. Mater. **20**, 4744 (2008)
27. Liang, J., Huang, Y., Zhang, L., Ma, Y., Wang, Y., Guo, T., Chen, Y.: Molecular-level dispersion of graphene into poly (vinyl alcohol) and effective reinforcement of their nanocomposites. Adv. Fun. Mater. **19**, 2302 (2009)
28. Ramanathan, T., Abdala, A.A., Stankovich, S., Dikin, D.A., Herrera-Alonso, M., Piner, R.D., Adamson, D.H., Schniepp, H.C., Chen, X.R., Ruoff, R.S., Nguyen, S.T.: Functionalized graphene sheets for polymer nanocomposites. Nat. Nanotechnol. **3**, 327 (2008)
29. Bao, C., Guo, Y., Song, L., Hu, Y.J.: Poly (vinyl alcohol) nanocomposites based on graphene and graphite oxide: a comparative investigation of property and mechanism. Mater. Chem. **21**, 13950 (2011)
30. Tiwari, S.K., Nimbalkar, A.S., Hong, C.K., Ha, S.K.: A green route for quick and kilogram production of reduced graphene oxide and their applications at low loadings in epoxy resins. ChemistrySelect **4**, 1274 (2019)
31. Jun, Y.S., Sy, S., Ahn, W., Zarrin, H., Rasen, L., Tjandra, R., Amoli, B.M., Zhao, B., Chiu, G., Yu, A.: Highly conductive interconnected graphene foam based polymer composite. Carbon **95**, 658 (2015)
32. Lan, Y., Liu, H., Cao, X., Zhao, S., Dai, K., Yan, X., Guo, Z.: Electrically conductive thermoplastic polyurethane/polypropylene nanocomposites with selectively distributed graphene. Polymer **97**, 19 (2016)
33. Kuilla, T., Bhadra, S., Yao, D., Kim, N.H., Bose, S., Lee, J.H.: Recent advances in graphene based polymer composites. Prog. Polym. Sci. **35**, 1375 (2010)
34. Hu, H., Wang, X., Wang, J., Wan, L., Liu, F., Zheng, H., Chen, R., Xu, C.: Preparation and properties of graphene nanosheets–polystyrene nanocomposites via in situ emulsion polymerization. Chem. Phys. Lett. **484**, 253 (2010)
35. Chen, L., Chai, S., Liu, K., Ning, N., Gao, J., Liu, Q., Chen, F., Fu, Q.: Enhanced epoxy/silica composites mechanical properties by introducing graphene oxide to the interface. ACS Appl. Mater. Interfaces **4**, 4404 (2012)
36. You, F., Wang, D., Li, X., Liu, M., Dang, Z.M., Hu, G.H.: Synthesis of polypropylene-grafted graphene and its compatibilization effect on polypropylene/polystyrene blends. J. Appl. Polym. Sci. **131**, 13 (2014)
37. Luo, F., Chen, L., Ning, N., Wang, K., Chen, F., Fu, Q.: Interfacial enhancement of maleated polypropylene/silica composites using graphene oxide. J. Appl. Polym. Sci. **125**, E357 (2012)
38. Cao, Y., Feng, J., Wu, P.: Polypropylene-grafted graphene oxide sheets as multifunctional compatibilizers for polyolefin-based polymer blends. J. Mater. Chem. **22**, 15005 (2012)

39. Mohan, V.B., Jayaraman, K., Bhattacharyya, D.: Hybridization of graphene-reinforced two polymer nanocomposites. Inter. J. Smart Nano Mat. **7**, 201 (2016)
40. Pan, Y.X., Yu, Z.Z., Ou, Y.C., Hu, G.H.: A new process of fabricating electrically conducting nylon 6/graphite nanocomposites via intercalation polymerization. J. Polym. Sci. Part B: Poly. Phys. **38**, 1633 (2000)
41. Wang, W.P., Liu, Y., Li, X.X., You, Y.Z.: Synthesis and characteristics of poly (methyl methacrylate)/expanded graphite nanocomposites. J. Appl. Polym. Sci. **100**, 1431 (2006)
42. Chieng, B., Ibrahim, N., Yunus, W., Hussein, M., Then, Y., Loo, Y.: Effects of graphene nanoplatelets and reduced graphene oxide on poly (lactic acid) and plasticized poly (lactic acid): a comparative study. Polymers **6**, 2246 (2014)
43. Yuan, B., Bao, C., Song, L., Hong, N., Liew, K.M., Hu, Y.: Preparation of functionalized graphene oxide/polypropylene nanocomposite with significantly improved thermal stability and studies on the crystallization behavior and mechanical properties. Chem. Eng. J. **237**, 420 (2014)
44. Tang, L.C., Wan, Y.J., Yan, D., Pei, Y.B., Zhao, L., Li, Y.B., Wu, L.B., Jiang, J.X., Lai, G.Q.: The effect of graphene dispersion on the mechanical properties of graphene/epoxy composites. Carbon **60**, 27 (2013)
45. Cai, D., Yusoh, K., Song, M.: The mechanical properties and morphology of a graphite oxide nanoplatelet/polyurethane composite. Nanotechnology **20**, 085712 (2009)
46. Yasmin, A., Luo, J.J., Daniel, I.M.: Processing of expanded graphite reinforced polymer nanocomposites. Compos. Sci. Technol. **66**, 1189 (2006)
47. Pullicino, E., Zou, W., Gresil, M., Soutis, C.: The effect of shear mixing speed and time on the mechanical properties of GNP/epoxy composites. Appl. Compos. Mater. **24**, 311 (2017)
48. Ma, H.L., Zhang, Y., Hu, Q.H., He, S., Li, X., Zhai, M., Yu, Z.Z.: Enhanced mechanical properties of poly (vinyl alcohol) nanocomposites with glucose-reduced graphene oxide. Mater. Lett. **102**, 18 (2013)
49. Kumar, S.K., Castro, M., Saiter, A., Delbreilh, L., Feller, J.F., Thomas, S., Grohens, Y.: Development of poly (isobutylene-co-isoprene)/reduced graphene oxide nanocomposites for barrier, dielectric and sensing applications. Mater. Lett. **96**, 112 (2013)
50. Wang, J., Hu, H., Wang, X., Xu, C., Zhang, M., Shang, X.: Preparation and mechanical and electrical properties of graphene nanosheets–poly (methyl methacrylate) nanocomposites via in situ suspension polymerization. J. Appl. Polym. Sci. **122**, 1871 (2011)
51. Guo, J., Ren, L., Wang, R., Zhang, C., Yang, Y., Liu, T.: Water dispersible graphene noncovalently functionalized with tryptophan and its poly (vinyl alcohol) nanocomposite. Comp. Part B: Eng. **42**, 2135 (2011)
52. Fang, M., Wang, K., Lu, H., Yang, Y., Nutt, S.: Covalent polymer functionalization of graphene nanosheets and mechanical properties of composites. J. Mater. Chem. **19**, 7105 (2009)
53. Aldosari, M., Othman, A., Alsharaeh, E.: Synthesis and characterization of the in situ bulk polymerization of PMMA containing graphene sheets using microwave irradiation. Molecules **18**, 3167 (2013)
54. Swain, S.: Synthesis and characterization of graphene based unsaturated polyester resin composites. Trans. Electr. Electr. Mater. **14**, 58 (2013)
55. Zhang, H.B., Zheng, W.G., Yan, Q., Yang, Y., Wang, J.W., Lu, Z.H., Ji, G.Y., Yu, Z.Z.: Electrically conductive polyethylene terephthalate/graphene nanocomposites prepared by melt compounding. Polymer **51**, 1196 (2010)
56. Yang, Y.K., He, C.E., Peng, R.G., Baji, A., Du, X.S., Huang, Y.L., Xie, X.L., Mai, Y.W.: Noncovalently modified graphene sheets by imidazolium ionic liquids for multifunctional polymer nanocomposites. J. Mater. Chem. **22**, 5675 (2012)
57. Potts, J.R., Lee, S.H., Alam, T.M., An, J., Stoller, M.D., Piner, R.D., Ruoff, R.S.: Thermomechanical properties of chemically modified graphene/poly (methyl methacrylate) composites made by in situ polymerization. Carbon **49**, 2623 (2011)
58. Quan, H., Zhang, B.Q., Zhao, Q., Yuen, R.K., Li, R.K.: Facile preparation and thermal degradation studies of graphite nanoplatelets (GNPs) filled thermoplastic polyurethane (TPU) nanocomposites. Compos. Pt A. Appl. Sci. Manuf. **40**, 1513 (2009)

59. Liu, C., Wang, Z., Huang, Y.A., Xie, H., Liu, Z., Chen, Y., Lei, W., Hu, L., Zhou, Y., Cheng, R.: One-pot preparation of unsaturated polyester nanocomposites containing functionalized graphene sheets via a novel solvent-exchange method. RSC Adv. **3**, 22388 (2013)
60. Mu, Q., Feng, S.: Thermal conductivity of graphite/silicone rubber prepared by solution intercalation. Thermochim. Acta **462**, 75 (2007)
61. Choi, W., Lahiri, I., Seelaboyina, R., Kang, Y.S.: Synthesis of graphene and its applications: a review. Crit. Rev. Solid State Mater. Sci. **35**, 71 (2010)
62. Lan, Y., Liu, H., Cao, X., Zhao, S., Dai, K., Yan, X., Zheng, G., Liu, C., Shen, C., Guo, Z.: Electrically conductive thermoplastic polyurethane/polypropylene nanocomposites with selectively distributed graphene. Polymer **97**, 19 (2016)
63. Zitolo, A., Goellner, V., Armel, V., Sougrati, M.T., Mineva, T., Stievano, L., Fonda, E., Jaouen, F.: Identification of catalytic sites for oxygen reduction in iron-and nitrogen-doped graphene materials. Nat. Mater. **14**, 937 (2015)
64. Hou, Y., Wen, Z., Cui, S., Ci, S., Mao, S., Chen, J.: An advanced nitrogen-doped graphene/cobalt-embedded porous carbon polyhedron hybrid for efficient catalysis of oxygen reduction and water splitting. Adv. Funct. Mater. **25**, 882 (2015)
65. Gnanasekaran, K., Heijmans, T., Van Bennekom, S., Woldhuis, H., Wijnia, S., de With, G., Friedrich, H.: 3D printing of CNT-and graphene-based conductive polymer nanocomposites by fused deposition modeling. App. Mater. Today. **9**, 28 (2017)
66. Wang, X., Jiang, M., Zhou, Z., Gou, J., Hui, D.: 3D printing of polymer matrix composites: a review and prospective. Comp. Pt B: Eng. **110**, 442–458 (2017)

Graphene and Its Derivatives for Secondary Battery Application

Anukul K. Thakur, Mandira Majumder and Shashi B. Singh

Abstract Graphene has prophesied itself as a potentially promising greenhorn with unique electronic properties. Attention toward graphene-based material is mainly attributed to its outstanding electrical, mechanical, thermal properties besides very large specific surface area and the tenability that can be achieved for various properties through functionalization and/or moderation. Due to the various unique properties possessed by the graphene sheets including the ease of synthesis and provision for surface functionalization, graphene and materials derived from graphene have been exhibiting great potential in the field of energy storage. This chapter accounts for a brief introduction to the graphene material followed by a brief discussion on the recent advances in the field of its derivatives. This chapter also accounts for the application of graphene and graphene-derived materials in the field of energy storage specifically batteries in various forms like lithium-ion, sodium-ion, lithium-air, and lithium-sulfur batteries.

Keywords Secondary batteries · Li-ion · Sodium-ion · Sulphur batteries

1 Introduction

The electricity production cost from the various renewable energy sources has significantly cut down the cost of electricity production. In recent years, the significance of the various sources of renewable energy has augmented eventually reducing the use of the conventional sources of energy [1–5]. Most of the renewable sources of energy like wind energy or the solar energy are irregular and hence need highly efficient storage devices to be used when required [5]. However, storage of the produced electric energy on a large scale still remains a major source of concern. Storage of

A. K. Thakur (✉) · S. B. Singh
Department of Physics, Indian Institute of Science Education and Research Berhampur, Berhampur 760010, India
e-mail: anukulphyiitd@gmail.com

M. Majumder
Nanostructured Composite Materials Laboratory, Department of Physics, Indian Institute of Technology (Indian School of Mines) Dhanbad, Dhanbad 826004, India

© Springer Nature Switzerland AG 2019
S. Sahoo et al. (eds.), *Surface Engineering of Graphene*, Carbon Nanostructures,
https://doi.org/10.1007/978-3-030-30207-8_3

energy is mainly subjugated by the electrochemical devices such as supercapacitors, batteries, and fuel cells [6–10]. Among these electrochemical energy devices, batteries are considered one of the most attractive energy storage and conversion devices due to their extremely high potential energy density [11–13]. Till date, advanced batteries have been occupying a significant position in meeting the energy storage requirements in the field of electronics. Mostly owed to their long cyclic life and substantially high energy density accompanied by a good coulombic efficiency, they have become indispensable [14–16]. In order to fulfill the ever-varying requirements related to energy storage, it becomes mandatory to keep on working on the various active materials as well as various parts of the battery like cathode and anode that would result in improved performance when used in a battery [17, 18]. Mostly, the present research on the battery is concentrated on the development of novel electrode materials possessing requisite properties such as high rate capability, high energy density, along with longer cycle life.

From the preliminary times of the application of batteries as an energy storage mode, mainly graphene has been executed as the primary electrode material. The properties like large Young's modulus, large theoretical surface area, high mobility, large thermal conductivity, and substantially good electrical conductivity are better than other carbon materials and proved to be very useful in making its application possible as an electrode material (see Table 1) [19–21]. Graphene is a laminar material which is typically one atom thick consisting of carbon atoms arranged in a hexagonal pattern with unique electronic and electrochemical properties. Since the very first report of its synthesis by the peel-off strategy using "Scotch tape" in 2004, research on graphene has developed as among the most active research

Table 1 Various properties of graphene compared to the other carbonaceous materials [19]

	Graphene	Carbon nanotube	Fullerene	Graphite
Dimensions	2	1	0	3
Hybridization	sp^2	Mostly sp^2	Mostly sp^2	sp^2
Hardness	Highest (for single layer)	High	High	High
Tenacity	Flexible, elastic	Flexible, elastic	Elastic	Elastic, non-elastic
Experimental surface area (m^2 g^{-1})	~1500	~1300	80–90	~10–20
Electrical conductivity (S cm^{-1})	~2000	Structure dependent	10^{-10}	Anisotropic: 2–3 $\times 10^{4a}$, 6[b]
Thermal conductivity (W m^{-1} K^l)	4840–5300	3500	0.4	Anisotropic: 1500–2000[a], 5–10[b]

Reprinted from Ref. [19], with permission from nature publication 2015
[a]A direction, [b]direction

fields in material science [22–24]. During the past few years, graphene has been intensively studied by the material scientists, chemists, physicists, and engineers. Graphene can be synthesized easily and duly modified by implementing different methods [22]. The modified graphene also is known as graphene derivative results in enhanced electrochemical properties and energy storage properties [25–27]. The improved properties established in case of the graphene and its various derivatives prove to be very promising as an electrode material for battery applications. Hence, the implementation of graphene and its various derivatives as environment-friendly, and as well as high-efficiency electrode materials for batteries is greatly favored. This chapter accounts for a focused discussion on the synthesis of graphene and its derivatives along with a discussion on its battery application.

2 Fundamentals of Battery

Apart from the combustion, chemical energy can be extracted by chemical reactions too. Batteries primarily rely on the chemical reactions to convert chemical energy to electrical energy. A battery mainly consists of three parts: cathode, anode (electrodes), and electrolyte. In case of an electrolytic cell, reaction between the electrolyte and the anode leads to deficiency of electrons in anode which disturbs the equilibrium and hence generation of potential difference. When the circuit is closed, that is the anode and the cathode are connected with externally, electrons flow from cathode to the anode, and current is generated. However, the flow of current stops as soon as the potential difference is eliminated. At this stage, either the cell is discarded forever and cannot be reused or these types of cells are termed as primary cell. And, the second case is where the reaction can be reversed by connecting the cell to an external source like solar panel, a current source from hydal/thermal/nuclear power plant, or any other external current source. These types of rechargeable batteries are termed as secondary batteries [6].

Also, in some cases, use of liquid in paste form and not liquid form enhances the portability of the battery to a large extent. All modern batteries like lithium ion, lithium polymer, and nickel-metal hydride cells use paste electrolyte and are termed as dry cells [14, 17].

Also, batteries are available in several sizes and shapes. Round pencil-shaped batteries are available in A, AA, and AAA sized. On the other hand, we also have the coin cell-type batteries used for small power gadgets.

The designing of complete circuits often has their specific requirement of power, and for that individual cells are often integrated in series or in parallel to each other to meet the power requirement.

Often in the commercial cells, the charge and discharge rates of a battery are declared by C-rates. C-rate is relatable to the capacity and the rate of charge discharge of the concerned cell. To explain, a battery rated at 1C and with capacity 1 A h should provide 1 A of current for 1 h continuously. Similarly, a battery rated as 0.5 C and

1 A h capacity would provide 500 mA of current for 2 h and a battery rated at 2 C with the same capacity would deliver 2 A of current for 30 min.

The performance of a battery and its capacity much depends on the electrode and the electrolyte material. A few types of batteries involving graphene and its derivatives as the electrode material are discussed in the following sections.

3   Brief History of Graphene

Graphene is composed of continuously linked honeycomb lattice consisting of two sub-lattices containing C–C σ-bonds, as illustrated in Fig. 1 [23, 24]. Each C atom in the lattice also contains a π-orbital majorly responsible for contributing delocalized electrons to the carbon network [23, 24, 28, 29]. The electronic properties of the graphene materials are solely based on the architectural arrangement and the number

Fig. 1 Graphene: the parent of all graphitic forms. Reprinted from Ref. [23], with permission from nature publication

of layers of the individual graphene sheets. Few-layered graphene possesses electronic structure and properties which significantly vary from that of the bulk graphite [23, 24, 30, 31]. At such a small thickness, the number of stacked layers and the order of the interlayers affect the chemical, physical, and the electronic properties of graphene. Single-layered graphene sheet behaves like a zero-band-gap semiconductor [30–33]. To get an insight into the phenomenon of widening of the band gap in case of the graphene sheets, a numerous number of computational means have been undertaken on the method of hydrogenation and halogenation of graphene to explore the possibilities of engineering the band structure [34, 35]. The electrons from the double-layered graphene follow energy dispersion of parabolic nature, and the material also renders itself as a zero-band-gap semiconductor. However, the tunable band gap of the double-layered graphene can easily be modified by inducing some asymmetric nature between the bi-layers [36, 37]. These studies propose that bi-layer graphene could act as a promising novel material for various applications in the future.

4   Techniques for Preparation of Graphene

Since the discovery of the graphene in the year 2004, it has attracted huge attention, and as a result, a substantial amount of work has been associated with the synthesis of graphene through various top-down and bottom-up routes [22]. The availability of graphene in the required form becomes a significant precondition for its application in energy storage, especially in battery applications. Till date, various captivating methods for the synthesis of mono-layer and a few-layer graphene have been reported and can be categorized as given below and shown in Scheme 1.

4.1   *Mechanical Exfoliation*

The year 2004 marks the discovery of graphene by mechanically unraveling separate sheets of mono-layer graphene from a sturdily bonded layered assembly of sheets termed as graphite [23, 24, 27]. This approach involved the micromechanical cleavage of the natural graphite flakes or well-oriented pyrolytic graphite by the means of a scotch tape. It is the very first method to be implemented to obtain graphene and is very apt for basic research purpose attributed to the great electronic quality, structural homogeneity, absence of any crystal defects, and exhibition of very high carrier mobility [28, 38]. Though, this technique allows reliable and easy preparation of graphene but suffers from a low yield [29]. Yet another route of mechanical exfoliation relates to the principle of anodic bonding. This approach is related to the fact that bulk graphite is bonded with glass made of borosilicate at a specific temperature followed by passing voltage. After that, it is peeled away resulting in a single-layer or a few-layer graphene sheet perching on the substrate [39, 40]. The

Scheme 1 Schematic diagram showing different common routes for fabricating graphene

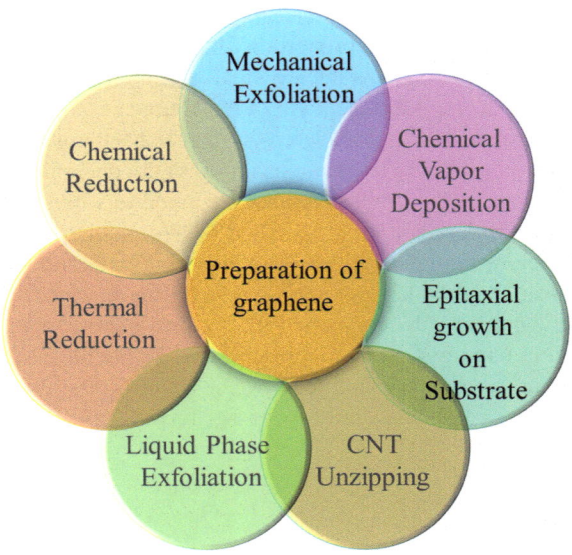

graphene sheets produced by the mechanical exfoliation result in graphene sheets with varying thicknesses. Some of the methods producing graphene depend on natural graphite specimen and ordered pyrolytic graphite [29–31].

4.2 Ultrasonic Cleavage

This is conceptually similar to the method of mechanical cleavage discussed earlier and is termed as ultrasonic cleavage. In this approach, the graphene precursors are generally suspended in an aqueous or organic solvent followed by its ultrasonic agitation acting as a source of energy to cleave the graphene precursors [22, 34, 41]. This idea is conceived solely after the observation made about successful exfoliation of the carbon nanotubes in organic solvents like N-methylpyrrolidine [42, 43]. The accomplishment of the proper synthesis of graphene through the method of ultrasonic cleavage significantly depends on the apt choice of solvents along with compatible surfactants associated with proper frequency, amplitude, and duration of sonication [22, 44].

4.3 Chemical Vapor Deposition

Synthesis of graphene on a large scale is made possible by another technique called chemical vapor deposition resulting in single or a few-layered graphene [22, 45, 46].

This method allows the synthesis of graphene sheets on a large scale by the means of utilizing a suitable metal surface treated at high temperature. A typical chemical vapor deposition process involves dissolution of carbon in a metal substrate followed by its precipitation on the substrate by cooling. A typical process may include placing a nickel substrate in a chemical vapor deposition chamber kept under vacuum and at a temperature maintained below 1000 °C, containing diluted hydrocarbon gas [46]. Apart from nickel, copper, and platinum, various other metal substrates with varied carbon solubility and catalytic effects, such as ruthenium and cobalt, are also used as substrate for chemical vapor deposition [46–49].

4.4   Chemical Vapor Deposition Through Plasma Induction

Chemical vapor deposition can also be achieved through plasma induction offering a substitute route for the synthesis of graphene at lower temperatures [50, 51]. Reports are available related to the synthesis of single and a few-layered graphene by the chemical vapor deposition associated in order to synthesize graphene layers on the various substrates involving a methane-hydrogen mixture [52, 53]. This method is advantageous as it includes very fast rate of deposition and relatively lower growth temperature of around 500–600 °C [50].

4.5   Unrolling Carbon Nanotube

Another interesting method of graphene sheet synthesis involves unrolling of the carbon nanotubes. Ammonia-solvated lithium cation ions are actually electrostatically attracted to the negatively charged multi-walled carbon nanotubes [54, 55]. The interlayer distance of the multi-walled carbon nanotubes increases from 3.35 to 6.62 Å as a result of the simultaneous intercalation of the lithium and ammonia into multi-walled carbon nanotubes [56]. This enhancement in the interlayer distance leads to the unwrapping and exfoliation as shown in Fig. 2 [57]. A strong oxidizing agent like potassium permanganate can also be used in order to unroll multi-walled carbon nanotubes by cutting them in the longitudinal direction [58]. The major advantage of this method is that it can be executed at room temperature.

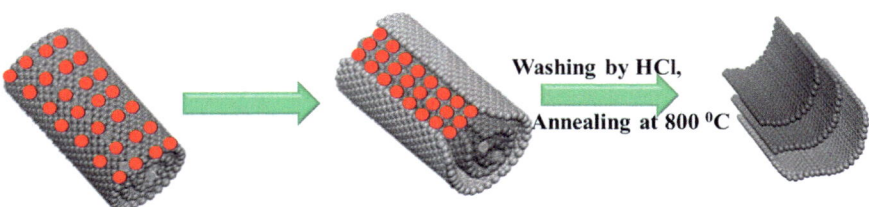

Fig. 2 Schematic illustration of unzipping the multi-walled carbon nanotubes [57]

4.6 Thermal Reduction

Another method involves thermal expansion of the graphene sheets with adequate oxidation of graphite by acid treatment and sufficient pressure at the time of the thermal heat treatment [59]. Critically, when the rate of decomposition of the oxide and the hydroxyl sites surpasses the rate of diffusion of the gases that are evolved, the existing pressure between the layers increases. At the point where the interlayer pressure outdoes the existing Van der Waals forces binding the graphene layers together, the graphite oxide finally splits into individual sheets of graphene [60].

4.7 Chemical Exfoliation

Graphene can also be produced by chemical exfoliation of the graphitic materials involving chemical oxidation, intercalation of ions, exfoliation, and/or finally reduction of the various graphene derivatives like natural graphite flakes, graphite oxide, graphite fluoride expandable graphite, other graphite intercalation compounds, and carbon nanotubes [61–63]. Especially, the yield of graphene obtained from the graphite is appreciable commercially. In addition to the reduction by heat treatment, reduction carried out by chemical reaction is another significant method for preparing graphene. In this method, first reduced graphene oxide is synthesized by increasing the interlayer spacing of graphite oxide obtained by Hummers' method [64, 65]. Synthesis of the mono-layer graphene oxide from multi-layered graphite oxide by the means of ultrasonication can serve as a classical example of chemical exfoliation route [22].

4.8 Graphene Material from Other Graphite Derivatives

Naturally occurring graphite, various graphite intercalated compounds, and/or graphites in expandable form are categorized as a specific type of graphite and are popularly used as precursors for obtaining mono-layer graphene sheets. Graphene synthesized through this method is generally highly pure [64]. Although the quality of the graphene obtained was very good, but the concentration of the suspension obtained is less. Various solvents those can be used for a making a homogeneous colloidal suspension of graphene can broadly be categorized according to the solvent cohesive energy and the surface energy of graphene [66, 67]. Another report regarding a uniform colloidal suspension consisting of graphene sheets could be synthesized by stirring potassium-based graphite intercalated compound in an N-methyl-2-pyrrolidone solvent [43, 64].

5 Graphene Derivatives

Experimentally and theoretically, novel carbon materials have been synthesized by using variously modified graphene, i.e., graphene oxide, graphane, hydrogenated graphene, fluorographene, graphyne/graphdiyne, and graphene nanoribbons commonly termed as graphene derivatives [25, 26]. Surface functionalization introduced in graphene results in improvement of the electronic, electrochemical properties, stability of the resulting material. This chapter accounts for a detailed overview of the various graphene-derived carbon materials, especially emphasizing on their exclusive properties and applications in the field of energy storage (battery applications).

5.1 Graphene Oxide

Graphite oxide is widely used as a reduction precursor for the synthesis of graphene, specifically a mono-layered graphite oxide, commonly termed as graphene oxide (GO) [68–70]. GO has attracted the tremendous attention of chemists as a significant oxide-functionalized derivative of graphene, as shown in Scheme 2 [71]. GO can be synthesized commercially by implementing a first-step chemical oxidation approach by using natural graphite followed by its exfoliation [72, 73]. Mostly, GO was synthesized by the synthesis process as proposed by Hummers and Offeman [74]. This method is originally imitated from the process of Staudenmaier, often termed as the Hummers' method [75]. Apart from this method, in some cases, the Brodie method was also implemented, in which the process of oxidation of the graphite was performed in fuming nitric acid along with potassium chlorate [76]. The basic steps in this method include oxidation of graphite in concentrated sulfuric acid with potassium permanganate and sodium nitrate. This step is followed by the elimination

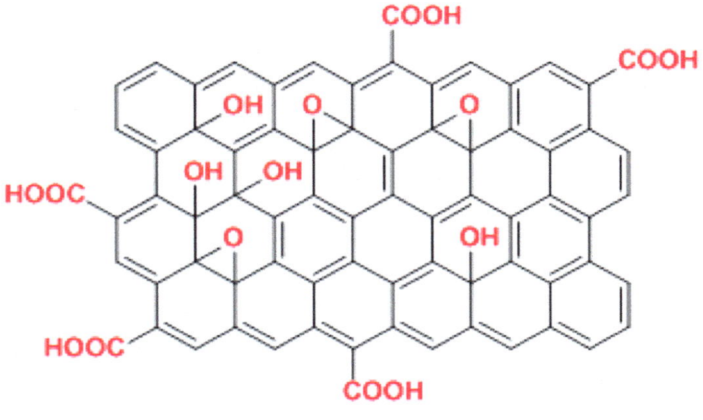

Scheme 2 Structure of graphene oxide [71]

of the excess potassium permanganate by reducing to water-soluble manganese(II) sulfate with the help of hydrogen peroxide followed by its washing by methanol [77, 78]. The oxidation reaction actually disrupts the existing long-range π-π conjugation interaction associated with the graphite surface. As a consequence, the sp^2 graphitic domains are formed, surrounded by disorderly arranged sp^3 hybridized domains with other functional groups containing oxygen like epoxides, carbonyls (–C=O), hydroxyls (–OH), ketone, quinine, and phenol. The interlayer forces are largely reduced due to the existence of these intervening functional groups containing oxygen, and in turn, results in possession of strongly hydrophilic character [79–81]. Further, GO can also be synthesized with the help of direct oxidation of the free-standing graphene by atomic oxygen under the condition of ultrahigh vacuum [81], by exposing to ozone [82] or molecular oxygen [83], by carrying out photochemical reaction in presence of oxygen and ultraviolet light and by the properly controlled electrochemical oxidation under potentio-static conditions in nitric acid medium [84, 85].

5.2 Graphane

Graphane is actually a derivative of graphene which is an incomplete hydrogenated form of graphene. It is having chemical stoichiometric formula CH as predicted by the total energy calculations from first principles [25, 26, 86]. In graphane, each carbon atom is sp^3 hybridized, and one hydrogen atom is attached to each carbon atom on the alternate sides of the carbon network. This type of arrangement is structurally similar to that of hydrogenated (111) sheet of diamond [87]. A substantial amount of theoretical studies has been dedicated for the understanding of such newly architected. Graphane can possess two configurations, as shown schematically Scheme 3, which include a conformer which is boat-shaped with the hydrogen atoms alternating in pairs, and the other is shaped in a chair-like fashion where the hydrogen atoms

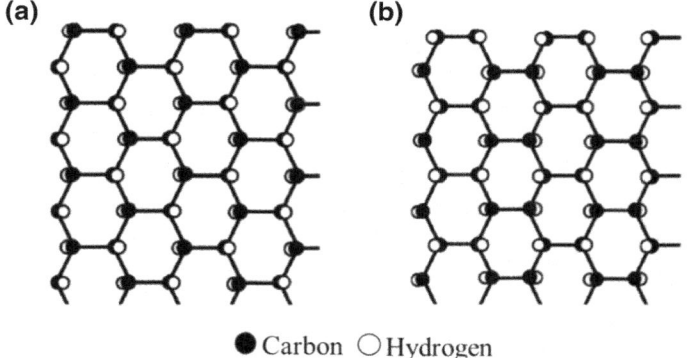

● Carbon ○ Hydrogen

Scheme 3 Structure of graphane **a** chair type and **b** boat type [25]

alternate on both sides of the carbon atom layer [26, 87, 88]. In case of graphane, C–C bond with sp³ hybridization is much longer as compared to the C–C bond of graphene. Graphane is assumed to be a stable structure associated with a binding energy similar to the other hydrocarbons like cyclohexane, benzene, and polyethylene [87–91].

5.3 Fluorographene

Besides graphane and graphene oxide, fluorographene which is basically fluorinated graphene possessing a stoichiometric formula of CF is another significant structural derivative for graphene. The study related to the reaction between graphite and fluorine has been studied long before in 1934 [25, 26, 92, 93]. Fluorographene is a monolayer of graphite fluoride which can be obtained by fluorination of bulk graphite at a very high temperature. Since graphite fluoride exhibits a weak interaction between the interlayers, it is characterized as an outstanding precursor for the synthesis of mono-layer graphene fluoride [26, 94]. The various methods for the synthesis of fluorographene are largely categorized into two major groups: exfoliation and fluorination approaches. Fluorination of graphene sheets mainly involves plasma fluorination, direct gas fluorination, photochemical/electrochemical synthesis, and hydrothermal fluorination [95]. Exfoliation can be achieved by using various methods such as modified Hummer's exfoliation, sonochemical exfoliation, and exfoliation through thermal treatment [22]. Fluorographene seems to be a very promising 2-D material characterized by the ease of synthesis and with exceptional properties. Fluorographene shows fairly ordered hexagonal structure and strongly insulating properties. In addition, fluorographene is fairly stable and also chemically inert along with high thermal stability up to 400 °C [96]. Moreover, fluorographene gets the outstanding stiffness and mechanical strength of the graphene, associated with the value of Young's modulus three times less than that of graphene and substantially sustaining elastic deformations [97, 98].

5.4 Graphyne and Graphdiyne

Carbon possesses a wide range of hybridized states such as sp, sp², and sp³ and can engage in various diverse bonds. Yet the well-known crystalline phases involve either completely sp³- or sp²-hybridized carbon atoms [25, 26, 99, 100]. Graphyne and graphdiyne have recently sought augmented attention as novel forms of artificial carbon allotropes, attributed to their exceptional structures and fascinating mechanical and electronic properties [25, 26]. Both of them possess a remarkable planar network of only carbon atom arranged in a hexagonal lattice-like benzene and/or alkyne units as can be seen in Scheme 4.

(a)   **(b)**

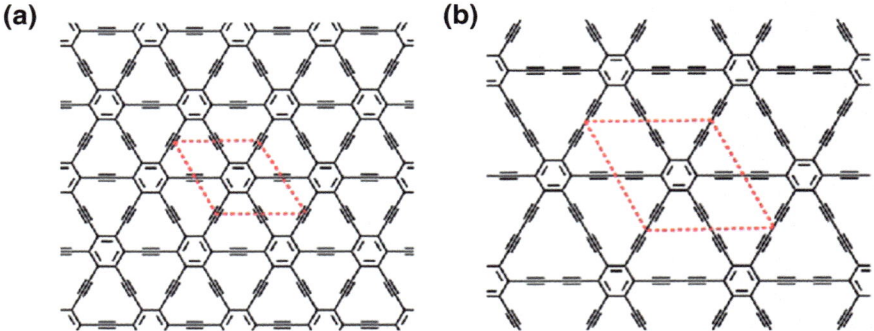

Scheme 4 Structure of **a** graphyne and **b** graphdiyne. The parallelogram is drawn with a red line represents a unit cell [25]

Graphyne was first assumed to possess a stable crystalline structure associated with a high possibility of synthesis in 1987 [100]. Graphyne consists of both the sp^2- and sp-hybridized carbon atoms, involving mostly aromatic benzene rings and feeble involvement of 12-membered antiaromatic rings [101]. The various graphene forms consist of mainly their kind of C–C bonds, viz. $C(sp^2)$–$C(sp^2)$ associated to the central aromatic ring, $C(sp^2)$–$C(sp)$ linking the adjacent C≡C and C=C bonds, and $C(sp)$–$C(sp)$ for the connected triple bonds. The graphyne layers are structurally similar to that of graphene in the sense that they also possess the similar hexagonal symmetry as graphene. Graphyne can be perceived to be built from graphene by the replacement of almost one-third of C–C bonds with the C≡C connections [102, 103]. Graphyne is assumed to have a good stability at high temperature and shows similar mechanical characteristics as graphite. From the point of view, the existing in-plane stiffness, the mono-layer graphyne is foreseen to be more lenient than graphene attributed to its comparatively lesser coordination number of bonds [25, 26, 104–106].

On the other hand, graphdiyne is composed of a structure where 1/3rd of C–C bonds are injected with two C≡C linkages [25, 26]. The additional alkyne unit enhanced the pore size of the material. The lattice length for the 2D graphdiyne can be optimized to reach a value which is larger than graphyne [107, 108]. In addition, the C–C bond length order in graphdiyne network also follows the same trend as exhibited by the graphyne, only with the inclusion of an additional $C(sp)$–$C(sp)$ single bond attaching two consecutive C≡C bonds. The existence of –C≡C– groups for the graphyne or graphdiyne imparts enhanced localization to the π-electron resulting to a non-zero gap [109, 110].

5.5   Graphane Nanoribbons

Keeping in mind the similarity to the graphene, graphane can be additionally dis-associated into one-dimensional hydrocarbons, termed as graphane nanoribbons.

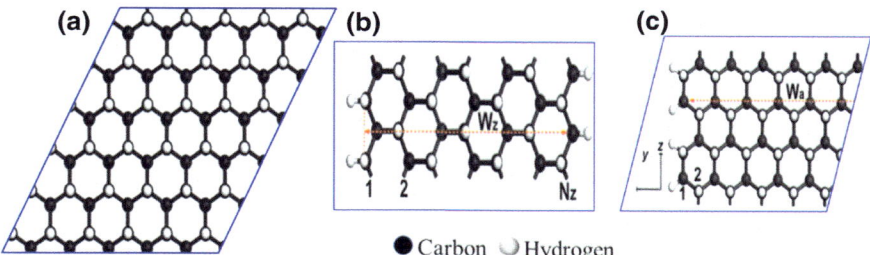

Scheme 5 **a** Upper and lower view of a 2D graphane layer, geometric structures of the **b** 7-zigzag, and **c** 13-armchair graphane nanoribbons [111]

These are completely hydrogenated graphene nanoribbons derivatives. Graphane nanoribbons can easily be synthesized by hydrogenation of the graphane nanoribbons in a hydrogen plasma medium [25, 110, 111]. Graphene can also be synthesized by graphane nanoribbons by unzipping hydrogenated carbon nanotubes [25]. Graphane nanoribbons exhibit completely dissimilar properties as compared to graphene nanoribbons [112]. As a result of the presence of sp^3 hybridized C atoms the delocalized π-electrons, the edge states disappear. For this reason, the armchair and zigzag graphane nanoribbons are generally wide-band-gap and non-magnetic semiconductors. As can be seen in the Scheme 5, the graphane nanoribbons are periodic in the z-direction. The width of the ribbons is denoted by Wa and Wz, respectively [111].

6 Graphene and Its Derivatives for Energy Storage: Battery Application

Batteries are enrapturing market as an essential constituent in several energy-efficient applications. They are being incorporated into various electrical, aerospace, textiles, automotive, electric vehicles, and stationary markets intensely refining our work efficiency [1–4]. Electrochemical energy storage modes, like batteries possessing large specific energy densities, are serving to eliminate the problems related to the renewable energy sources like solar energy, wind energy, tidal energy, and wave energy [5]. Attributed to their compact structure and size, batteries are considered to be promising to provide with all energy storage facility for our daily work. For the electrochemical energy storage modes, specifically, lithium-ion batteries constitute the most widely utilized power sources attributed to the large energy density, long cycle life, safety, and eco-friendliness [113–115]. Other electrochemical energy storage modes like lithium-air batteries, the sodium-ion batteries, and the lithium-sulfur batteries have also attracted augmented attention attributed to their distinguishing features [14]. Figure 3 demonstrates the evolution of the various solid-state electrolyte batteries over years. However, all these categories of batteries exhibit their own

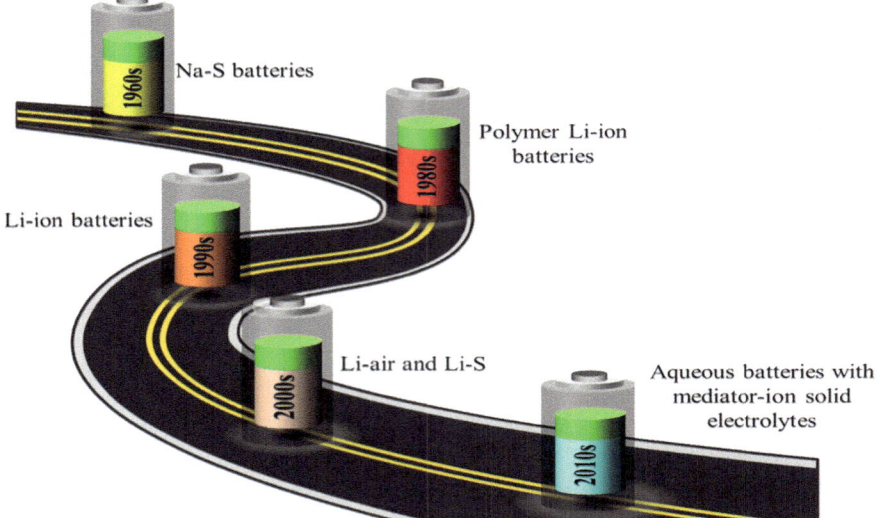

Fig. 3 The timeline shows the key developments in solid-state electrolyte batteries [15]

advantages/disadvantages which leave a room for further enhancement. Graphene and its derivatives are very attractive as battery material, and there can exist several forms of graphene derivatives and applications in batteries, which are shown in Fig. 4.

6.1 Lithium-Ion Batteries

Lithium-ion batteries (LIBs), used for the commercial purposes, were originally designed by Sony Co., Japan, 28 years ago. These commercialized batteries utilize laminar stacked-type compounds, as the cathode material like lithium cobalt oxide and the anode is made of graphite material. When the charge-discharge process goes on, the lithium ions pass through a liquid lithium ion electrolyte residing in-between the electrodes [116, 117]. The working of such type of batteries mechanism is illustrated in Fig. 5. During the discharge process, ions are expelled from the anode which is charged negatively and injected in the cathode charged positively. On the other hand, ions are mainly involved in circulating current between the negative and the positive electrode through the polymeric separator and electrolyte. Graphene and its derivatives are engaged in several applications related to in LIBs as anode and cathode material and attributed to their exclusive structural, other mechanical and chemical properties [113, 116].

Graphite exhibiting a theoretical capacity of value 372 mAh g^{-1} assumed according to the fact that a single Li ion can interact with six carbon atoms. This is the

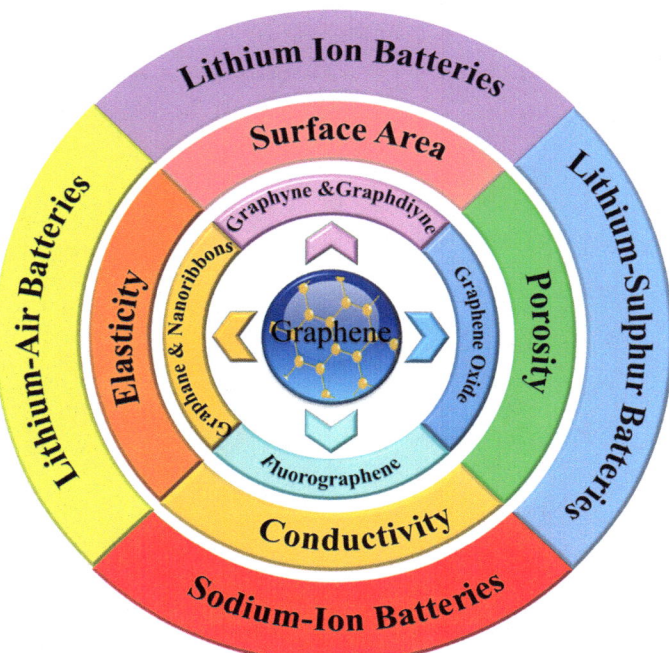

Fig. 4 Schematic illustration for the unique properties and applications in batteries for graphene and its derivatives

mostly used anode material in LIBs attributed to its outstanding reversibility for the Li ions and very elongated cycling life [14, 117]. On the other hand, graphite continued to be the leading anode material used commercially for a long time. Theoretically, graphene can house Li ions on each of its sides making its theoretical capacity double than that of the other carbon materials. Calculating theoretically, it has been concluded that graphene can exhibit a maximum capacity of 740 mAh g^{-1} according to the double-layer adsorption assumption [14, 118].

However, for achieving the practical values that would enable LIBs applicable in electric vehicles, designing a suitable electrode material is still a challenge. On the other hand, attributed to its electrochemical activeness and conductive and elastic nature, graphene has successfully overcome various issues like low conductivity, volume expansion, rate capability, and the quick fading of capacity [1–4]. Till now, both experimental and theoretical results have indicated that augmentation in the electron transportability and its electrochemical performance of the graphene derivatives is still possible [14, 25]. Graphene derivatives: graphene oxide which can be considered as a soft material becomes the second choice for LIBs. Compared to the pure form of graphene, graphene oxide is mechanically less strong, but the strength is still considered good, and it still shows potential in the synthesis of composite materials

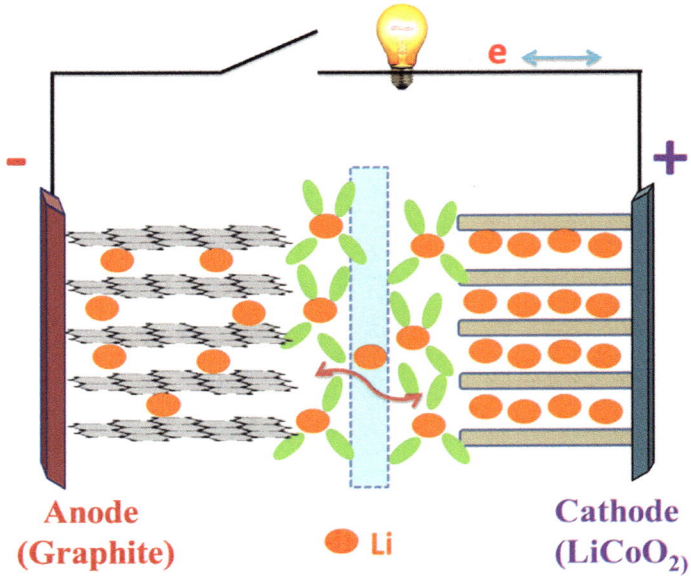

Fig. 5 Schematic diagram for the mechanism of ion intercalation into a LIB. Reprinted from Ref. [116], with permission from nature publication

[27, 28, 119–121]. The various electronic properties of graphene can be controlled which makes it suitable for application in LIBs.

In recent times, chemical modification of graphene is wide spreading in order to synthesize novel functional materials [121, 122]. Among the various functionalized materials, hydrogenated graphene possessing a graphene-like structure is a special and interesting one as it possesses a large hydrogen content. In this case, mostly all the carbon atoms are in sp^3 hybridized state rendering this material as very suitable for LIBs fields [123, 124]. Hydrogenated graphene synthesized by using gamma-ray radiation for application in LIBs shows good capacity, i.e., ~680 mAh g^{-1} (see Fig. 6) [123].

On the other side, fluorographene exhibits way more varying properties as compared to that of graphene as it lacks the association with the $\prod$-conjugated electrons. Fluorographene exhibits a theoretical specific capacity of 865 mAh g^{-1} for Li, which is considerably high as compared to the pure graphene [125, 126]. This material also finds its application as cathode material in the primary and secondary LIBs batteries, leading to the net enhancement in the shelf-life characteristics and discharge characteristics of the battery cells [123].

The captivating electronic and structural properties render graphyne/graphdiyne as very interesting candidates as anode material for application in LIBs [127, 128]. The first principle calculations were employed to investigate two-dimensional carbon allotropes dispersed by lithium, viz. graphdiyne and graphyne for exploring their potential applications in lithium storage [127]. The largest Li storage capacity for

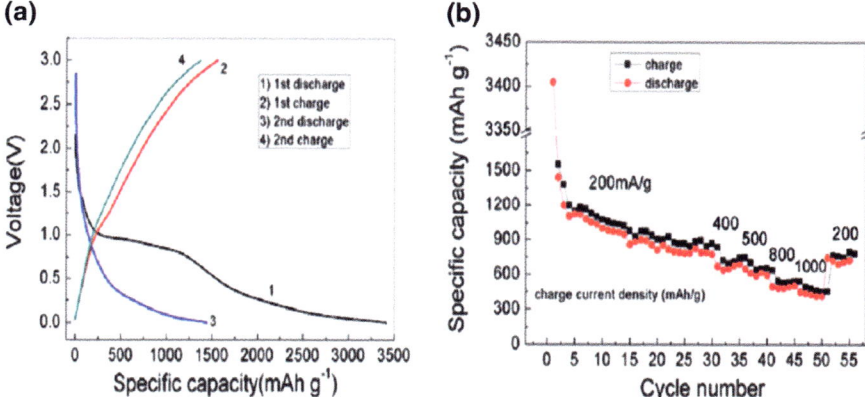

Fig. 6 a Discharge/charge profiles and b rate performance of a half-cell composed of highly hydrogenated graphene and Li. Reprinted from Ref. [123], with permission from nature publication

the graphyne is greater than the storage capacity in graphite. The intercalation of Li has a small effect on the spacing between two consecutive graphyne layers, and this is very helpful for the charging processes. The high Li storage capacity and large mobility enable graphyne to render itself as a promising material to be used as the anode for battery applications [127].

Apart from these, graphene nanoribbons are also developing as a promising candidate for LIBs. The mechanical strength and large surface area of graphane nanoribbons at the time of lithiation indicate graphane nanoribbons can act as an outstanding anode material from the point of view of capacity and durability [19, 25]. Peralta et al. [129] have reported the lithium storage capacity of graphene nanoribbons on the basis of density functional theory and exposed that the strength of interaction between graphane nanoribbons and lithium is 50% stronger as compared to the graphene nanosheet. This augmented interaction leads to the better performance of LIBs as anode materials. Besides, the lithium diffusion coefficient and the lithium storage capacity have also been reported to be enhanced up to two orders as compared to that of the graphene nanosheet [129–132].

6.2 Sodium-Ion Batteries

In the recent few decades, LIBs have been widely used as a rechargeable energy source. However, the continuously growing necessity of the LIBs on a large scale puts forward growing concerns regarding elevated cost and inadequate terrestrial stock of lithium [133–135]. Therefore, sodium-ion batteries (SIBs), involving abundantly available inexpensive sodium belonging to the alkaline earth metal family, have recently gained increasing attention as a cheaper alternative to LIBs. SIBs were proposed long before as an alternative to the LIBs; however, they were neglected for long and not put to the practical applications attributed to their substandard performance. As compared to the lithium, sodium is better as far as the availability is concerned, but it is not as great from the point of view of the redox potential and the ionic radius [134–136]. However, still, SIBs prove to be the most promising alternatives for LIBs in case of the application on a large scale, attributed to its low cost and greater availability [136]. Furthermore, the physical and the chemical characteristics of sodium are very much similar to those of lithium, as shown in Fig. 7 [134, 135]. Sodium ions are passed among anode and the cathode electrodes via. an aqueous or a nonaqueous. Apart from the sodium, graphene and its various derivates also prove to be of great importance in designing SIBs exhibiting a better rate capability, high specific capacity, and elongated cycle life as compared to the bare graphene [137]. Till date, graphene and its derivates have shown enhanced energy density and gravimetric capacity when compared to the other materials attributed to their high diffusion ability, high capacity, short ion-diffusion distance, and a huge number of active sites for ion storage. Enthused by the preceding works of rendering graphene and its various derivatives as the outstanding anode material for application in the LIBs, graphene and various derivatives seem to be an effective material for [137–141].

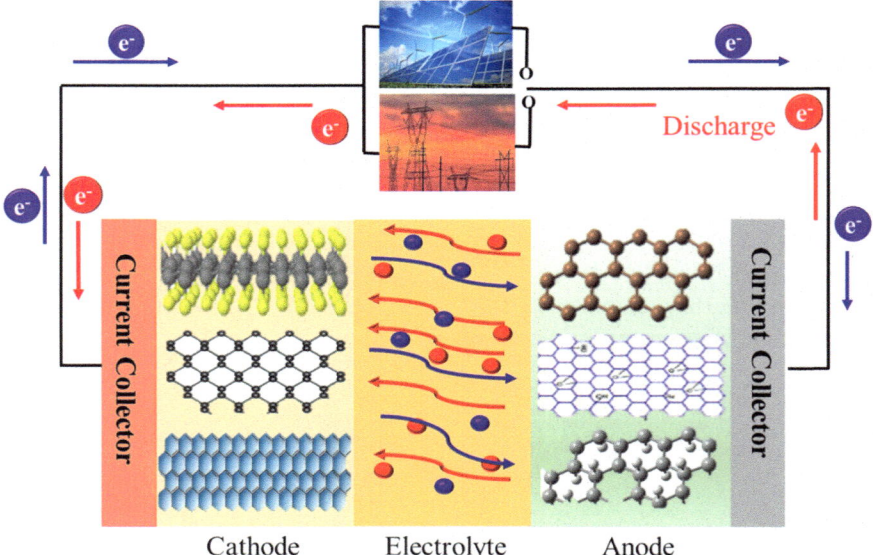

Fig. 7 Schematic diagram illustrating the mechanism of ion intercalation into a SIB

6.3 Lithium-Air Batteries

The lithium-air storage systems have captured increased attention recently in the year 2009 as a likely alternative to LIBs for electric vehicle impulsion applications. The lithium-air battery (Li-air) proves to be a progressive energy conversion as well as energy storage technology [142–144]. The first report regarding the rechargeable lithium-oxygen battery was published in 1987 which described a design very similar to a solid oxide fuel cell [143]. It is known to convert chemical energy possessed by oxygen (cathode) and lithium (anode) into the electric energy during the discharge, process, and, in turn, store the electric energy by the means of the splitting of the $Li-O_2$ discharge products while charging process by using electricity. Generally, two varieties of Li-air batteries are mostly studied: aqueous systems and nonaqueous systems, both of which are electrically rechargeable [144]. Li-air batteries are generally based on a large operating voltage related to the lithium-oxygen electrochemical couple and can result in an exceptionally large theoretical energy density of ~11,680 Wh kg^{-1}. This value is very close to that of petrol having a value of 13,000 Wh kg^{-1}, as shown in Fig. 8 [143]. According to the estimation, a practical Li-air battery can possibly provide a specific energy density as large as around 800 Wh kg^{-1}, and this value is about four times as compared to the Li-ion batteries [143]. Figure 8 illustrates comparatively the obtained gravimetric energy densities (Wh kg^{-1}) for several kinds of rechargeable batteries with that of petrol.

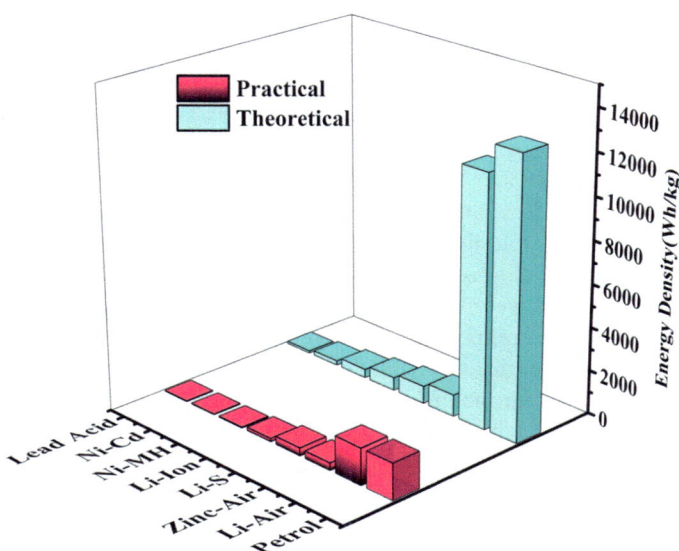

Fig. 8 Comparative illustration of the gravimetric energy densities with various types of rechargeable batteries with that of petrol [143]

Graphene as a basic component is used widely as air-electrode in Li-air systems, where an organic electrolyte along with an aqueous electrolyte are used together at the cathode and anode, respectively [145, 146]. In recent periods, graphene has emerged as an outstanding cathode material to be used with the Li-air batteries attributed to its very high discharge capacity and large round-trip efficiency [141]. The detailed mechanisms related to oxygen reduction and the oxygen evolution reactions for graphene associated with Li-air batteries are not clear so far. The usage of graphene nanosheets as an electrode in Li-air batteries exhibited 8705.9 mAh g carbon^{-1} of discharge capacity obtained at the 75 mA g carbon^{-1} of current density. This obtained value was substantially higher as compared to that obtained for the electrode fabricated with the commercial carbon [147, 148]. Further, the electrode based on graphene nanosheets exhibited a discharge capacity of 2000 mAh g carbon^{-1} obtained at 2.80 V of discharge voltage and a 3.90 V of charge voltage with a corresponding current density of 50 mA g carbon^{-1} [148]. From the various reports available, it is implied that the Li-air battery associated with the graphene electrode is very much promising. In addition, graphene derivatives are also considered to be very promising and indicated huge future possibilities to be used as electrode material in Li-air batteries due to Li storage capacity [149, 150]. However, reports are very few for Li-air batteries associated with derivatives of graphene as an electrode material.

6.4 Lithium-Sulfur Batteries

Modern generation batteries involving Li electrodes are well known attributed to much-promised safety and high energy density. However, the batteries in the market are failing to fulfill the requisite terms associated with long-term periods storage due to low energy densities. To eradicate this problem yet another class of lithium-based batteries, termed as "lithium-sulfur" (Li–S) batteries, has emerged seeking much attention attributed to its lesser weight, high energy density, and larger safety [14, 15]. The batteries obtained as a result of replacement of the bulky anodes of the LIBs by the lightly weighted sulfur is termed as lithium-sulfur batteries. As a result of their large theoretical capacity, sulfur shows much potential to act as an effective replacement for the conventional electrodes used in the LIBs [14, 151]. Yet other advantages of sulfur include cost-effectiveness, natural availability and environmental benignity, which render it very much appealing as an electrode material for second-generation LIBs for application in day to day life. However, the electrical conductivity of sulfur is very low which is considered as the reason for low contact and meager electrochemical properties exhibited by the electrode based on sulfur, leading to low capacity and quickly fading capacity of the electrode [152, 153]. These drawbacks are resolved by the integrating highly conductive substrates such as graphene and graphene derivatives in order to increase the electrical conductivity of the sulfur electrode. In recent years, various sulfur-graphene hybrids have been designed and studied in Li–S batteries [153–155]. Recently, the utilization of fluorographene was further extended as a fascinating electrode separator for application in Li–sulfur batteries leading to the prevention of the migration of the polysulfides toward the Li anode [156]. However, the performance for these was inferior to that obtained for the carbon-sulfur hybrids, indicating their further advances in case of the Li–S batteries necessitates advancement for the existing sulfur/graphene and graphene derivatives interface in order to achieve better performance [157, 158].

6.5 Other Battery Technologies

Battery technologies like sodium-air (Na-air) and sodium-sulfur (Na–S) batteries have also attracted attention as potential heirs of the other battery technologies.

Na-air has attracted increased attention attributed to its cost-effectiveness and several other exceptional characteristics, like large theoretical energy density, lesser cost, and negligible environmental impact [159]. However, Na-air battery faces the problem of low power density, short cyclability, and large overpotential. Furthermore, Na-air batteries need highly pure oxygen stocked within a tank to supply oxygen, which further enhances the price of the entire setup and also increases the inconvenience [14].

On the other hand, Na-S batteries exhibit various advantages, such as large theoretical energy density (760 Wh kg^{-1}), high efficiency, long life, and low material cost [14]. All these advantages project the Na–S as promising storage applications, e.g., load-leveling and seamless power supply [14, 160]. However, as a result of less sodium-ion conductivity, the cell has to be functioning at an elevated temperature to attain a significant amount of current flow, leading to challenges related to safety. Graphene has been anticipated as an encapsulating carbon matrix and a conductive agent for the Li-S battery, but such actions are yet not identified for the Na-S battery [161]. In spite of the fact that the above discussed two-battery technologies, i.e., Na-air and Na-S battery technologies are still lying in the initial stage of development and hence requires more extensive association with research.

7 Conclusions and Future Perspectives

So, as we can see that in order to smoothly drive our day to day life, it is very necessary to further develop the charge storage modes. Carbon forms the premier electrode material in the energy storage field. In case of batteries also, carbon materials have been widely used as electrode. Among the various carbon materials, graphene forms a very interesting electrode material owed to its various unique properties. However, the process of obtaining graphene in mono and multi-layered forms has been extensively studied by the researchers to develop efficient and cost-effective ways of graphene synthesis. Various synthesis methods are developed in order to synthesize graphene in bulk. Graphene and various graphene-derived materials including graphene oxide, hydrogenated graphene, graphane, graphyne/graphdiyne, fluorographene, and graphene nanoribbons have boosted the urge to study various scientific and technical properties of these materials. All these graphene derivatives exhibit distinct and independent properties different from that of graphene, which expands the scope of applications for these materials. Forming a very significant part of the family of two-dimensional materials, these graphene derivatives provide promising aspects compelling to carry out research on their structure, properties, chemistry, and various possible applications. However, the difficulties faced to produce these materials on a commercial scale with the requisite properties are still a challenge, and there still lies a lot of unexplored about these materials. Undoubtedly, graphene and various graphene derivatives are prophesied to have an exciting future for application in varied fields. Further, theoretical as well as experimental determinations are requisite to explore the physical properties, chemical properties, stability features, facile synthesis routes, and various applications of the graphene derivatives. To our strong belief, graphene-related derivatives would bring revolution in the performance of the energy storage modes acting as promising electrodes in battery applications.

References

1. Goodwin, S., Darren, A.W.: Closed bipolar electrodes for spatial separation of H_2 and O_2 evolution during water electrolysis and the development of high-voltage fuel cells. ACS Appl. Mater. Interfaces **9**, 23654 (2017)
2. Schafzahl, L., Mahne, N., Schafzahl, B., et al.: Singlet oxygen during cycling of the aprotic sodium–O_2 battery. Angew. Chem. Int. Ed. **56**, 15728 (2017)
3. Hassoun, J., Panero, S., Reale, P., et al.: A new, safe, high-rate and high-energy polymer lithium-ion battery. Adv. Mater. **21**, 4807 (2009)
4. Fakharuddin, A., Jose, R., Brown, T.M., et al.: A perspective on the production of dye-sensitized solar modules. Energ. Environ. Sci. **7**, 3952 (2014)
5. Yang, Z., Zhang, J., Kintner-Meyer, C.W., et al.: Electrochemical energy storage for green grid. Chem. Rev. **111**, 3577 (2011)
6. Conway, B.E.: Electrochemical Supercapacitors: Scientific Fundamentals and Technological Applications. Springer Science & Business Media (2013)
7. Reddy, M.V., Rao, G.V.S., Chowdari, B.V.R.: Metal oxides and oxysalts as anode materials for Li ion batteries. Chem. Rev. **113**, 5364 (2013)
8. Wang, G., Zhang, L., Zhang, J.: A review of electrode materials for electrochemical supercapacitors. Chem. Soc. Rev. **41**, 797 (2012)
9. Wang, Y., Chen, K.S., Mishler, J., et al.: A review of polymer electrolyte membrane fuel cells: technology, applications, and needs on fundamental research. Appl. Energy **88**, 981 (2011)
10. Yang, Y., Bremner, S., Menictas, C., et al.: Battery energy storage system size determination in renewable energy systems: a review. Renew. Sust. Energ. Rev. **91**, 109 (2018)
11. Harks, P.P.R.M.L., Mulder, F.M., Notten, P.H.L.: In situ methods for Li-ion battery research: a review of recent developments. J. Power Sources **288**, 92 (2015)
12. Wakihara, M., Yamamoto, O. (eds): Li-Ion Batteries. Wiley-VCH, New York (1998)
13. Nykvist, B., Nilsson, M.: Rapidly falling costs of battery packs for electric vehicles. Nat. Clim. Change **5**, 329 (2015)

14. Peng, L., Zhu, Y., Chen, D.: Two-dimensional materials for beyond-lithium-ion batteries. Adv. Energ. Mater. **6**, 1600025 (2016)
15. Manthiram, A., Yu, X., Wang, S.: Lithium battery chemistries enabled by solid-state electrolytes. Nat. Rev. Mater. **2**, 16103 (2017)
16. Kiehne, H.A.: Battery Technology Handbook, vol. 118. CRC Press (2003)
17. Mekonnen, Y., Sundararajan, A., Arif, I.S.: A review of cathode and anode materials for lithium-ion batteries. In: Southeast Conference IEEE, pp 1–6 (2016)
18. Lecerf, A., Lubin, F., Broussely, M.: Rechargeable electrochemical battery including a lithium anode. U.S. Patent 4,975,346, issued 4 Dec 1990
19. Raccichini, R., Varzi, A., Passerini, S., et al.: The role of graphene for electrochemical energy storage. Nat. Mater. **14**, 271 (2015)
20. Zhu, J., Yang, D., Yin, Z., et al.: Graphene and graphene-based materials for energy storage applications. Small **10**, 3480 (2014)
21. Wang, C., Li, D., Too, O., et al.: Electrochemical properties of graphene paper electrodes used in lithium batteries. Chem. Mater. **21**, 2604 (2009)
22. Bhuyan, M.S.A., Uddin, M.N., Islam, M.M., et al.: Synthesis of graphene. Int. Nano Lett. **6**, 65 (2016)
23. Geim, A.K., Novoselov, S.: The rise of graphene. A collection of reviews. Nat. J. **11** (2010)
24. Geim, A.K.: Graphene: status and prospects. Science **324**, 1530 (2009)
25. Tang, Q., Zhou, Z., Chen, Z.: Graphene-related nanomaterials: tuning properties by functionalization. Nanoscale **5**, 4541 (2013)
26. Inagaki, M., Kang, F.: Graphene derivatives: graphane, fluorographene, graphene oxide, graphyne and graphdiyne. J. Mater. Chem. A **2**, 13193 (2014)
27. Lonkar, S.P., Yogesh, S.D., Ahmed, A.A.: Recent advances in chemical modifications of graphene. Nano Res. **8**, 1039 (2015)
28. Allen, M.J., Tung, C., Kaner, B.: Honeycomb carbon: a review of graphene. Chem. Rev. **110**, 132 (2009)
29. Jia, X., Campos, D.J., Terrones, M., et al.: Graphene edges: a review of their fabrication and characterization. Nanoscale **3**, 86 (2011)
30. Rao, C., Sood, A., Subrahmanyam, K., et al.: Graphene: the new two-dimensional nanomaterial. Angew. Int. Ed. **48**, 7752 (2009)
31. Ruoff, R.: Graphene: Calling all chemists. Nat. Nanotech. **3**, 10 (2008)
32. Stankovich, S., Dikin, A., Dommett, H.B., et al.: Graphene-based composite materials. Nature **442**, 282 (2006)
33. Sluiter, M.H.F., Kawazoe, Y., et al.: Cluster expansion method for adsorption: application to hydrogen chemisorption on graphene. Phys. Rev. B **68**, 085410 (2003)
34. Sofo, J.O., Chaudhari, A.S., Barber, G.D.: Graphane: a two-dimensional hydrocarbon. Phys. Rev. B **75**, 153401 (2007)
35. Zeng, Q., Wang, H., Fu, W.: Band engineering for novel two-dimensional atomic layers. Small **11**, 1868 (2015)
36. Sovoselov, N.K., Fal, V.I., Colombo, L., et al.: A roadmap for graphene. Nature **490**, 192 (2012)
37. Gao, W.: The chemistry of graphene oxide. In: Gao W. (eds) Graphene Oxide. Springer, Cham pp. 61–95 (2015)
38. Yi, M., Shen, Z.: A review on mechanical exfoliation for the scalable production of graphene. J. Mater. Chem. A **3**, 11700 (2015)
39. Moldt, T., Eckmann, A., Klar, P., et al.: High-yield production and transfer of graphene flakes obtained by anodic bonding. ACS Nano **5**, 7700 (2011)
40. Balan, A., Kumar, R., Boukhicha, M., et al.: Anodic bonded graphene. J. Phys. D: Appl. Phys. **43**, 374013 (2010)
41. Zhu, Y., Murali, S., Cai, W., et al.: Graphene and graphene oxide: synthesis, properties, and applications. Adv. Mater. **22**, 3906 (2010)
42. Dumonteil, S., Demortier, A., Detriche, S., et al.: Dispersion of carbon nanotubes using organic solvents. J. Nanosci. Nanotechnol. **6**, 1315 (2006)

43. Hasan, T., Scardaci, V., Tan, P.H.: Stabilization and "debundling" of single-wall carbon nanotube dispersions in N-methyl-2-pyrrolidone [NMP] by polyvinylpyrrolidone [PVP]. J. Phys. Chem. C **111**, 12594 (2007)
44. Jun, Z.: Graphene production: new solutions to a new problem. Nat. Nanotechnol. **3**, 528 (2008)
45. Reina, A., Jia, X., Ho, J., et al.: Large area, few-layer graphene films on arbitrary substrates by chemical vapor deposition. Nano Lett. **9**, 30 (2008)
46. Zhang, Y.I., Zhang, L., Zhou, C.: Review of chemical vapor deposition of graphene and related applications. Acc. Chem. Res. **46**, 2329 (2013)
47. Bhaviripudi, S., Jia, X., Dresselhaus, S., et al.: Role of kinetic factors in chemical vapor deposition synthesis of uniform large area graphene using copper catalyst. Nano Lett. **10**, 4128 (2010)
48. Yu, H.K., Balasubramanian, K., Kim, K., et al.: Chemical vapor deposition of graphene on a "peeled-off" epitaxial cu [111] foil: a simple approach to improved properties. ACS Nano **8**, 8636 (2014)
49. Green, M.L., Gross, M.E., Papa, L.E., et al.: Chemical vapor deposition of ruthenium and ruthenium dioxide films. J. Electrochem. Soc. **132**, 2677 (1985)
50. Yue, D.W., Ra, C.H., Liu, X.C.: Edge contacts of graphene formed by using a controlled plasma treatment Nanoscale **7**, 825 (2015)
51. Shah, J., Lopez-Mercado, J., Carreon, M.G., et al.: Plasma synthesis of graphene from mango peel. ACS Omega **3**, 455 (2018)
52. Ho, G.W., Wee, A.T.S., Lin, J.: Synthesis of well-aligned multiwalled carbon nanotubes on Ni catalyst using radio frequency plasma-enhanced chemical vapor deposition. Thin Solid Films **388**, 73 (2001)
53. Deng, J., Zheng, R., Yang, Y., et al.: Excellent field emission characteristics from few-layer graphene–carbon nanotube hybrids synthesized using radio frequency hydrogen plasma sputtering deposition. Carbon **50**, 4732 (2012)
54. Peng, C., Chen, B., Qin, Y., et al.: Facile ultrasonic synthesis of CoO quantum dot/graphene nanosheet composites with high lithium storage capacity. ACS Nano **6**, 1074 (2012)
55. Wang, G., Yang, J., Park, J., et al.: Facile synthesis and characterization of graphene nanosheets. J. Phys. Chem. C **112**, 8192 (2008)
56. Zhao, X., Liu, Y., Inoue, S., et al.: Smallest carbon nanotube is 3 Å in diameter. Phys. Rev. Lett. **92**, 125502 (2004)
57. Shinde, D.B., Majumder, M., Pillai, K.V.: Counter-ion dependent, longitudinal unzipping of multi-walled carbon nanotubes to highly conductive and transparent graphene nanoribbons. Sci. Rep. **4**, 4363 (2014)
58. Aitchison, T.J., Ginic-Markovic, M., Matisons, J.G., et al.: Purification, cutting, and sidewall functionalization of multiwalled carbon nanotubes using potassium permanganate solutions. J. Phys. Chem. C **111**, 2440 (2007)
59. Zhu, Y., Stoller, M.D., Cai, W., et al.: Exfoliation of graphite oxide in propylene carbonate and thermal reduction of the resulting graphene oxide platelets. ACS Nano **4**, 1227 (2010)
60. McAllister, M.J., Li, J.L., Adamson, D.H., et al.: Single sheet functionalized graphene by oxidation and thermal expansion of graphite. Chem. Mater. **19**, 4396 (2007)
61. Cai, M., Thorpe, D., Adamson, D.H., et al.: Methods of graphite exfoliation. J. Mater. Chem. **22**, 24992 (2012)
62. Jiang, H., Liu, B., Huang, Y., et al.: Thermal expansion of single wall carbon nanotubes. J. Eng. Mater. Technol. **126**, 265 (2004)
63. Li, X., Wang, X., Zhang, L., et al.: Chemically derived, ultrasmooth graphene nanoribbon semiconductors. Science **319**, 1229 (2008)
64. Park, S., Ruoff, S.R.: Chemical methods for the production of graphenes. Nat. Nanotechnol. **4**, 217 (2009)
65. Hernandez, Y., Nicolosi, V., Lotya, M., et al.: High-yield production of graphene by liquid-phase exfoliation of graphite. Nat. Nanotechnol. **3**, 563 (2008)

66. Akhavan, O.: The effect of heat treatment on formation of graphene thin films from graphene oxide nanosheets. Carbon **48**, 509 (2010)
67. Valles, C., Drummond, C., Saadaoui, H., et al.: Solutions of negatively charged graphene sheets and ribbons. J. Am. Chem. Soc. **130**, 15802 (2008)
68. Mahmood, N., Zhang, C., Yin, H., et al.: Graphene-based nanocomposites for energy storage and conversion in lithium batteries, supercapacitors and fuel cells. J. Mater. Chem. A **2**, 15 (2014)
69. El-Kady, M.F., Shao, Y., Kaner, R.B.: Graphene for batteries, supercapacitors and beyond. Nat. Rev. Mater. **1**, 16033 (2016)
70. Singh, R.K., Kumar, R., Singh, D.P.: Graphene oxide: strategies for synthesis, reduction and frontier applications. RSC Adv. **6**, 64993 (2016)
71. Kinoshita, H., Nishina, Y., Alias, A.A., et al.: Tribological properties of monolayer graphene oxide sheets as water-based lubricant additives. Carbon **66**, 720 (2014)
72. Yamamoto, S., Kinoshita, H., Hashimoto, H., et al.: Facile preparation of Pd nanoparticles supported on single-layer graphene oxide and application for the Suzuki-Miyaura cross-coupling reaction. Nanoscale **6**, 6501 (2014)
73. Stankovich, S., Dikin, D.A., Piner, R.D., et al.: Synthesis of graphene-based nanosheets via chemical reduction of exfoliated graphite oxide. Carbon **45**, 1558 (2007)
74. Marcano, D.C., Kosynkin, D.V., Berlin, J.M.: Improved synthesis of graphene oxide. ACS Nano **4**, 4806 (2010)
75. Hummers, J., William, S., Offeman, R.E.: Preparation of graphitic oxide. J. Am. Chem. Soc. **80**, 1339 (1958)
76. Botas, C., Álvarez, P., Blanco, P., et al.: Graphene materials with different structures prepared from the same graphite by the Hummers and Brodie methods. Carbon **65**, 156 (2013)
77. Yan, J., Chou, M.Y.: Oxidation functional groups on graphene: structural and electronic properties. Phys. Rev. B **82**, 125403 (2010)
78. Zhang, L., Xia, J., Zhao, Q., et al.: Functional graphene oxide as a nanocarrier for controlled loading and targeted delivery of mixed anticancer drugs. Small **6**, 537 (2010)
79. Wilson, N.R., Pandey, P.A., Beanland, R., et al.: Graphene oxide: structural analysis and application as a highly transparent support for electron microscopy. ACS Nano **3**, 2547 (2009)
80. Hansora, D.P., Mishra, S.: Graphene Nanomaterials: Fabrication, Properties, and Applications. Pan Stanford (2017)
81. Songfeng, P., Cheng, H.M.: The reduction of graphene oxide. Carbon **50**, 3210 (2012)
82. Yuan, J., Ma, L.P., Pei, S., et al.: Tuning the electrical and optical properties of graphene by ozone treatment for patterning monolithic transparent electrodes. ACS Nano **7**, 4233 (2013)
83. Ishii, Y., Sakaguchi, S., Iwahama, T.: Innovation of hydrocarbon oxidation with molecular oxygen and related reactions. Adv. Synth. Catal. **343**, 393 (2001)
84. Cheng, Y.C., Kaloni, T.P., Zhu, Z.Y.: Oxidation of graphene in ozone under ultraviolet light. Appl. Phys. Lett. **10**, 073110 (2012)
85. Zhou, M., Wang, Y., Zhai, Y., et al.: Controlled synthesis of large-area and patterned electrochemically reduced graphene oxide films. Chem.: Eur. J. **15**, 6116 (2009)
86. Zhou, C., Chen, S., Lou, J., et al.: Graphene's cousin: the present and future of graphane. Nanoscale Res. Lett. **9**, 26 (2014)
87. Sahin, H., Leenaerts, O., Singh, S.K., et al.: Graphane. Wiley Interdiscip. Rev. Comput. Mol. Sci **5**, 255 (2015)
88. Elias, D.C., Nair, R.R., Mohiuddin, T.M.G., et al.: Control of graphene's properties by reversible hydrogenation: evidence for graphane. Science **323**, 613 (2009)
89. Umadevi, D., Sastry, G.N.: Graphane versus graphene: a computational investigation of the interaction of nucleobases, aminoacids, heterocycles, small molecules [CO_2, H_2O, NH_3, CH_4, H_2], metal ions and onium ions. Phys. Chem. Chem. Phys. **17**, 30260 (2015)
90. Reshak, A.H., Auluck, S.: Electronic and optical properties of chair-like and boat-like graphane. RSC Adv. **4**, 37411 (2014)
91. Wen, X.D., Yang, T., Hoffmann, R., et al.: Graphane nanotubes. ACS Nano **6**, 7142 (2012)

92. Feng, W., Long, P., Feng, Y., et al.: Two-dimensional fluorinated graphene: synthesis, structures, properties and applications. Adv. Sci. **3**, 1500413 (2016)
93. Samarakoon, D.K., Chen, Z., Nicolas, C., et al.: Structural and electronic properties of fluorographene. Small **7**, 965 (2011)
94. Paupitz, R., Autreto, P.A.S., Legoas, S.B., et al.: Graphene to fluorographene and fluorographane: a theoretical study. Nanotechnology **24**, 035706 (2012)
95. Chronopoulos, D.D., Bakandritsos, A., Pykal, M., et al.: Chemistry, properties, and applications of fluorographene. Appl. Mater. Today **9**, 60 (2017)
96. Grayfer, E.D., Makotchenko, V.G., Kibis, L.S., et al.: Synthesis, Properties, and Dispersion of Few-Layer Graphene Fluoride. Chem. Asian J. **8**, 2015 (2013)
97. Nair, R.R., Ren, W., Jalil, R., et al.: Fluorographene: a two-dimensional counterpart of Teflon. Small **6**, 2877 (2010)
98. Yuan, S., Rösner, M., Schulz, A., et al.: Electronic structures and optical properties of partially and fully fluorinated graphene. Phys. Rev. Lett. **114**, 047403 (2015)
99. Sturala, J., Luxa, J., Pumera, M., et al.: Chemistry of graphene derivatives: synthesis, applications, and perspectives. Chem. Eur. J. **24**, 5992 (2018)
100. Baughman, R.H., Eckhardt, H., Kertesz, M.: Structure-property predictions for new planar forms of carbon: layered phases containing sp^2 and sp atoms. J. Chem. Phys. **87**, 6687 (1987)
101. Enyashin, A.N., Ivanovskii, A.L.: Graphene allotropes. Phys. Status Solidi (b) **248**, 1879 (2011)
102. Xu, Z., Lv, X., Li, J., et al.: A promising anode material for sodium-ion battery with high capacity and high diffusion ability: graphyne and graphdiyne. RSC Adv. **6**, 25594 (2016)
103. Srinivasu, K., Ghosh, S.K.: Graphyne and graphdiyne: promising materials for nanoelectronics and energy storage applications. J. Phys. Chem. C **116**, 5951 (2012)
104. Kim, B.G., Choi, H.J.: Graphyne: hexagonal network of carbon with versatile Dirac cones. Phys. Rev. B **86**, 115435 (2012)
105. Coluci, V.R., Galvao, D.S., Baughman, R.H.: Theoretical investigation of electromechanical effects for graphyne carbon nanotubes. J. Chem. Phys. **121**, 3228 (2004)
106. Coluci, V.R., Braga, S.F., Legoas, S.B., et al.: New families of carbon nanotubes based on graphyne motifs. Nanotechnology **15**: S142 (2004)
107. Sun, L., Jiang, P.H., Liu, H.J. Graphdiyne: a two-dimensional thermoelectric material with high figure of merit. Carbon **90**, 255 (2015)
108. Long, M., Tang, L., Wang, D., et al.: Electronic structure and carrier mobility in graphdiyne sheet and nanoribbons: theoretical predictions. ACS Nano **5**, 2593 (2011)
109. Zhong, J., Wang, J., Zhou, J.G., et al.: Electronic structure of graphdiyne probed by X-ray absorption spectroscopy and scanning transmission X-ray microscopy. J. Phys. Chem. C **117**, 5931 (2013)
110. Pan, L.D., Zhang, L.Z., Song, B.Q., et al.: Graphyne-and graphdiyne-based nanoribbons: density functional theory calculations of electronic structures. Appl. Phys. Lett. **98**, 173102 (2011)
111. Li, Y., Zhou, Z., Shen, P., et al.: Structural and electronic properties of graphane nanoribbons. J. Phys. Chem. C **113**, 15043 (2009)
112. Terrones, M., Botello-Méndez, A.R., Campos-Delgado, J., et al.: Graphene and graphite nanoribbons: morphology, properties, synthesis, defects and applications. Nano Today **5**, 351 (2010)
113. Kucinskis, G., Bajars, G., Kleperis, J.: Graphene in lithium ion battery cathode materials: a review. J. Power Sources **240**, 66 (2013)
114. Yoo, E.J., Kim, J., Hosono, E.: Large reversible Li storage of graphene nanosheet families for use in rechargeable lithium ion batteries. Nano Lett. **8**, 2277 (2008)
115. Marom, R., Francis, A.S., Leifer, N., et al.: A review of advanced and practical lithium battery materials. J. Mater. Chem. **21**, 9938 (2011)
116. Subrahmanyam, G., Miele, E., Angelis, F.D., et al.: Review on recent progress of nanostructured anode materials for Li-ion batteries. J. Power Sources **257**, 421 (2014)

117. Dengyu, P., Song, W., Bing, Z., et al.: Li storage properties of disordered graphene nanosheets. Chem. Mater. **21**, 3136 (2009)
118. Ali, A.T., Ullah, H., Sudhagar, P., et al.: The application of graphene and its derivatives to energy conversion, storage, and environmental and biosensing devices. Chem. Rec. **16**, 1591 (2016)
119. Xu, C., Xu, B., Gu, Y., et al.: Graphene-based electrodes for electrochemical energy storage. Energy Environ. Sci. **6**, 1388 (2013)
120. Zhu, X., Zhu, Y., Murali, S., et al.: Nanostructured reduced graphene oxide/Fe_2O_3 composite as a high-performance anode material for lithium ion batteries. ACS Nano **5**, 3333 (2011)
121. Shuvo, M.A.I., Khan, M.A.R., Karim, H., et al.: Investigation of modified graphene for energy storage applications. ACS Appl. Mater. Int. **5**, 7881 (2013)
122. Georgakilas, H., Otyepka, M., Bourlinos, A.B., et al.: Functionalization of graphene: covalent and non-covalent approaches, derivatives and applications. Chem. Rev. **112**, 6156 (2012)
123. Chen, W., Zhu, Z., Li, S., et al.: Efficient preparation of highly hydrogenated graphene and its application as a high-performance anode material for lithium ion batteries. Nanoscale **4**, 2124 (2012)
124. Zhu, S., Li, T.: Hydrogenation-assisted graphene origami and its application in programmable molecular mass uptake, storage, and release. ACS Nano **8**, 2864 (2014)
125. Sun, C., Feng, Y., Li, Y., et al.: Solvothermally exfoliated fluorographene for high-performance lithium primary batteries. Nanoscale **6**, 2634 (2014)
126. Amini, M.N., Leenaerts, O., Partoens, B., et al.: Graphane-and fluorographene-based quantum dots. J. Phys. Chem. C **117**, 16242 (2013)
127. Zhang, H., Zhao, M., He, X., et al.: High mobility and high storage capacity of lithium in $sp-sp^2$ hybridized carbon network: the case of graphyne. J. Phys. Chem. C **115**, 8845 (2011)
128. Becton, M., Zhang, L., Wang, X., et al.: Mechanics of graphyne crumpling. Phys. Chem. Chem. Phys. **16**, 18233 (2014)
129. Uthaisar, C., Barone, V., Peralta, J.E.: Lithium adsorption on zigzag graphene nanoribbons. J. Appl. Phys. **106**, 113715 (2009)
130. Li, L., Raji, A.R.O., Tour, J.M.: Graphene-wrapped MnO_2–graphene nanoribbons as anode materials for high-performance lithium ion batteries. Adv. Mater. **25**, 6298 (2013)
131. Xiao, B., Li, X., Li, X., et al.: Graphene nanoribbons derived from the unzipping of carbon nanotubes: controlled synthesis and superior lithium storage performance. J. Phys. Chem. C **118**, 881 (2013)
132. Lin, J., Peng, Z., Xiang, C., et al.: Graphene nanoribbon and nanostructured SnO_2 composite anodes for lithium ion batteries. ACS Nano **7**, 6001 (2013)
133. Pan, H., Hu, Y.S., Chen, L.: Room-temperature stationary sodium-ion batteries for large-scale electric energy storage. Energy Environ. Sci. **6**, 2338 (2013)
134. Slater, M.D., Kim, D., Lee, E., et al.: Sodium-ion batteries. Adv. Funct. Mater. **23**, 947 (2013)
135. Hwang, J.Y., Myung, S.T., Sun, Y. K.: Sodium-ion batteries: present and future. Chem. Soc. Rev. **46**, 3529 (2017)
136. Kundu, D., Talaie, E., Duffort, V., et al.: The emerging chemistry of sodium ion batteries for electrochemical energy storage. Angew. Chem. Int. Ed. **54**, 3431 (2015)
137. Balogun, M.S., Luo, Y., Qiu, W., et al.: A review of carbon materials and their composites with alloy metals for sodium ion battery anodes. Carbon **98**, 162 (2016)
138. He, J., Wang, N., Cui, Z., et al.: Hydrogen substituted graphdiyne as carbon-rich flexible electrode for lithium and sodium ion batteries. Nat. Commun. **8**, 1172 (2017)
139. Zhang, S., Liu, H., Huang, C., et al.: Bulk graphdiyne powder applied for highly efficient lithium storage. Chem. Commun. **51**, 1834 (2015)
140. An, H., Li, Y., Gao, Y., et al.: Free-standing fluorine and nitrogen co-doped graphene paper as a high-performance electrode for flexible sodium-ion batteries. Carbon **116**, 338 (2017)
141. Liu, Y., Yang, Y., Wang, X., et al.: Flexible paper-like free-standing electrodes by anchoring ultrafine SnS_2 nanocrystals on graphene nanoribbons for high-performance sodium ion batteries. ACS Appl. Mater. Int. **9**, 15484 (2017)

142. Shao, Y., Park, S., Xiao, J.: Electrocatalysts for nonaqueous lithium–air batteries: status, challenges, and perspective. ACS Catal. **2**, 844 (2012)
143. Girishkumar, G., McCloskey, B., Luntz, A.C., et al.: Lithium−air battery: promise and challenges. J. Phys. Chem. Lett. **1**, 2193 (2010)
144. Park, M., Sun, H., Lee, H., et al.: Lithium-air batteries: survey on the current status and perspectives towards automotive applications from a battery industry standpoint. Adv. Energy Mater. **2**, 780 (2012)
145. Luntz, A.C., McCloskey, B.D.: Nonaqueous Li–air batteries: a status report. Nonaqueous. Chem. Rev. **114**, 11721 (2014)
146. Ma, Z., Yuan, X., Li, L., et al.: A review of cathode materials and structures for rechargeable lithium–air batteries. Energy Environ. Sci. **8**, 2144 (2015)
147. Ottakam, T., Muhammed, M., Stefan, A., et al.: The carbon electrode in nonaqueous Li–O$_2$ cells. J. Am. Chem. Soc. **135**, 494 (2012)
148. Li, Y., Jiajun, W., Xifei, L., et al.: Superior energy capacity of graphene nanosheets for a nonaqueous lithium-oxygen battery. Chem. Commun. **47**, 9438 (2011)
149. Wang, L., Ara, M., Wadumesthrige, K., Salley, S., Ng, K.Y.S.: Graphene nanosheet supported bifunctional catalyst for high cycle life Li-air batteries. J. Power Sources **234**, 8 (2013)
150. Kun, W., Wang, N., He, J., et al.: Graphdiyne nanowalls as anode for lithium−ion batteries and capacitors exhibit superior cyclic stability. Electrochim. Acta **253**, 506 (2015)
151. Zhang, Y., Gao, Z., Song, N., et al. Graphene and its derivatives in lithium–sulfur batteries. Mater. Today. Energy **9**, 319 (2018)
152. Li, L., Ruan, G., Peng, Z., et al.: Enhanced cycling stability of lithium sulfur batteries using sulfur–polyaniline–graphene nanoribbon composite cathodes. ACS Appl. Mater. Interfaces. **6**, 15033 (2014)
153. Zu, C., Manthiram, A.: Hydroxylated graphene–sulfur nanocomposites for high-rate lithium–sulfur batteries. Adv. Energy Mater. **3**, 1008 (2013)
154. Zhao, M.Q., Zhang, Q., Huang, J.Q., et al.: Unstacked double-layer templated graphene for high-rate lithium–sulphur batteries. Nature Commun. **5**, 3410 (2014)
155. Lu, S., Chen, Y., Wu, X., et al.: Three-dimensional sulfur/graphene multifunctional hybrid sponges for lithium-sulfur batteries with large areal mass loading. Sci. Rep. **4**, 4629 (2014)
156. Liu, Z., Li, J., Xiang, J., et al.: Hierarchical nitrogen-doped porous graphene/reduced fluorographene/sulfur hybrids for high-performance lithium–sulfur batteries. Phys. Chem. Chem. Phys. **19**, 2567 (2017)
157. Yu, M., Li, R., Wu, M., et al.: Graphene materials for lithium–sulfur batteries. Energy Storage Mater. **1**, 51 (2015)
158. Wu, S., Ge, R., Lu, M., et al.: Graphene-based nano-materials for lithium–sulfur battery and sodium-ion battery. Nano Energy **15**, 379 (2015)
159. Yin, W.W., Fu, Z.W.: The potential of Na-air batteries. Chem. Cat. Chem. **9**, 1545 (2017)
160. Wang, J., Yang, J., Nuli, Y., et al.: Room temperature Na/S batteries with sulfur composite cathode materials. Electrochem. Commun. **9**, 31 (2007)
161. Phil, K.: Batteries included. Electrical Connection Autumn **52** (2018)

Recent Progress in Graphene Research for the Solar Cell Application

Raju Nandi, Soumyadeep Sinha, Jaeyeong Heo, Soo-Hyun Kim
and Dip K. Nandi

Abstract In the past few years, tremendous efforts have been devoted to the synthesis and application of graphene and its derivatives toward the development of graphene-based solar photovoltaics. With their extraordinary electrical, mechanical, and thermal properties graphene-based materials are considered as an ideal candidate for the fabrication of low-cost and scalable photovoltaic devices. In fact, graphene-based materials have been successfully implemented in all types of photovoltaics including Si-based Schottky junction solar cells to the perovskite solar cells. Although, graphene-based solar cells have not yet been commercially applied and most of them are still limited to the research and development phase, however, it has a great potential to replace conventional transparent conducting oxides. This chapter provides a comprehensive overview of the applications of graphene and its derivatives, namely graphene oxide and reduced graphene oxide in the field of organic, perovskite, and dye-sensitized solar cells. The key challenges of the graphene-based solar cells are also addressed along with their promising future in flexible photovoltaics.

Keywords Graphene and its derivatives · Solar cells · Power conversion efficiency

R. Nandi
School of Advanced Materials Engineering, Chonbuk National University, Baekje-Daero 567,
Jeonju 54896, Republic of Korea
e-mail: rajunandi2008@gmail.com

S. Sinha · J. Heo
Department of Materials Science and Engineering, Optoelectronics Convergence Research
Center, Chonnam National University, Gwangju 61186, Republic of Korea
e-mail: soumyadeep20@gmail.com

S.-H. Kim · D. K. Nandi (✉)
School of Materials Science and Engineering, Yeungnam University, Gyeongsan, Gyeongbuk
38541, Republic of Korea
e-mail: dpnandi@gmail.com

© Springer Nature Switzerland AG 2019
S. Sahoo et al. (eds.), *Surface Engineering of Graphene*, Carbon Nanostructures,
https://doi.org/10.1007/978-3-030-30207-8_4

1   Introduction

The worldwide ever-growing energy demand puts forth numerous efforts for harvesting it from the renewable sources such as sunlight, wind, and ocean waves [1–3]. The Sun serves as a primary source of energy in the form of sunlight and heat, both of which are considered as clean and apparently infinite reservoir. Therefore, the Sun as an energy resource is not depletable in nature and as a result is apprehended in the center of green or renewable energy technology. Solar cells or solar photovoltaics (PVs) are such devices which produce electricity directly from sunlight and have been proven to be one of the most promising replacements to the conventional energy sources (coal, oil, and natural gas) [4–7]. The installation of PV devices, either in stand-alone mode or in large scale (solar PV power plants) is increasing rapidly and expected to play a pivotal role in electrical power generation in near future. In parallel to the practical implementation of the solar panels (a solar panel includes several cells in series and/or in parallel to achieve a desired voltage and current), the research and development are rigorously being carried out on different types of solar cells considering their fundamental working principles. We are currently in the era of developing fourth generation of solar cells. The first-generation cells are based on crystalline silicon, which not only exhibit highest power conversion efficiency (PCE) but also are commercially predominant in PV technology. Second-generation cells are thin-film solar cells (TFSCs) that include amorphous or polycrystalline silicon, CdTe and $CuInGaSe_2/CuZnSnS_2$ (CIGS/CZTS) cells and are also commercially being used. The third-generation cells are based on organic (or polymer) solar cells (OSCs), perovskite solar cells (PSCs), and dye- or semiconductor-sensitized solar cells (DSSCs). The fourth generation of PV technology is the hybridization of low-cost/flexible organic thin films with novel inorganic nanostructured materials in order to improve the performance of the devices. However, third- and fourth-generation solar cells have not yet been commercially applied and most of them are still limited to the research and development phase.

In this chapter, we will describe the major and emerging applications of graphene and its derivatives, namely graphene oxide (GO) and reduced graphene oxide (rGO) in the field of all categorized solar cells (mainly in OSCs, PSCs, and DSSCs). The superior optical and electrical properties of graphene and its derivatives have already displayed very promising results in solar PV research. In addition, graphene and its derivatives are also well known for its high thermal and chemical stability [8]. Graphene possesses extremely high carrier mobility (~20,000 cm^2 V^{-1} s^{-1}) and optical transmittance in the near-infrared (NIR) to visible spectrum of the light (~2.3% light gets absorbed in this range) [9]. However, the transmittance ($T\%$) and the sheet resistance (R_s) of graphene and its derivatives depend on the number of graphene layers, doping and the methods of synthesis process, as shown in Fig. 1. In addition, Fig. 1b also depicts the expected values for T (%) and R_s that could satisfy the industrial applications of these materials. These two properties are enabling this group of materials as a potential transparent conducting electrode (TCE) in several types of solar cells.

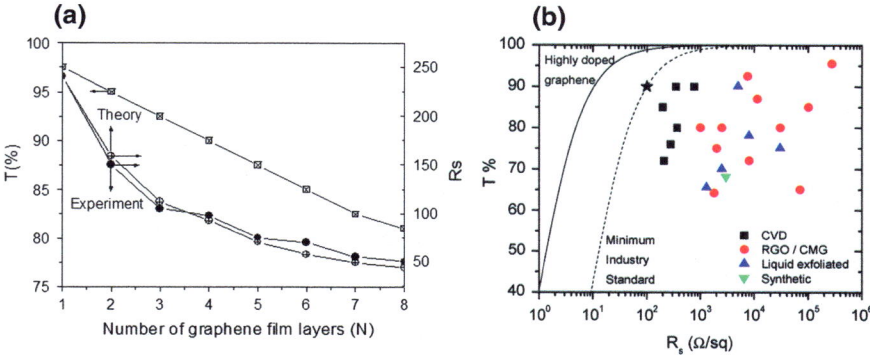

Fig. 1 a Variation of transmittance and sheet resistance of graphene as a function of number of layers [10] and **b** transmittance and sheet resistance values of graphene prepared by different methods [11]

In recent years, graphene-based materials have been successfully applied in all types of photovoltaics including Si-based Schottky junction solar cells to the newest member of this family, the perovskite solar cells [12–18]. Though the success is still restricted to laboratory-based research scale, it has a great potential to replace conventional transparent conducting oxide (TCO) such as indium (In)-doped tin oxide (ITO) or fluorine (F)-doped tin oxide (FTO). The abundance of carbon will easily make graphene and its derivatives economically beneficial over ITO/FTO and large-scale synthesis can further enable industrial process of graphene-based PV technology. Moreover, the scarcity of In or the safety hazards associated with F both could be avoided once graphene-based materials are up to the mark for large-scale commercialization. The similar work function (ϕ) of both pristine graphene and ITO/FTO (~4.5 eV) clearly indicates that the graphene could serve the same purpose. Figure 2 shows a schematic illustration of band alignments (Fermi level, valance and conduction band edge or the band-gaps) of graphene and GO along with

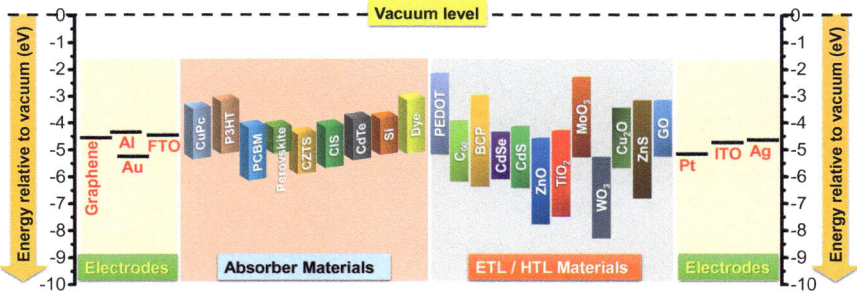

Fig. 2 Schematic diagram of energy levels of graphene-based materials with other materials commonly used in solar cells

the other major components (electrodes, absorbers, electron and hole transport layers (ETL/HTL) used in the field of solar PVs [12–18]. Furthermore, graphene and its derivatives exhibit excellent mechanical flexibility and bending durability and are thus a suitable candidate for flexible solar cells. The following sections deal with several specific types of solar cells where graphene and graphene-based composites were successfully implemented as a potential candidate.

2 Fundamentals of a Solar Cell

Solar cell is a large-area semiconductor device with a p–n junction that produces electricity under sunlight. However, this definition directly suits for the first-generation crystalline Si-based and second-generation thin-film-based solar PV technologies. As mentioned above, the commercialization of PV mostly occurs with the first generation of solar cells and to some extent with the second generation as well, we will mainly concentrate on the working principle of p–n junction photovoltaic device. Each photon (of incident light) with equal or larger energy than the band-gap ($h\nu \geq E_g$) of the semiconductor (Si or thin-film absorber) in principle should generate an electron-hole (e–h) pair. Such photo-generated e–h pairs are then spontaneously separated with the help of an electric field (built-in potential, ξ_B) that exists across the p–n junction. The separated electrons are thereafter collected through the external circuit producing electric energy and then will recombine with the holes at the opposite side of the p–n junction. A simple schematic illustration is shown in Fig. 3 to depict the fundamental working principle of a p–n junction Si solar cell. However,

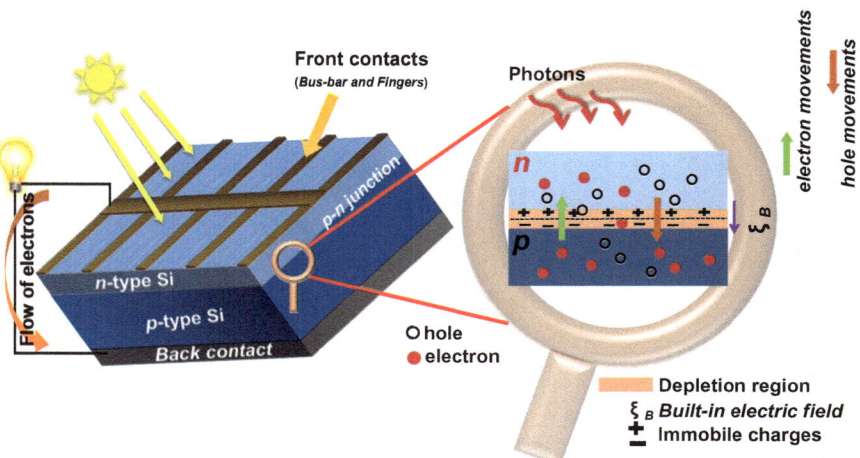

Fig. 3 Schematic illustration of a p–n junction Si solar cell showing the fundamental working principle under illumination

these photo-generated carriers once separated also develop a forward bias across the junction that results in a flow of diode current (I_D) opposite to the current under illumination (I_L). The simplistic current–voltage (I–V) equation of a solar cell therefore can be written as follows:

$$I = I_L - I_D \tag{1}$$

$$I = I_L - I_0\left[e^{qV/nkT} - 1\right] \tag{2}$$

where, I_0 is reverse saturation current, q is charge of an electron, n is diode ideality factor, k is Boltzmann's constant, and T is temperature in Kelvin.

In reality, every solar cell suffers from two different kinds of parasitic resistances known as series resistance (R_s) and shunt resistance (R_{sh}). The separated e–h pairs encounter resistance inside the bulk of the semiconductor as well as at the semiconductor-metal junction (at the front and rear contacts of the solar cell) before they are finally extracted to the external circuit. This resistance is defined as R_s for a solar cell, and its value should be as low as possible. On the other hand, any less resistive path across the p–n junction will lead some of the separated e–h pairs to recombine and this resistance is termed as R_{sh}. As R_{sh} provides an alternative path to flow the charge carriers opposite to the desired path, the magnitude of this resistance should be as high as possible. Any kind of defects at the junction generally gives rise to this resistance. Therefore, considering these two types of parasitic resistances, the more realistic I–V relation of a solar cell will differ from its ideal Eq. (2) and can be rewritten as follows:

$$I = I_L - I_0\left[e^{q[V+IR_s]/\eta kT} - 1\right] - [V + IR_s]/R_{sh} \tag{3}$$

The electrical equivalent circuits corresponding to an ideal solar cell and a practical cell with parasitic resistances (corresponding to Eqs. 2 and 3, respectively) are shown in Fig. 4. While the maximum voltage (under open circuit condition) that can be achieved from a solar cell is defined as open-circuit voltage (V_{oc}), the maximum current (with zero external load) drawn from the cell is defined as short-circuit current (I_{sc}). The other two major fundamental parameters to comprehend the potential of a solar cell are power conversion efficiency (PCE) and fill factor (FF). PCE is the ratio of maximum power output (P_{max}) of a cell to the input power of the incident radiation (P_{in}). On the other hand, FF is quantified as the ratio of P_{max} to the multiplication of V_{oc} and I_{sc}. The FF and PCE are therefore could be formulated as follows:

$$FF = P_{max}/[I_{sc} \times V_{oc}] = \left[I_{mp} \times V_{mp}\right]/[I_{sc} \times V_{oc}] \tag{4}$$

$$PCE = P_{max}/P_{in} = [FF \times I_{sc} \times V_{oc}]/P_{in} \tag{5}$$

where, I_{mp} and V_{mp} correspond to the current and voltage at maximum power, respectively.

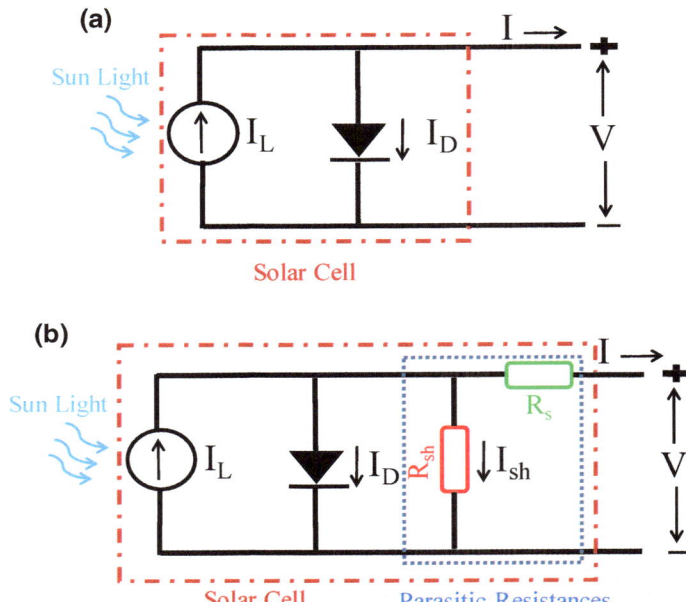

Fig. 4 Schematic illustrations of the equivalent electrical circuits (single diode model) of a solar cell **a** without (ideal) and **b** with parasitic resistances

While p- and n-type Si are used in crystalline Si-based solar PV, a hetero-junction (e.g., p-type CIGS/CZTS and n-type CdS/ZnOS junction or p-type CdTe and n-type CdS junction) is used in TFSCs. On the other hand, third- and fourth-generation solar cells generally consist of three active layers that include a low band-gap light-harvesting (absorber) layer sandwiched between electron and hole collectors. The absorber layer includes organic polymers, dye-sensitizers, or a blend of organic–inorganic material (like perovskites) and the electron/hole collectors are usually large band-gap inorganic and/or organic compound semiconductors. The common properties of an absorber layer include high absorption coefficient and carrier lifetime (or diffusion length), low-defect sites. Furthermore, the electron-hole collectors must have proper band alignments with the absorber and should also provide least recombination paths to the separated e–h pairs. Above all, a solar cell must be stable in the ambient condition and should withstand the harsh weather conditions over a reasonably long lifetime.

The metallic contact (both front and rear, also known as electrodes) is another important component of a solar cell which plays a significant role to extract the photo-generated carriers to the external circuit. In this regard, TCO (like ITO and FTO) and several metals (including expensive noble metals such as Ag, Au, or Pt) are most commonly used electrodes. While ITO and FTO restrict the flexibility of the device, best-performing metal electrodes (e.g., Ag or Au) limit the device to be widely applied due to the huge production cost. Additionally, the issues like scarcity and hazardousness associated with In and F make these electrodes an infeasible choice for large-scale fabrication as well as long-term utilization of the device. Graphene, with all its advantages, has great potential to address most of the above concerns and could become a game-changer to the existing PV technology in imminent future.

3   Graphene-Based Materials in Different Solar Cells

Due to their favorable opto-electronic properties, graphene-based materials have been and are being extensively used in various types of solar cells, including organic, perovskite, dye-sensitized, and inorganic solar cells. Pristine and functionalized graphene and its derivatives like GO or rGO are mainly used for this purpose. These materials are found to be potentially applicable in several aspects of a solar cell such as electrodes (anode and cathode), carrier (both electron and hole) transport layers, additives to active layers, and protecting layers. The following subsections briefly describe the research and developments in each of these categories.

3.1   Graphene-Based Materials for Organic Solar Cells (OSCs)

Organic–polymer solar cells have attracted enormous attention due to their flexibility, lightweight, solution-synthesis process, and low cost. More importantly, the flexibility of these devices opened a room to install them on any kind of surface with random geometry. Recently, graphene-based materials have been applied extensively in OSCs because of their high transparency, conductivity, chemical stability, and flexibility. A typical device structure and working principle of organic solar cells are schematically shown in Fig. 5.

When a photon is absorbed by polymers or their composites (also known as donor), an electron is excited from the highest occupied molecular orbital (HOMO) to the lowest unoccupied molecular orbital (LUMO), leading to a formation of bound e-h pair (exciton). These photo-generated charge carriers can then be separated by establishing a hetero-junction between the donor and the acceptor material. The most commonly used hetero-junction is termed as bulk hetero-junction (BHJ). Several polymers and their composites have been successfully used so far in BHJ-based OSCs. However, most commonly used polymers are poly (3-hexylthiophene) (P3HT) and/or poly (3-octythiophene) (P3OT) as donor and (6, 6)-phenyl C61-butyric acid methyl ester (PCBM) as electron acceptor. Moreover, the highest PCEs also have been achieved in BHJ-based OSCs, fabricated using these materials. Graphene materials used in OSCs are investigated as additives to the donor or donor–acceptor material, and as conductive electrodes (anode or cathode).

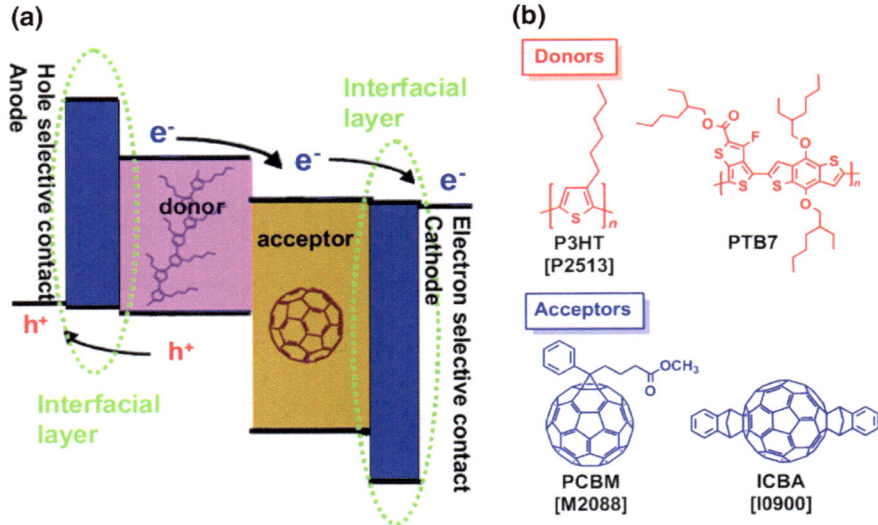

Fig. 5 **a** A typical schematic diagram of an OSC showing the different parts and movements of the photo-generated *e–h* inside the cell [19] and **b** chemical structure of commonly used donors and acceptors in OSCs [20]

3.1.1 Graphene as Additive Materials

Graphene in many forms (pristine graphene, GO, rGO, fullerene or graphene quantum dots) has been blended or grafted with commonly used donor polymer P3HT (or P3OT) in order to improve the performance of OSCs [21–23]. It has been shown that the incorporation of graphene materials in P3HT results in the enhancement of both V_{oc} and short-circuit current density (J_{sc}). The variation of photovoltaic parameters as a function of graphene concentration in P3HT and P3HT: PCBM is typically shown in Fig. 6 [24]. The increase in V_{oc} is primarily attributed to the modification of HOMO level of polymer, and the enhanced J_{sc} is associated with the improved electron transport due to its superior electrical conductivity. In a similar context, fullerene-grafted graphene nanosheets have been used as electron acceptors in P3HT-based bulk hetero-junction OSCs. The solar cell based on fullerene-grafted: P3HT active layer displayed a 2.5-fold increase in PCE compared to that of its P3HT counterpart [22]. Similarly, Gupta et al. [25] investigated aniline ($C_6H_5NH_2$) functionalized graphene quantum dots (ANI-GQDs) as additive to P3HT acceptor materials in OSCs. It was observed that the overall PCE monotonously increased from 0.77 to 1.14% with increase in the ANI-GQDs incorporation in P3HT from 0.5 to 1 wt%. Further increase in the ANI-GQDs resulted in a rapid decrease in the overall PCE (0.12% with 5 wt% ANI-GQDs). In this context, the best result was obtained from the OSCs fabricated with poly (2-methoxy-5-(30,70-dimethyloctyloxy)-1,4-phenylenevinylene) (MDMO-PPV) doped GO-grafted: P3HT as active layer [21]. Incorporation of MDMO-PPV doped GO into P3HT not only resulted in substantial

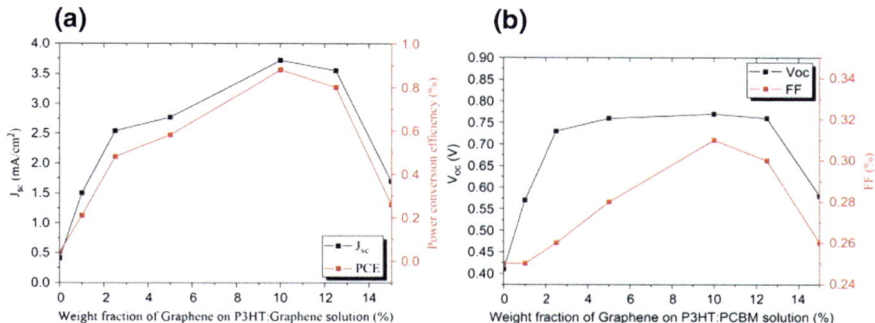

Fig. 6 Dependence of the **a** J_{sc}, PCE and **b** V_{oc}, FF on the different graphene concentrations in P3HT and P3HT: PCBM, respectively [24]

increase in the absorption range but also significantly improved the photovoltaic performance. The OSCs exhibited highest PCE of 1.5% with V_{oc} of 0.88 V, J_{sc} of 4.39 mA cm^{-2} and FF ~39%.

3.1.2 Graphene as Electrode Materials

Electrodes in any photovoltaic devices are expected to be highly transparent and conductive. As mentioned above, ITO and FTO are the most commonly used electrodes for this purpose. However, these electrodes have limitations in OSCs, such as sensitive to pH, high cost of preparation, difficulty in patterning, and its brittle nature which is not suitable for flexible devices. To address some of these issues, graphene materials have been investigated as alternative electrodes in OSCs. In the last few years, graphene materials are not only applied as anode for the substitution of conventional TCOs but also as cathode for which low work function metals are frequently used. Chemical vapor-deposited graphene films and solution-processed GO or rGO films are the most studied electrode materials in OSCs. The schematic diagrams of photovoltaic devices using aforementioned materials as electrodes (anode and cathode) are shown in Figs. 7 and 8. As an anode, it has been found out that the sheet resistance of graphene film primarily controls the efficiency of OSCs. In order to lower the sheet resistance, different strategies such as controlling the thickness, functionalization, post-deposition treatment, and doping of graphene film have been adopted [26–29]. The thickness of graphene in the form of thin film has strong influence on the sheet resistance and optical transmittance which substantially affects the performance of OSCs, particularly J_{sc} of the device. For example, Yin et al. [30] have demonstrated that the V_{oc} of the device was nearly independent on the thickness of rGO film, however, the J_{sc} was drastically influenced with the thickness (V_{oc}~ 0.56 V, J_{sc} 1.74, 3.31, 4.39 and 4.24 mA cm^{-2}, for thickness of 4, 10, 16, 21 nm, respectively). The highest PCE (0.78%) of OSCs was obtained from an optimum thickness of 16 nm of the rGO film. The sheet resistance of graphene films can also be

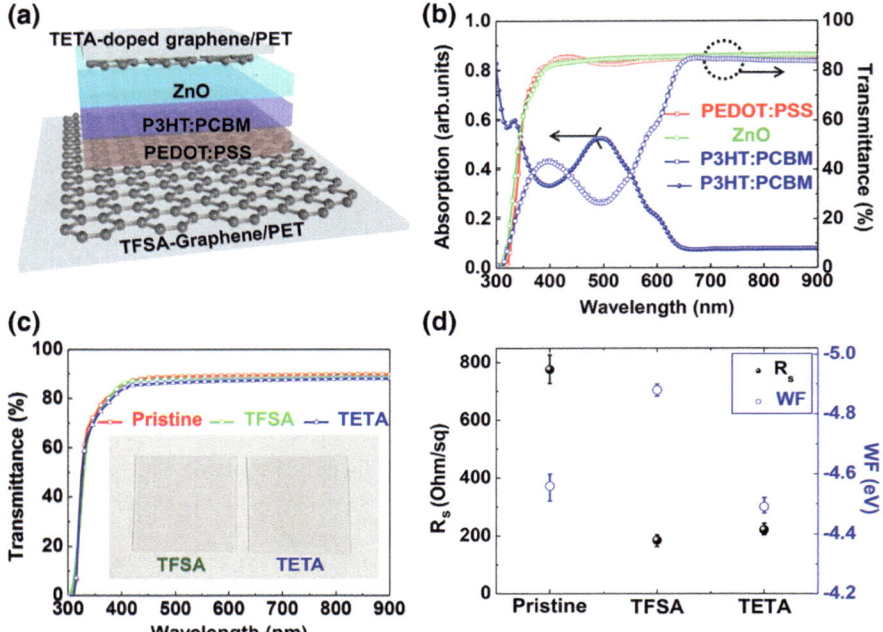

Fig. 7 **a** Schematic of the device architecture for typical OSCs with graphene electrodes, **b** transmission spectra of PEDOT: PSS, ZnO, and P3HT: PCBM and absorption spectrum of P3HT: PCBM, **c** transmittance spectra of pristine graphene, triethylenetetramine-doped graphene, and TFSA-doped graphene and **d** sheet resistances and work functions of pristine graphene, triethylenetetramine-doped graphene, and TFSA-doped graphene [27]

reduced by chemical functionalization without sacrificing the optical transmittance. Butylamine-functionalized graphene films displayed twofold decrease in the sheet resistance [28]. Subsequently, the OSCs fabricated with butylamine functionalized graphene as anode resulted in drastic increase in PCE from 0.38 to 0.74% without affecting the V_{oc} (0.53 V) of the device. The photovoltaic performance of OSCs has also been considerably enhanced by post-deposition treatment of graphene films. In this regard, the highest cell efficiency was obtained by the HNO_3-(nitric acid-) treated multilayer graphene films as anode materials exhibiting PCE of 2.86% [31]. Among all of the above-mentioned methods, doping has been proven to be the most beneficial method for improving the performance of OSCs [27]. The best cell efficiency was obtained by trifluoromethanesulfonylamide (TFSA)-doped graphene film as anode materials exhibiting PCE of 3.3% with V_{oc} of 0.62 V, J_{sc} of 9.84 mA cm^{-2} and FF of 54.5%.

Although several studies have reported on the graphene-based electrode as anode in OSCs, however, the use of the graphene materials as cathode is still very limited. In this case, the control of the work function of graphene is the major strategy to improve the overall PCE of OSCs. To tune the work function of graphene materials,

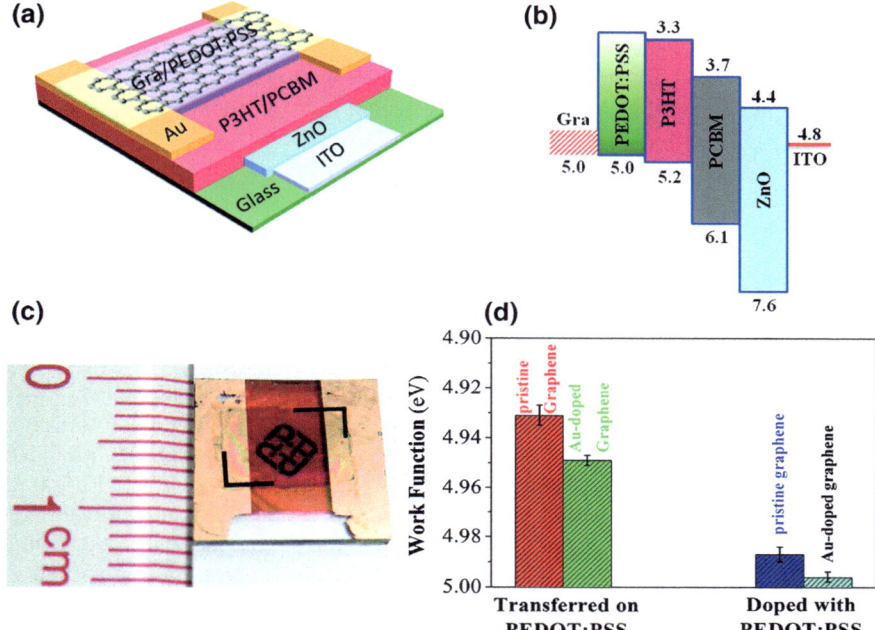

Fig. 8 **a** Schematic diagram of a semitransparent OSC with a structure of glass/ITO/ZnO/P3HT: PCBM/PEDOT: PSS/graphene, **b** band structure of the cell, **c** semitransparent cell with graphene as top electrode, and **d** work functions of graphene/PEDOT: PSS electrodes prepared at different conditions [35]

different procedures, for example, using interfacial layers or doping of graphene films with different kinds of elements have been implemented [32–34]. Interfacial engineered and/or doped graphene films exhibited reduced work function and sheet resistance which resulted in an increase in built-in potential as well as enhancement of charge extraction in OSCs. Jo et al. [34] have demonstrated that the work function of multilayer graphene films can be reduced by introducing an interfacial dipole layer poly (9,9-bis((6′(N,N,N-trimethyl ammonium) hexyl)-2,7-fluorene)-alt-(9,9-bis(2-(2-(2-methoxyethoxy) ethoxy)ethyl)-9-fluorene)) dibromide (WPF-6-oxy-F). The highest PCE obtained from OSCs fabricated with WPF-6-oxy-F interfacial-engineered graphene films as cathode material was 1.23%. Like anode materials, doping has also been revealed as a more effective approach for the modification of graphene-based cathode materials. The OSCs fabricated with Au nanoparticles' doped graphene as cathode materials have displayed maximum PCE of 2.7% with J_{sc} of 10.58 mA cm^{-2}, V_{oc} of 0.59 V, and FF of 43% [35].

In conclusion, the graphene and its derivatives have found to be explored mostly as electrode materials and the enhancement of the device performance of such graphene-based OSCs could be considered worthwhile. Although the graphene-based electrodes still restrict the PCE of the devices in comparison with the conventional TCOs like FTO, further research and development would lead to a successful replacement of TCOs in OSCs owing to the additional advantages of graphene such as the flexibility, compatibility, and cost-effectiveness. However, the extremely poor PCEs obtained using graphene-based materials as additives seem to be an inefficient approach for modifying OSCs.

3.2 Graphene-Based Materials for Perovskite Solar Cells (PSCs)

Perovskite solar cells (PSCs) are one of the major breakthroughs in the field of solar PV. The efforts of a decade or so makes this cell probably the best alternative to well-established and commercialized Si-based solar cells. Perovskite, which has been so far explored as an absorber layer in PSCs, is the organic–inorganic halide-based materials with a common structure of ABX_3 (where A is CH_3NH_3, B is Pb/Sn and X is I, Cl and Br). The potential of this extraordinary material results in extremely high PCE (over 20%) of these PSCs in almost no time in comparison with any other kind of solar cells, what so ever. Among several types of similar perovskite materials, methyl ammonium lead iodide ($CH_3NH_3PbI_3$, $MAPbI_3$) is the most commonly studied absorber material in PSCs. It is also envisioned that the improved stability with large-scale production of these PSCs could be a real game-changer to current solar PV market. In this context, several strategies such as electrode modification, surface modification of perovskite layer, introduction of plasmonic nanoparticles, and graphene-based materials into the device layers have been employed to improve the performance and stability of PSCs. In particular, introduction of graphene materials into PSCs has attracted intensive attention due to its flexibility, low cost of production, high chemical stability, and appropriate work function.

Perovskite materials are the intrinsic (neither p-type nor n-type) semiconductors with low exciton binding energy (2–55 meV) [36]. Different configurations of PSCs in direct and inverted architectures with planar and mesoscopic structures are schematically shown in Fig. 9. The photo-generated e–h pairs in the perovskite are quickly collected to the electron and hole-transporting materials (ETM and HTM) due to slow charge carrier recombination. Finally, these charge carriers are extracted to the external circuit with the help of cathode/anode. Graphene materials in PSCs were investigated as conductive electrodes, carrier-transporting materials (hole and electron) and stabilizer material.

3.2.1 Graphene as Conductive Electrodes

Graphene films have been successfully used as top or bottom conductive electrode for PSC applications [37–40]. Typical schematics of graphene materials as transparent electrodes in PSCs are shown in Fig. 10. In general, Au is used as the best-performing

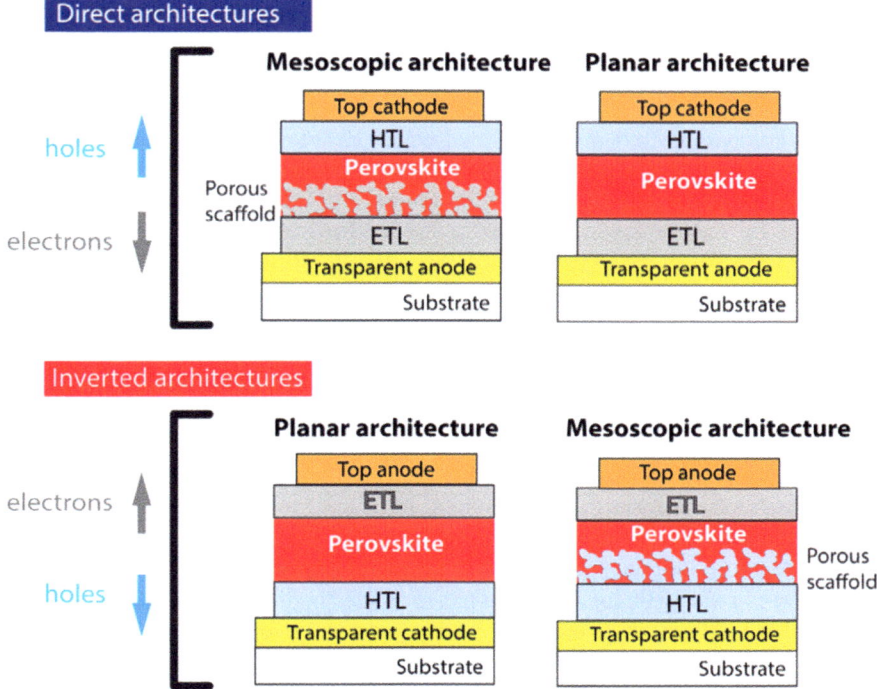

Fig. 9 Direct and inverted PSCs device architectures with planar and mesoporous scaffolds [15]

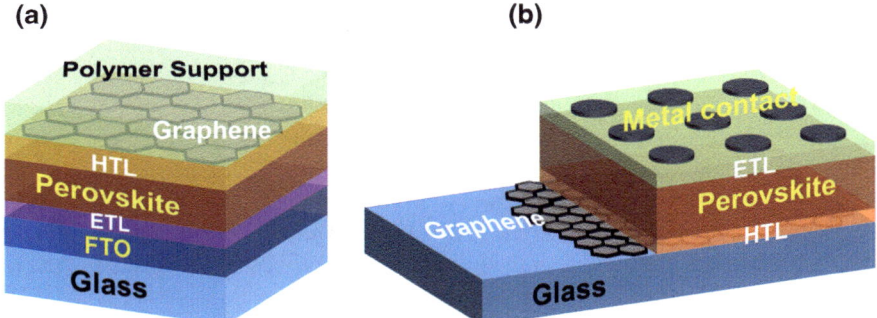

Fig. 10 **a** Schematic of a MAPbI$_3$-based PSC with transparent CVD-grown graphene as top contact [44] and **b** schematic structure of the inverted FTO-free PSC utilizing graphene as a transparent anode [39]

top electrode in PSCs. As a replacement, graphene film is first grown on Cu foil by chemical vapor deposition (CVD) and then mechanically transferred on top of the HTM (usually spiro-OMeTAD). In this context, Lang et al. [41] have demonstrated that the photovoltaic performance of the PSCs fabricated with single-layer graphene is nearly comparable to that of a reference device based on Au electrode. In addition, the incorporation of graphene avoiding Au results in a semitransparent device which also shows a reasonable PCE of 8.3%. However, it is to be noticed that the sheet resistance of a single-layer graphene is high and it is also poorly adherent to the spiro-OMeTAD. In order to overcome these issues, a further study has shown that the conductivity and adhesion of the graphene could be improved by introducing a thin layer (~20 nm) of PEDOT: PSS on spiro-OMeTAD [40]. In this case, the optimized device with double-layer graphene films achieved a PCE of 12.37%, which is comparable to the Au-electrode-based device but relatively higher with respect to that of the TCO-free PSCs [42, 43]. On the other hand, Sung et al. [39] investigated graphene-coated glass substrate as transparent bottom anode in PSCs with inverted architecture. Unfortunately, it was difficult to deposit HTM (PESOT: PSS or spiro-OMeTAD) uniformly on the graphene film owing to its hydrophobic nature and thus the device performance could not be evaluated. To address this problem, a thin (~2 nm) layer of molybdenum trioxide (MoO$_3$) HTM has been additionally incorporated on top of the graphene surface, which provided hydrophilicity and formation of desirable band alignment between the MoO$_3$/graphene electrode and HTM. Through such interfacial engineering, the device performance was significantly improved to a maximum PCE of 17.1% with J_{sc} of 21.9 mA cm^{-2}, V_{oc} of 1.03 V, and FF of 72% [39].

In view of flexible PV applications, PSCs have also been fabricated on polyethylene naphthalate (PEN) and polyethylene terephthalate (PET) substrates with graphene as bottom electrode [38, 45]. While the PCE of the flexible PSCs fabricated on PEN substrate retains ~85% of its initial PCE after 5000 bending cycles, the performance of PSCs fabricated on PET degraded by ~14% only after 500 cycles. Nevertheless, these results are very promising compared to the brittle ITO-based flexible PSCs, which showed a rapid decrease in PCE only after 250 bending cycles [46, 47]. These studies certainly set a good example that the brittleness of ITO/FTO is not a practical solution for flexible solar cells. Therefore, utilization of graphene, with further research and development, seems to be the most appropriate method for future flexible PV technology.

3.2.2   Graphene as Carrier-Transporting Materials

One of the most relevant application of graphene materials for PSCs is related to both electron and hole transport layers. Spiro-OMeTAD is a well-accepted and the best HTM for PSCs because of its suitable work function, high solubility, and large hole mobility. However, this HTM is prone to oxidation and suffers from photodoping mechanisms which result in significant processing issues. Another critical issue of spiro-OMeTAD is associated with its considerably short device lifetime [46,

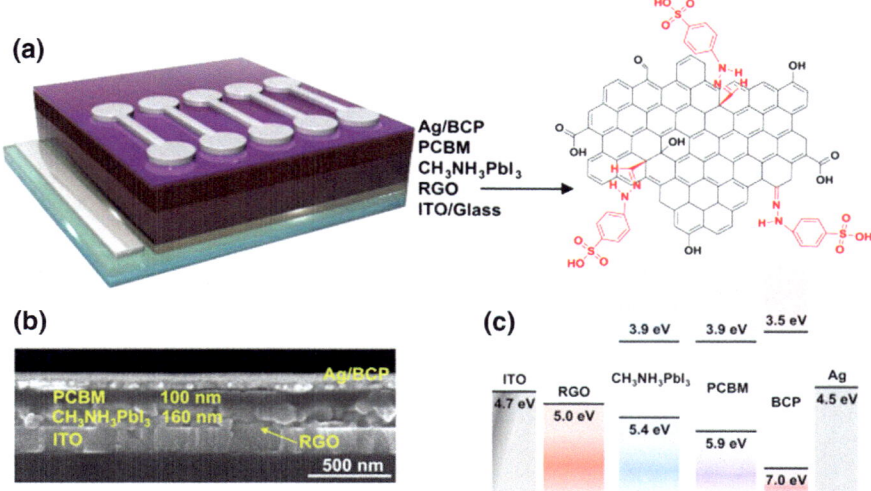

Fig. 11 a Schematic PSC architecture employing rGO as HTM and the chemical structure of rGO used in this study, b the cross-sectional SEM image of the resultant device, and c the corresponding energy-level diagram of each layer [48]

47]. In this regard, graphene derivatives particularly, GO and rGO were identified as promising alternatives to spiro-OMeTAD and other organic or inorganic HTMs (typical device structure is schematically shown in Fig. 11) [48]. Graphene oxide film was first employed as HTM in an inverted planar hetero-junction PSCs based on the iodide/chloride mixed halide perovskite absorber [49]. The multilayer GO film was deposited by spin-coating from an aqueous suspension, followed by a low temperature annealing. The PSCs fabricated with this GO film as HTM displayed superior performance (PCE ~11.1%) compared to the PEDOT: PSS-based device (PCE ~9.3%). It has been shown that rGO also has similar effects on the performance of PSCs [49]. The improved PV characteristics are attributed to the formation of preferentially oriented larger perovskite domains on top of the GO or rGO films. As mentioned above, doping has a great influence in the sheet resistance and work function of GO films. In this context, GO film doped with both PEDOT: PSS and silver trifluoromethane sulfonate (AgOTf) were found to be beneficial for PSCs with inverted device architecture and showed a slightly better performance (compared to PEDOT: PSS HTM) with maximum PCE of 11.9% [50]. Furthermore, GO or rGO films stacked with spiro-OMeTAD also has been explored as HTM in PSCs. Due to the presence of poly-aromatic rings at its surface and oxygenated carboxylic groups on the edges of amphiphilic GO film, a much stronger wettability of the spiro-OMeTAD could be achieved on the perovskite layer. As a result, device performance was significantly improved, with a PCE increasing from 10.0% (without GO) to 14.5% (with GO) [51]. The photovoltaic characteristics of the PSCs fabricated with different types of rGO are shown in Fig. 12.

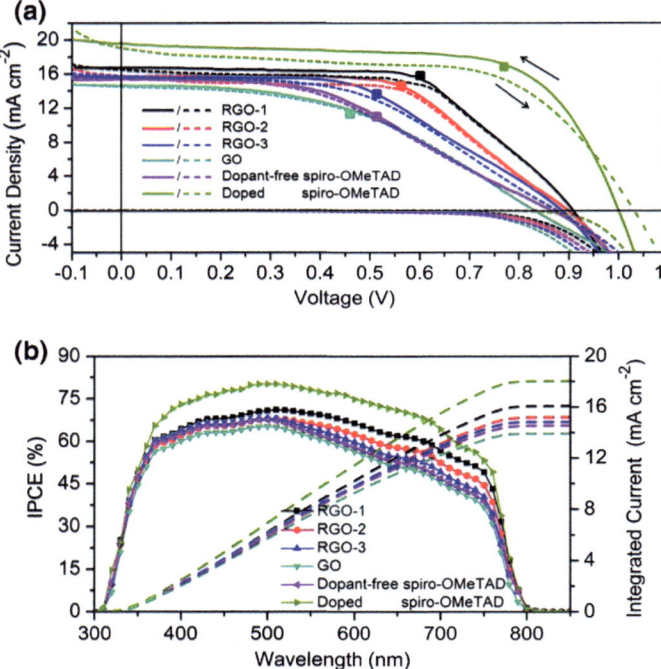

Fig. 12 **a** *J-V* curves of perovskite solar cells with different HTMs (rGO-1 (GO reduced at pH 1), rGO-2 (GO reduced at pH 2), rGO-3 (GO reduced at pH 3), GO, dopant-free spiro-OMeTAD and doped spiro-OMeTAD, respectively), **b** the incident photon to current efficiency spectra of perovskite solar cells with different hole transport layers and integrated current density [51]

Due to its ambipolar nature, graphene materials have been further explored to improve electron extraction efficiency as ETM in PSCs (schematics shown in Fig. 13). In order to overcome the intrinsic limitation of direct device architectures based on TiO_2 (dense and/or mesoporous) ETM, graphene/TiO_2 nanocomposites (or graphene

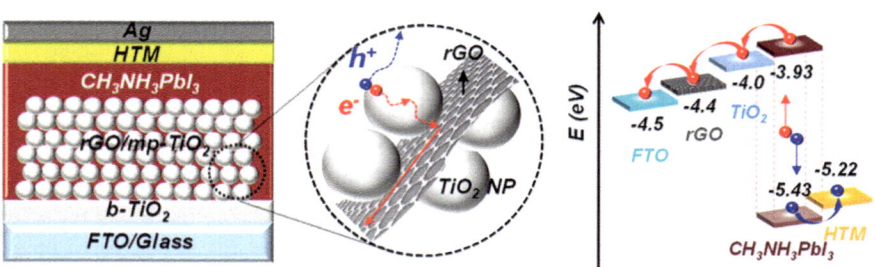

Fig. 13 Schematic representation of rGO/mesoporous-TiO_2 nanocomposite as ETM in PSCs along with energy-level diagram of each layer [53]

materials as interlayer between TiO_2 and perovskite) have been investigated. It is observed that the introduction of graphene materials in TiO_2 significantly influences the electrical properties of the fabricated device. The intimate interfacial contact between TiO_2 and graphene materials not only resulted in substantial reduction in series resistance but also decreased e-h recombination in the active layer, improving the device performance considerably. For example, Umeyama et al. [52] demonstrated that the overall PCE was enhanced by 40% for the rGO-embedded dense TiO_2 (~9.3%), compared to a reference cell fabricated only with TiO_2 (PCE 6.6%) as an ETM. The device performance could be further improved significantly by embedding rGO with mesoporous TiO_2 nanoparticles which delivered a highest PCE of ~14.5% [53]. However, ETM layer consists of pristine graphene nanoflakes and mesoporous TiO_2 nanoparticles exhibited the relatively better performance of the device with a PCE of 15.6% [54]. More promising results were recently obtained by introducing three-dimensional scaffold rGO between perovskite and TiO_2 films. Compared with conventional devices, rGO scaffold-based PSCs showed better charge transfer and lower carrier recombination, which resulted in 27% improvement in the device performance. The rGO-based device displayed PCE as high as 17.2%, with J_{sc} of 22.8 mA cm^{-2}, V_{oc} of 1.05 V, and FF of 72%. In addition, an increase in the external quantum efficiency (EQE) of the device is also observed, which is associated with light-trapping property and larger surface area of the perovskite film. Moreover, the rGO-based device showed lower hysteresis and superior stability over TiO_2-based devices (Fig. 14) [55].

3.2.3 Graphene as Stabilizer Material

It is well known that the instability of perovskite thin films in ambient environment is the biggest hindrance to its commercial applications in solar PVs. The perovskite is easily hydrolyzed and decomposed from dark brown into yellowish thin films under humid environment [56]. To solve this problem, an additional layer of graphene and its derivatives have been deposited as an encapsulation on top of the perovskite layer [57–59]. Pristine graphene is intrinsically hydrophobic in nature and hence an appropriate candidate as a barrier for moisture which has been also employed in hybrid or organic solar cells. The presence of a graphene or rGO layer in PSCs exhibited an increased device lifetime, through a better stability not only with respect to moisture but also by preventing photo-oxidation [60, 61]. In addition, the diffusion of halide ions from the perovskite layer to the top metal electrode can also be prevented by such modification. In this regard, edge-functionalization of graphene nanoplatelets (EFGnPs) with different elements like H and F were further studied [62]. However, edge-functionalization with F (EFGnPs-F) has displayed excellent stability over a long period of time (for 30 days) compared to H-functionalized or pure graphene nanoplatelets owing to the enhanced hydrophobicity of the PSC. Figure 15 depicts a schematic of such edge-fucntionalization of graphene used in PSCs which resulted in a significant enhancement in the device stability without compromising the device performance.

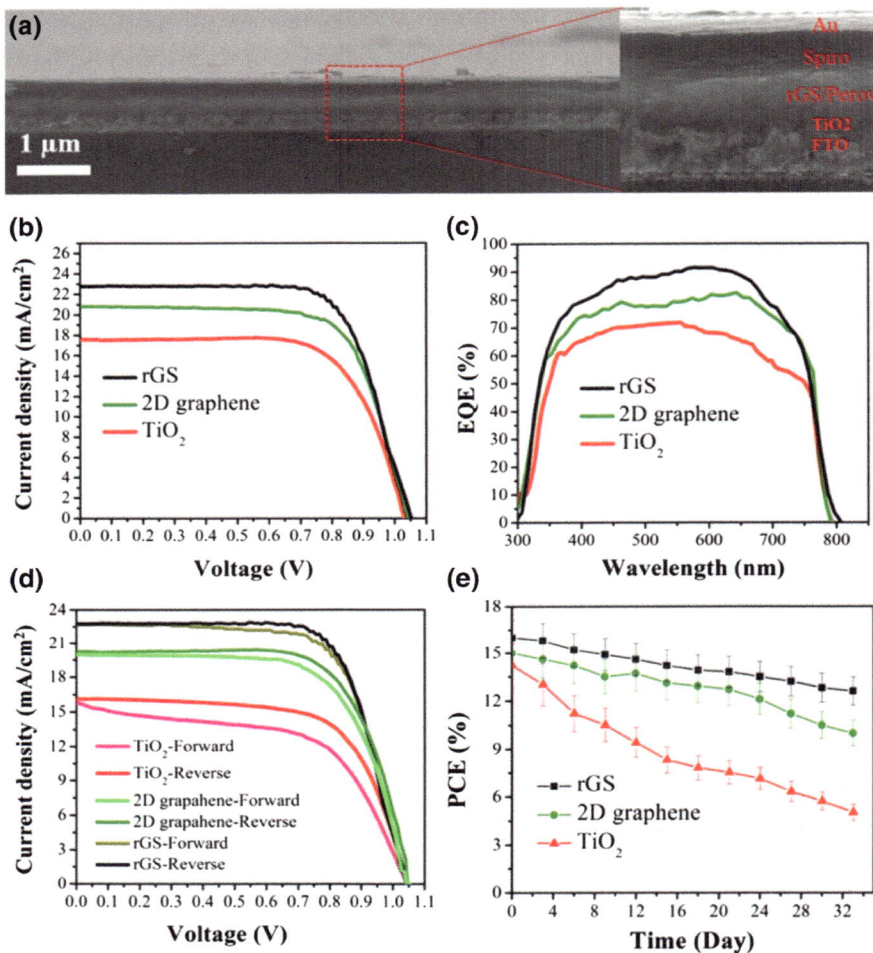

Fig. 14 PV characteristics of PSCs based on different interfacial layers such as rGO scaffold (rGS), two-dimensional rGO (2D graphene). **a** Cross-sectional SEM image of the device on the rGO scaffold, **b** J–V curves under 1.5 AMG indicate the performance of the devices, **c** EQE spectra, **d** hysteresis in the J–V measurements, and **e** stability of the devices in an ambient environment with ~65% humidity [55]

Among all existing types of solar cells, graphene and its derivatives displayed extremely high PCEs for PSCs. The overwhelming success of this latest category of solar cells is primarily attributed to the inherent capabilities associated with the perovskite material itself as an absorber. However, the successful incorporation of graphene-based materials as electrodes, carrier transport layers and stabilizer/protection layer paved the way for commercialization of PSCs. The recent studies of PSCs using graphene materials are truly remarkable in this direction to

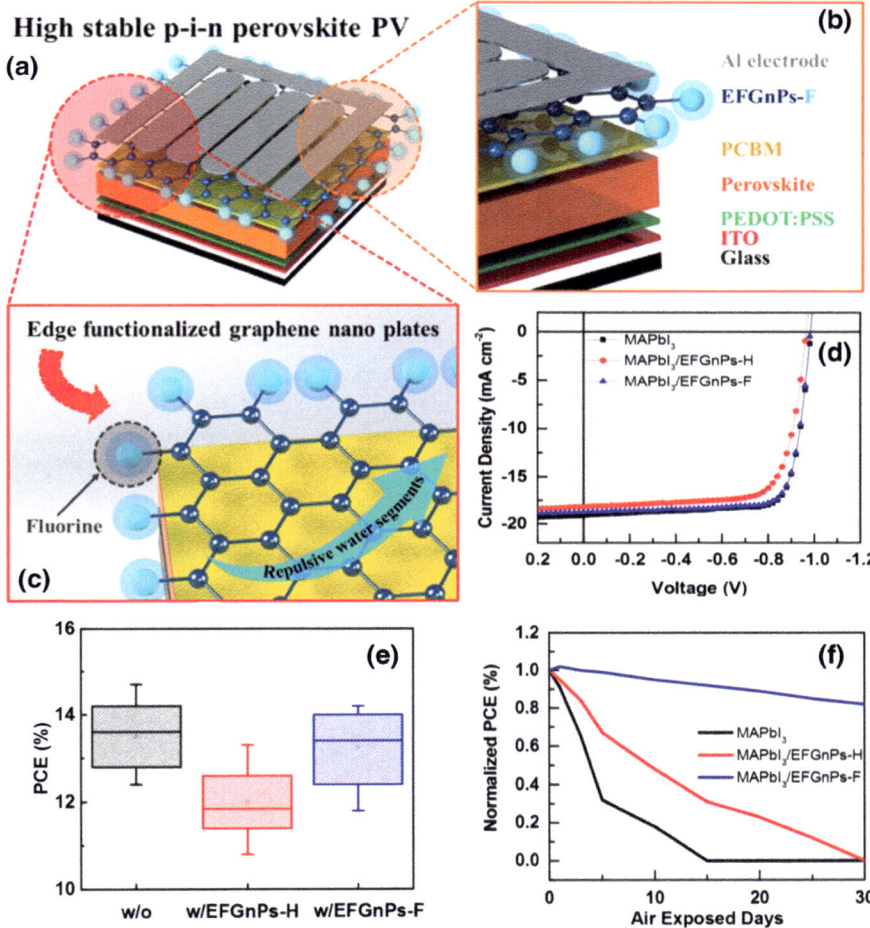

Fig. 15 **a–c** Schematic illustrations of functionalized graphene platelets used as a protecting layer in MAPbI$_3$-based PSC and its effects on stability and the performance of the cell, **d** $J-V$ characteristics of MAPbI$_3$ devices with EFGnPs-H and EFGnPs-F layers, **e** PCE distribution of MAPbI$_3$ devices with EFGnPs-H and EFGnPs-F layers and **f** stability of MAPbI$_3$ devices with EFGnPs-H and EFGnPs-F layers over 30 days [62]

resolve the critical stability issue without sacrificing the device performance. The flexible and stable PSCs including graphene and/or its derivatives possess significant potential to revolutionize the solar PV industry in imminent future.

3.3 Graphene-Based Materials for Dye-Sensitized Solar Cells (DSSCs)

Dye-sensitized solar cells (DSSCs) have drawn considerable interest from researchers as a promising low-cost thin-film solar cell technology. The DSSCs typically consist of five components: [1] a mechanical support coated with TCOs; [2] a semiconductor thin film or nanostructures, usually TiO_2; [3] a sensitizer (S, usually Ruthenium complexes) adsorbed onto the surface of the semiconductor; [4] an electrolyte containing a redox mediator (usually triiodide/iodine); and [5] a counter-electrode capable of regenerating the redox mediator (usually platinum) [12, 63, 64]. A schematic representation of typical DSSC structure and the incorporation of graphene-based materials into several parts of it are shown in Fig. 16. Under illumination, when a photon is absorbed by the sensitizer, it is excited to a higher energy state (S*) followed by the injection of an electron into the conduction band of the semiconductor that diffuses further to the current collector, leaving behind the sensitizer in the oxidized state. The injected electron arrives at the counter-electrode through the external load where it is transferred to electrolyte or hole collector to reduce the redox mediator which in turn regenerates the sensitizer and completes the electrical circuit. Graphene materials, with their superior electrical, optical, and mechanical properties, have been investigated as photo-anode, electrolyte, and counter-electrode in DSSCs.

3.3.1 Graphene as Photo-anode

The photo-anode of a DSSC is composed of a semiconducting layer sensitized by a dye on a transparent conducting material deposited on glass or plastic substrate. The

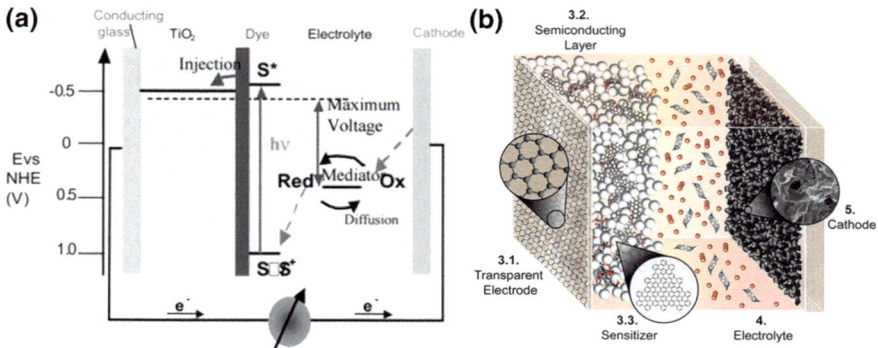

Fig. 16 **a** Schematic representation and working principle of a typical DSSC [65] and **b** schematic of a DSSC incorporating graphene materials in different parts of the device [66]

graphene-based materials in photo-anode of a DSSC have been used as transparent conductors, semiconducting layer, and sensitizer.

As transparent conductors Graphene has been used as transparent conducting electrode for DSSCs as early as in 2008, even though further studies for such implementations are limited. The graphene films used in this case were obtained by the exfoliation of graphite oxide followed by thermal reduction [67]. However, the device performance and resultant PCE (0.26%) were hindered by the relatively high sheet resistance compared to conventional TCO. In a similar context, metal oxides have been electrodeposited on mechanically exfoliated graphene as well as on chemically reduced GO films [68, 69]. Although these films were not implemented in solar cells, the approaches were adopted to develop flexible electrodes for DSSCs.

As semiconductor layer Graphene-based materials have also been incorporated in semiconductor layer of DSSCs to prevent the recombination between window electrode and electrolyte as well as to facilitate the electron transport leading to an enhanced photocurrent density of the device. In this regard, both rGO and GO provide a suitable energy barrier against recombination of photo-generated electrons between FTO and triiodide (mechanism schematically shown in Fig. 17), but it would also create an insulating barrier. Therefore, a very thin layer of GO/rGO is essential

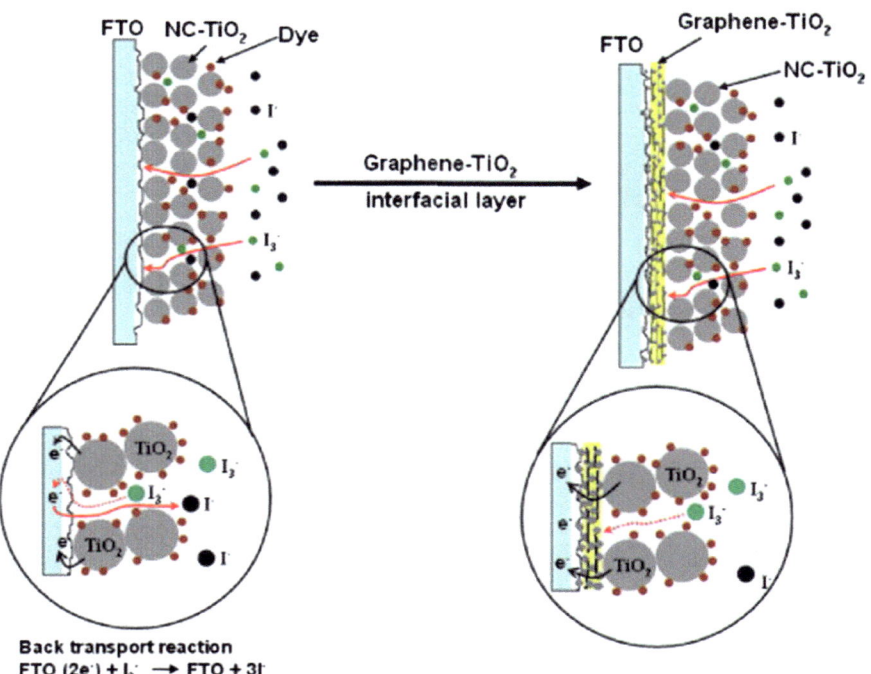

Fig. 17 Schematic representation and mechanism of graphene–TiO_2 interfacial layer to prevent recombination of electrons [70]

for such implementations. For example, photocatalytically reduced GO–TiO$_2$ interfacial layer used in DSSC displayed an approximately 7.6% relative improvement in PCE over a cell without GO layer [70]. Similarly, a significant enhancement in the overall performance is observed for thermally reduced rGO acting as a blocking layer, resulting in 8.1% efficient cell [71].

Another aspect of utilization of graphene materials in semiconductor layer is to increase the collection efficiency, i.e., photocurrent in the photo-anode. The atomically thin and high aspect ratio of graphene offers an advantage for low percolation threshold in DSSCs. Several studies have shown that the inclusion of 0.5–1.0 wt% graphene materials provides the best results in terms of overall PCE. In this context, Yang et al., demonstrated an increase in PCE from 5 to 7% by incorporating 0.6 wt% chemically reduced GO into the TiO$_2$ film followed by a heat treatment of the composite at 450 °C to sinter the TiO$_2$ and further reduce the GO [72]. The improvement in the device performance was achieved solely from the increased photocurrent. A similar approach has also shown an enhancement in J_{sc} and the PCE by 52.4 and 55.3%, respectively [73]. In addition to this effect, incorporation of graphene materials into the TiO$_2$ layer also has a significant role in dye absorption. In other words, adding graphene materials to the TiO$_2$ layer resulted in an increase in dye absorption, thus enhanced the light-harvesting efficiency leading to an increase in PCE [74–76].

As sensitizer Graphene materials have received further interest as a possible sensitizing material in DSSCs owing to the broad and strong absorption profile that absorbs about 2.3% of incident light by each monolayer [77]. A few experimental studies added credence to this possibility of graphene as a photosensitizer along with its ability to efficient charge injection from graphene materials to TiO$_2$ [78–80]. However, graphene materials have so far shown poor or mediocre performance in such application, much improvement needed for future implications.

3.3.2   Graphene as Electrolyte

Graphene materials have been combined with the electrolyte of a DSSC to address some limitations of conventional electrolyte, such as electrochemical instability, inefficient charge transport, and potential losses in electrolyte. In this aspect, the application of graphene materials can be divided into two main categories as: (a) minor additive and (b) main component (>1 wt%) in electrolyte. There are few studies where researchers used graphene materials in high concentrations as conductive fillers to decrease the electrolyte resistance [81, 82]. However, due to the high light absorption and catalytic activity of graphene materials, it is unlikely to impact on the performance of DSSCs [66]. On the other side, graphene materials using as a minor component in the electrolyte have shown to be more beneficial in the photovoltaic characteristics of DSSCs. Inclusion of graphene materials resulted in the gelation as well as the bleaching of the electrolyte, which would be favorable for photovoltaic applications. Velten et al. [83] found that an extremely low mass loading of graphene nanoribbons (~0.005 wt%) could lead to a substantial decrease in the optical absorption of iodide/triiodide which in turn resulted in a 22% increase in PCE for devices with an inverted architecture (from $\eta = 5.8$ to 7.0%). These results are shown in Fig. 18. In a similar approach, Gun et al. [84] observed that the addition of comparatively high mass loading of GO (~1 wt%) could yield a gel formation in a range of solvents containing ionic liquids. Although no long-term device performance was studied for this case, initial results showed that the devices with GO performed better ($\eta = 7.5\%$) than those with the pure liquid electrolyte ($\eta = 6.9\%$). It has been also found that the size of the GO sheets, which can be tuned by sonication, is inversely related to the ionic conductivity of the gels. DSSCs fabricated using well-ultrasonicated GO as the gelator have shown to retain the performance over a month of period, though the actual PCE of the device was considerably low [85, 86].

Fig. 18 **a** *J–V* curves of front- and back (inverted)-illuminated DSSCs containing electrolytes with and without graphene nanoribbons (GNR) and **b** optical images of iodide/triiodide electrolyte without (left) and with (right) GNR under illumination [83]

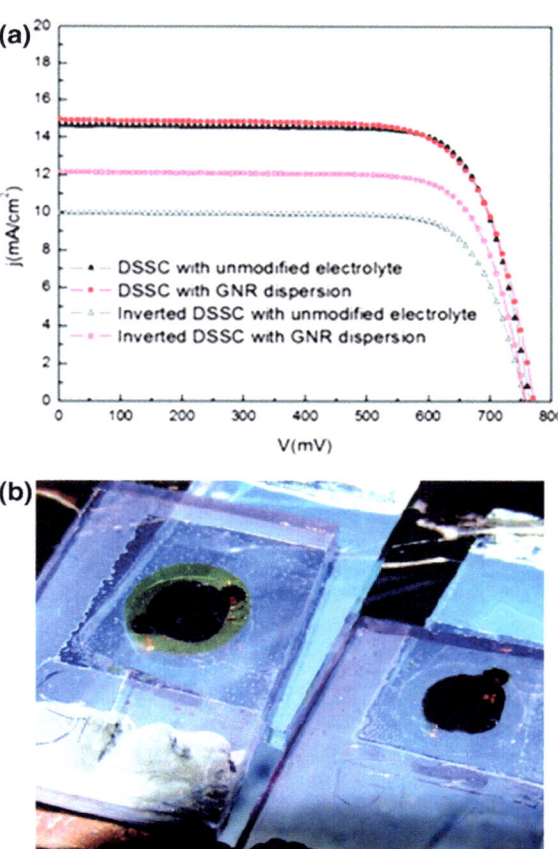

3.3.3 Graphene as Counter-Electrode

The counter-electrodes composed of platinum (Pt) nanoparticles deposited on FTO have been widely used in DSSCs to collect holes. However, it is undesired to use such expensive metal (platinum) in a supposed low-cost photovoltaic device. Therefore, graphene materials have been alternatively explored as counter-electrodes in DSSCs due to their high conductivity and electrocatalytic properties. Xu et al. [87] first studied the performance of DSSCs utilizing graphene materials as the counter-cathode. The fabricated device showed a better PCE ($\eta = 2.2\%$) than bare FTO ($\eta = 0.05\%$), however, it was poorer compared to the conventional platinized FTO ($\eta = 4.0\%$). Since then several approaches, such as reduction process, improvements in the morphology (by increasing the surface area or pore size) and increasing the intrinsic activity of the material by chemical modification have been implemented to enhance the performance of DSSCs. For example, Jang et al. [88] reported a systematic study on the effect of thermal annealing of GO films (~4 nm thick) in the performance of DSSCs. It was observed that a strong increase in performance with increased temperature treatment ($\eta = 0.50\%$, 0.51%, 2.9%, and 3.6% for GO films and thermally reduced GO films at 150, 250, and 350 °C, respectively). To increase the activity and achieve the high surface area, Wu and Zheng developed both vertically and horizontally oriented chemically reduced GO using spin coating and electrophoretic deposition, respectively [89]. They showed that the vertically oriented GO film has greater activity, indicating that ion mobility and assessable surface area were higher in this case. In another study, Zheng et al. showed that grinding rGO in polyethylene glycol and then thermolyzing the polymer which led to films with larger pores and hence resulting in DSSCs with higher efficiencies ($\eta = 7.2\%$) than those obtained from ultrasonicating rGO in the polymer ($\eta = 5.2\%$) [90]. The optimal PCE of DSSCs with the rGO counter-electrode is 7.19%, which is only slightly lower than that (7.76%) of control DSSCs with Pt as the counter-electrode. In addition, graphene materials also have been doped with nitrogen, functionalized with CF_4, treated with HNO_3 and integrated with carbon nanotubes to improve the intrinsic activity [91–94]. However, more promising results were obtained by changing iodide/triiodide redox electrolyte to cobalt-based and sulfur-based mediators. Kavan et al. have shown using graphene nanoplatelet films as electrode in DSSCs can outperform platinized FTO with both the Co(bpy)3(II/III) and the Co(L)2 redox couples, where L is 6-(1H-pyrazol-1-yl)-bpy [95, 96]. DSSCs fabricated using these electrolytes and graphene nanoplatelet cathodes exhibited PCE over 9%, and with V_{oc} greater than 1 V. These devices displayed 12 and 15% higher relative efficiencies, respectively, than those DSSCs using platinized FTO cathodes (Fig. 19).

Graphene-based materials have been incorporated successfully in DSSCs in every aspect of the device to improve the performance to a reasonable extent. It is observed that the chemical activity, light absorption, and charge transfer were greatly increased and concurrently the carrier recombination was substantially reduced by introducing graphene-based materials. Most importantly, recent studies showed that it has a great potential to replace expensive Pt metal which is conventionally used as a counter-electrode in DSSCs.

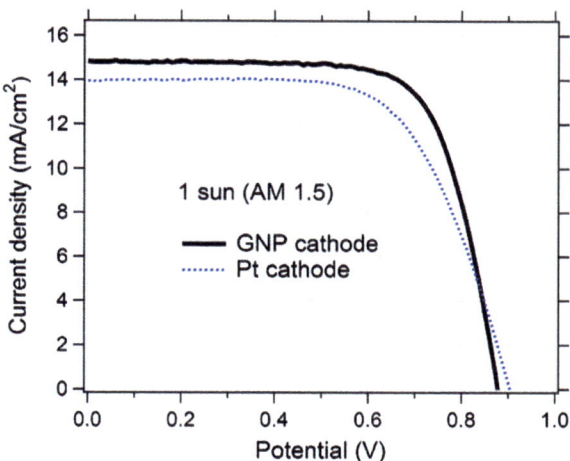

Fig. 19 J–V characteristics of DSSCs with platinized FTO counter-electrode and graphene nanoplatelets' FTO counter-electrode [96]

3.4 Graphene Materials in Other Solar Cells

Apart from heretofore-mentioned solar cells, graphene materials have also been used either individually or as a hybrid material in other types of solar cells. However, the number of such attempts are significantly less and the obtained results are also not very overwhelming. Li et al. used graphene sheets to fabricate a Schottky junction solar cells with n-Si and achieved a highest PCE of 1.65% for a 0.1 cm² cell area under air mass 1.5 (AM 1.5) illumination [14]. Similar attempts were adopted to use graphene nanowhiskers on amorphous carbon or carbon nanotubes and graphene in silicon Schottky junction solar cells which resulted in nearly comparable properties [97, 98]. Though the overall performance of these types of solar cells is still low, it represents a development of new type of solar cells. More recently, CVD-grown graphene films have been used in CIGS thin-film solar cells as window electrode as well as hole-transporting electrodes [13, 99]. Using graphene film as window electrode for the replacement of conventional Al-doped ZnO films in CIGS solar cell showed a high PCE of ~13.5% for a large active area up to 45 mm². On the other hand, employing graphene as hole-transporting electrode in CIGS solar cell, Sim et al. demonstrated that the fabricated device can achieve PCE of ~10% with a fill factor of 64.75%, which are substantially higher compared to reference cell fabricated using conventional Mo/stainless steel electrode [13]. High open-circuit voltage together with substantially large fill factor is primarily responsible for high cell efficiency of graphene/Cu-foil-based device. In addition, graphene and its derivatives also have shown good potential for CdS, CdTe, and TiO₂ quantum dot-sensitized solar cells [100–102]. Despite of these interesting results, further investigation is needed for the improvement in the performance as well as practical application of all these graphene-based solar cells.

4   Conclusions

A significant amount of research on graphene-based materials has already been carried out for the development of solar PV. The extremely high transmittance of graphene and its derivatives with high conductivity inherently establish this group of materials as one of the most suitable candidate for opto-electronic devices which eventually satisfy the vital aspects of a solar cell as well. Interestingly, the comparable work-functions of these materials with that of conventional electrodes provide a plausible room to replace ITO/FTO that is used commonly in solar cells. The excellent physical and chemical properties of graphene and its derivatives are indeed reflected through the device performance of several types of solar cells that mostly include organic/polymer, dye-sensitized and perovskite solar cells. Nevertheless, few attempts were also made in other PV technologies like Si-based solar cells and thin-film solar cells. The success of graphene-based materials can be best realized in perovskite solar cells compared to other types of solar cells. Next to perovskite, dye-sensitized and organic solar cells encountered a significant number of attempts utilizing graphene-based materials. In general, graphene, GO, and rGO have been applied as electrodes, as electron and hole transport materials as well as additives in absorber or semiconducting layer. Additionally, functionalizations and/or doping of these materials also facilitated the overall performance and more importantly, the stability of the device over a considerable period. Moreover, flexibility with desired mechanical stability of graphene-based materials opened a window for flexible solar PV technology avoiding conventional TCOs which are rigid and brittle in nature.

However, the major challenge of graphene-based materials is associated with the efficient large-scale synthesis in an economically feasible method. In view of preparation, two well-known methods, namely chemical vapor deposition and Hummers' method have been widely used to synthesize graphene-based materials. An easier fabrication of these materials using an inkjet printer with an appropriate solvent might provide a cost-effective technique for large-scale synthesis. The other limitations related to its transmittance and conductivity also need to be considered. These properties strongly depend on number of layers, doping, functionalization of graphene, GO, or rGO, which strongly affects the overall performance of the solar cells. Therefore, in spite of promising potential of graphene and its derivatives, more fundamental research and development are needed for scaling up and commercialization of solar PV using these materials.

References

1. Ellabban, O., Abu-Rub, H., Blaabjerg, F.: Renewable energy resources: Current status, future prospects and their enabling technology. Renew. Sustain. Energy Rev. **39**, 748 (2014)
2. Mathiesen, B.V., Lund, H., Connolly, D., Wenzel, H., Ostergaard, P.A., Möller, B., Nielsen, S., Ridjan, I., KarnOe, P., Sperling, K., Hvelplund, F.K.: Smart energy systems for coherent 100% renewable energy and transport solutions. Appl. Energy **145**, 139 (2015)

3. Panwar, N.L., Kaushik, S.C., Kothari, S.: Role of renewable energy sources in environmental protection: A review. Renew. Sustain. Energy Rev. **15**, 1513 (2011)
4. Branker, K., Pathak, M.J.M., Pearce, J.M.: A review of solar photovoltaic levelized cost of electricity. Renew. Sustain. Energy Rev. **15**, 4470 (2011)
5. Yan, J., Saunders, B.R.: Third-generation solar cells: A review and comparison of polymer: fullerene, hybrid polymer and perovskite solar cells. RSC Adv. **4**, 43286 (2014)
6. Yamaguchi, M., Lee, K., Araki, K.: A review of recent progress in heterogeneous silicon tandem solar cells. J. Phys. D Appl. Phys. **51**, 133002 (2018)
7. Chopra, K.L., Paulson, P.D., Dutta, V.: Thin-film solar cells: an overview. Prog. Photovoltaics Res. Appl. **12**, 69 (2004)
8. Matthew, R.B.K., Allen, J., Tung, V.C.: Honeycomb carbon: a review of graphene. Chem. Rev. **110**, 132 (2010)
9. Soldano, C., Mahmood, A., Dujardin, E.: Production, properties and potential of graphene. Carbon **48**, 2127 (2010)
10. Sun, Y., Zhang, W., Chi, H., Liu, Y., Hou, C.L., Fang, D.: Recent development of graphene materials applied in polymer solar cell. Renew. Sustain. Energy Rev. **43**, 973 (2015)
11. De, S., Coleman, J.N.: Are there fundamental limitations on the sheet resistance and transmittance of thin graphene films? ACS Nano **4**, 2713 (2010)
12. Ubani, C.A., Ibrahim, M.A., Teridi, M.A.M., Sopian, K., Ali, J., Chaudhary, K.T.: Application of graphene in dye and quantum dots sensitized solar cell. Sol. Energy **137**, 531 (2016)
13. Sim, J.K., Kang, S., Nandi, R., Jo, J.Y., Jeong, K.U., Lee, C.R.: Implementation of graphene as hole transport electrode in flexible CIGS solar cells fabricated on Cu foil. Sol. Energy **162**, 357 (2018)
14. Li, X., Zhu, H., Wang, K., Cao, A., Wei, J., Li, C., Jia, Y., Li, Z., Li, X., Wu, D.: Graphene-on-silicon schottky junction solar cells. Adv. Mater. **22**, 2743 (2010)
15. Bouclé, J., Herlin-Boime, N.: The benefits of graphene for hybrid perovskite solar cells. Synth. Met. **222**, 3 (2016)
16. Mahmoudi, T., Wang, Y., Hahn, Y.B.: Graphene and its derivatives for solar cells application. Nano Energy **47**, 51 (2018)
17. Acik, M., Darling, S.B.: Graphene in perovskite solar cells: Device design, characterization and implementation. J. Mater. Chem. A **4**, 6185 (2016)
18. Iwan, A., Chuchmała, A.: Perspectives of applied graphene: Prog. Polym. Sci. **37**, 1805 (2012)
19. Steim, R., Kogler, F.R., Brabec, C.J.: Interface materials for organic solar cells. J. Mater. Chem. **20**, 2499 (2010)
20. Wang, F., Xu, Q., Tan, Z., Li, L., Li, S., Hou, X., Sun, G., Tu, X., Hou, J., Li, Y.: Efficient polymer solar cells with a solution- processed and thermal annealing-free RuO2 anode buffer layer. J. Mater. Chem. A **2**, 1318 (2014)
21. Wang, J., Wang, Y., He, D., Liu, Z., Wu, H., Wang, H., Zhou, P., Fu, M.: Polymer bulk heterojunction photovoltaic devices based on complex donors and solution-processable functionalized graphene oxide. Sol. Energy Mater. Sol. Cells **96**, 58 (2012)
22. Yu, D., Park, K., Durstock, M., Dai, L.: Fullerene-grafted graphene for efficient bulk heterojunction polymer photovoltaic devices. J. Phys. Chem. Lett. **2**, 1113 (2011)
23. Wang, S., Goh, B.M., Manga, K.K., Bao, Q., Yang, P., Loh, K.P.: Graphene as atomic template and structural scaffold in the synthesis of graphene-organic hybrid wire with photovoltaic properties. ACS Nano **4**, 6180 (2010)
24. Liu, Z., He, D., Wang, Y., Wu, H., Wang, J.: Solution-processable functionalized graphene in donor/acceptor-type organic photovoltaic cells. Sol. Energy Mater. Sol. Cells **94**, 1196 (2010)
25. Gupta, V., Chaudhary, N., Srivastava, R., Sharma, G.D., Bhardwaj, R., Chand, S.: Luminscent graphene quantum dots for organic photovoltaic devices. J. Am. Chem. Soc. **133**, 9960 (2011)
26. Wang, X., Zhi, L., Tsao, N., Tomović, Ž., Li, J., Müllen, K.: Transparent carbon films as electrodes in organic solar cells. Angew. Chemie-Int. Ed. **47**, 2990 (2008)
27. Shin, D.H., Jang, C.W., Lee, H.S., Seo, S.W., Choi, S.H.: Semitransparent flexible organic solar cells employing doped-graphene layers as anode and cathode electrodes. ACS Appl. Mater. Interfaces **10**, 3596 (2018)

28. Valentini, L., Cardinali, M., Bittolo Bon, S., Bagnis, D., Verdejo, R., Lopez-Manchado, M.A., Kenny, J.M.: Use of butylamine modified graphene sheets in polymer solar cells. J. Mater. Chem. **20**, 995 (2010)

29. Yu, D., Yang, Y., Durstock, M., Baek, J., Dai, L.: Soluble P3HT-grafted graphene for efficient bilayer−heterojunction photovoltaic devices. ACS Nano **4**, 5633 (2010)

30. Yin, Z., Sun, S., Salim, T., Wu, S., Huang, X., He, Q., Lam, Y.M., Zhang, H.: Organic photovoltaic devices using highly flexible reduced graphene oxide films as transparent electrodes. ACS Nano **4**, 5263 (2010)

31. Un Jung, Y., Na, S.-I., Kim, H.-K., Jun Kang, S.: Organic photovoltaic devices with low resistance multilayer graphene transparent electrodes. J. Vac. Sci. Technol. A **30**, 050604 (2012)

32. Huang, J.H., Fang, J.H., Liu, C.C., Chu, C.W.: Effective work function modulation of graphene/carbon nanotube composite films as transparent cathodes for organic optoelectronics. ACS Nano **5**, 6262 (2011)

33. Lee, Y.Y., Tu, K.H., Yu, C.C., Li, S.S., Hwang, J.Y., Lin, C.C., Chen, K.H., Chen, L.C., Chen, H.L., Chen, C.W.: Top laminated graphene electrode in a semitransparent polymer solar cell by simultaneous thermal annealing/releasing method. ACS Nano **5**, 6564 (2011)

34. Jo, G., Na, S.I., Oh, S.H., Lee, S., Kim, T.S., Wang, G., Choe, M., Park, W., Yoon, J., Kim, D.Y., Kahng, Y.H., Lee, T.: Tuning of a graphene-electrode work function to enhance the efficiency of organic bulk heterojunction photovoltaic cells with an inverted structure. Appl. Phys. Lett. **97**, 213301 (2010)

35. Liu, Z., Li, J., Sun, Z.-H., Tai, G., Lau, S.-P., Yan, F.: Direct measurement of the exciton binding energy and effective masses for charge carriers in organic-inorganic tri-halide perovskites. ACS Nano **6**, 810 (2012)

36. Miyata, A., Mitioglu, A., Plochocka, P., Portugall, O., Wang, J.T.W., Stranks, S.D., Snaith, H.J., Nicholas, R.J.: Direct measurement of the exciton binding energy and effective masses for charge carriers in organic–inorganic tri-halide perovskites. Nat. Phys. **11**, 582 (2015)

37. Heo, J.H., Shin, D.H., Kim, S., Jang, M.H., Lee, M.H., Seo, S.W., Choi, S.H., Im, S.H.: Highly efficient $CH_3NH_3PbI_3$ perovskite solar cells prepared by $AuCl_3$-doped graphene transparent conducting electrodes. Chem. Eng. J. **323**, 153 (2017)

38. Yoon, J., Sung, H., Lee, G., Cho, W., Ahn, N., Jung, H.S., Choi, M.: Superflexible, high-efficiency perovskite solar cells utilizing graphene electrodes: Towards future foldable power sources. Energy Environ. Sci. **10**, 337 (2017)

39. Sung, H., Ahn, N., Jang, M.S., Lee, J.K., Yoon, H., Park, N.G., Choi, M.: Transparent conductive oxide-free graphene-based perovskite solar cells with over 17% efficiency. Adv. Energy Mater. **6**, 2 (2016)

40. You, P., Liu, Z., Tai, Q., Liu, S., Yan, F.: Efficient semitransparent perovskite solar cells with graphene electrodes. Adv. Mater. **27**, 3632 (2015)

41. Lang, F., Gluba, M.A., Albrecht, S., Rappich, J., Korte, L., Rech, B., Nickel, N.H.: Perovskite solar cells with large-area CVD-graphene for tandem solar cells. J. Phys. Chem. Lett. **6**, 2745 (2015)

42. Guo, F., Azimi, H., Hou, Y., Przybilla, T., Hu, M., Bronnbauer, C., Langner, S., Spiecker, E., Forberich, K., Brabec, C.J.: High-performance semitransparent perovskite solar cells with solution-processed silver nanowires as top electrodes. Nanoscale **7**, 1642 (2015)

43. Li, Z., Kulkarni, S.A., Boix, P.P., Shi, E., Cao, A., Fu, K., Batabyal, S.K., Zhang, J., Xiong, Q., Wong, L.H., Mathews, N., Mhaisalkar, S.G.: Laminated carbon nanotube networks for metal electrode-free efficient perovskite solar cells. ACS Nano **8**, 6797 (2014)

44. Lang, F., Gluba, M.A., Albrecht, S., Shargaieva, O., Rappich, J., Korte, L., Rech, B., Nickel, N.H.: In situ graphene doping as a route toward efficient perovskite tandem solar cells. Phys. Status Solidi Appl. Mater. Sci. **213**, 1989 (2016)

45. Kim, H., Lim, K.-G., Lee, T.-W.: Planar heterojunction organometal halide perovskite solar cells: roles of interfacial layers. Energy Environ. Sci. **9**, 12 (2016)

46. Kim, B.J., Kim, D.H., Lee, Y.Y., Shin, H.W., Han, G.S., Hong, J.S., Mahmood, K., Ahn, T.K., Joo, Y.C., Hong, K.S., Park, N.G., Lee, S., Jung, H.S.: Highly efficient and bending durable perovskite solar cells: Toward a wearable power source. Energy Environ. Sci. **8**, 916 (2015)

47. Heo, J.H., Lee, M.H., Han, H.J., Patil, B.R., Yu, J.S., Im, S.H.: Highly efficient low temperature solution processable planar type $CH_3NH_3PbI_3$ perovskite flexible solar cells. J. Mater. Chem. A **4**, 1572 (2016)

48. Yeo, J.S., Kang, R., Lee, S., Jeon, Y.J., Myoung, N.S., Lee, C.L., Kim, D.Y., Yun, J.M., Seo, Y.H., Kim, S.S., Na, S.I.: Highly efficient and stable planar perovskite solar cells with reduced graphene oxide nanosheets as electrode interlayer. Nano Energy **12**, 96 (2015)

49. Wu, Z., Bai, S., Xiang, J., Yuan, Z., Yang, Y., Cui, W., Gao, X., Liu, Z., Jin, Y., Sun, B.: Efficient planar heterojunction perovskite solar cells employing graphene oxide as hole conductor. Nanoscale **6**, 10505 (2014)

50. Liu, T., Kim, D., Han, H., Bin Mohd Yusoff, A.R., Jang, J.: Fine-tuning optical and electronic properties of graphene oxide for highly efficient perovskite solar cells. Nanoscale **7**, 10708 (2015)

51. Luo, Q., Zhang, Y., Liu, C., Li, J., Wang, N., Lin, H.: Iodide-reduced graphene oxide with dopant-free spiro-OMeTAD for ambient stable and high-efficiency perovskite solar cells. J. Mater. Chem. A **3**, 15996 (2015)

52. Umeyama, T., Matano, D., Baek, J., Gupta, S., Ito, S., (Ravi) Subramanian, V., Imahori, H.: Boosting of the performance of perovskite solar cells through systematic introduction of reduced graphene oxide in TiO_2 layers. Chem. Lett. **44**, 1410 (2015)

53. Han, G.S., Song, Y.H., Jin, Y.U., Lee, J.-W., Park, N.-G., Kang, B.K., Lee, J.-K., Cho, I.S., Yoon, D.H., Jung, H.S.: Reduced graphene oxide/mesoporous TiO_2 nanocomposite based perovskite solar cells. ACS Appl. Mater. Interfaces **7**, 23521 (2015)

54. Wang, J.T.W., Ball, J.M., Barea, E.M., Abate, A., Alexander-Webber, J.A., Huang, J., Saliba, M., Mora-Sero, I., Bisquert, J., Snaith, H.J., Nicholas, R.J.: Low-temperature processed electron collection layers of graphene/TiO2 nanocomposites in thin film perovskite solar cells. Nano Lett. **14**, 724 (2014)

55. Tavakoli, M.M., Tavakoli, R., Hasanzadeh, S., Mirfasih, M.H.: Interface engineering of perovskite solar cell using a reduced-graphene scaffold. J. Phys. Chem. C **120**, 19531 (2016)

56. Zhao, J., Cai, B., Luo, Z., Dong, Y., Zhang, Y., Xu, H., Hong, B., Yang, Y., Li, L., Zhang, W., Gao, C.: Investigation of the hydrolysis of perovskite organometallic halide $CH_3NH_3PbI_3$ in humidity environment. Sci. Rep. **6**, 1 (2016)

57. Hu, X., Jiang, H., Li, J., Ma, J., Yang, D., Liu, Z., Gao, F., Liu, S.: Air and thermally stable perovskite solar cells with CVD-graphene as the blocking layer. Nanoscale **9**, 8274 (2017)

58. Cao, J., Liu, Y.M., Jing, X., Yin, J., Li, J., Xu, B., Tan, Y.Z., Zheng, N.: Well-defined thiolated nanographene as hole-transporting material for efficient and stable perovskite solar cells. J. Am. Chem. Soc. **137**, 10914 (2015)

59. Bi, E., Chen, H., Xie, F., Wu, Y., Chen, W., Su, Y., Islam, A., Grätzel, M., Yang, X., Han, L.: Diffusion engineering of ions and charge carriers for stable efficient perovskite solar cells. Nat. Commun. **8**, 1 (2017)

60. Kim, T., Kang, J.H., Yang, S.J., Sung, S.J., Kim, Y.S., Park, C.R.: Facile preparation of reduced graphene oxide-based gas barrier films for organic photovoltaic devices. Energy Environ. Sci. **7**, 3403 (2014)

61. Jiao, Y., Ma, F., Gao, G., Wang, H., Bell, J., Frauenheim, T., Du, A.: Graphene-covered perovskites: an effective strategy to enhance light absorption and resist moisture degradation. RSC Adv. **5**, 82346 (2015)

62. Kim, G.H., Jang, H., Yoon, Y.J., Jeong, J., Park, S.Y., Walker, B., Jeon, I.Y., Jo, Y., Yoon, H., Kim, M., Baek, J.B., Kim, D.S., Kim, J.Y.: Fluorine functionalized graphene nano platelets for highly stable inverted perovskite solar cells. Nano Lett. **17**, 6385 (2017)

63. Zhang, Y., Li, H., Kuo, L., Dong, P., Yan, F.: Recent applications of draphene in dye-sensitized solar cells. Curr. Opin. Colloid Interface Sci. **20**, 406 (2015)

64. Nazeeruddin, M.K., Baranoff, E., Grätzel, M.: Dye-sensitized solar cells: A brief overview. Sol. Energy **85**, 1172 (2011)

65. Grätzel, M.: Dye-sensitized solar cells. J. Photochem. Photobiol. C Photochem. Rev. **4**, 145 (2003)

66. Roy-Mayhew, J.D., Aksay, I.A.: Graphene materials and their use in dye-sensitized solar cells. Chem. Rev. **114**, 6323 (2014)
67. Wang, X., Zhi, L., Müllen, K.: Transparent, conductive graphene electrodes for dye-sensitized solar cells. Nano Lett. **8**, 323 (2008)
68. Cottineau, T., Albrecht, A., Janowska, I., MacHer, N., Bégin, D., Ledoux, M.J., Pronkin, S., Savinova, E., Keller, N., Keller, V., Pham-Huu, C.: Synthesis of transparent vertically aligned TiO_2 nanotubes on a few-layer graphene (FLG) film. Chem. Commun. **48**, 1224 (2012)
69. Wu, S., Yin, Z., He, Q., Huang, X., Zhou, X., Zhang, H.: Electrochemical deposition of semiconductor oxides on reduced graphene oxide-based flexible, transparent, and conductive electrodes. J. Phys. Chem. C **114**, 11816 (2010)
70. Kim, S.R., Parvez, M.K., Chhowalla, M.: UV-reduction of graphene oxide an its application as an interfacial layer to reduce the back-transport reactions in dye-sensitized solar cells. Chem. Phys. Lett. **483**, 124 (2009)
71. Chen, T., Hu, W., Song, J., Guai, G.H., Li, C.M.: Interface functionalization of photoelectrodes with graphene for high performance dye-sensitized solar cells. Adv. Funct. Mater. **22**, 5245 (2012)
72. Yang, N.L., Zhai, J., Wang, D., Chen, Y.S., Jiang, L.: Two-dimensional graphene bridges enhanced photoinduced charge transport in dye-sensitized solar cells. ACS Nano **4**, 887 (2010)
73. Wang, H., Leonard, S.L., Hu, Y.H.: Promoting effect of graphene on dye-sensitized solar cells. Ind. Eng. Chem. Res. **51**, 10613 (2012)
74. Tsai, T., Chiou, S., Chen, S.: Enhancement of dye-sensitized solar cells by using graphene-TiO2 composites as photoelectrochemical working electrode. Int. J. Electrochem. Sci. **6**, 3333 (2011)
75. Peining, Z., Nair, A.S., Shengjie, P., Shengyuan, Y., Ramakrishna, S.: Facile fabrication of TiO_2-graphene composite with enhanced photovoltaic and photocatalytic properties by electrospinning. ACS Appl. Mater. Interfaces **4**, 581 (2012)
76. Tang, Y., Lee, C., Xu, J., Liu, Z., Chen, Z., He, Z., Cao, Y., Yuan, G., Song, H., Chen, L., Luo, L., Cheng, H., Zhang, W., Bello, I., Lee, S.: Incorporation of graphenes in nanostructured TiO2 films via molecular grafting for dye-sensitized solar cell application. ACS Nano **4**, 3482 (2010)
77. Mak, K.F., Sfeir, M.Y., Wu, Y., Lui, C.H., Misewich, J.A., Heinz, T.F.: Measurement of the opptical conductivity of graphene. Phys. Rev. Lett. **101**, 2 (2008)
78. Long, R., English, N.J., Prezhdo, O.V.: Photo-induced charge separation across the graphene-TiO2interface is faster than energy losses: A time-domain ab initio analysis. J. Am. Chem. Soc. **134**, 14238 (2012)
79. Zhang, Y., Zhang, N., Tang, Z.R., Xu, Y.J.: Graphene transforms wide band gap ZnS to a visible light photocatalyst. the new role of graphene as a macromolecular photosensitizer. ACS Nano **6**, 9777 (2012)
80. Williams, K.J., Nelson, C.A., Yan, X., Li, L.S., Zhu, X.: Hot electron injection from graphene quantum dots to TiO_2. ACS Nano **7**, 1388 (2013)
81. Ahmad, I., Khan, U., Gun'ko, Y.K.: Graphene, carbon nanotube and ionic liquid mixtures: Towards new quasi-solid state electrolytes for dye sensitised solar cells. J. Mater. Chem. **21**, 16990 (2011)
82. Jung, M.H., Kang, M.G., Chu, M.J.: Iodide-functionalized graphene electrolyte for highly efficient dye-sensitized solar cells. J. Mater. Chem. **22**, 16477 (2012)
83. Velten, J.A., Carretero-González, J., Castillo-Martínez, E., Bykova, J., Cook, A., Baughman, R., Zakhidov, A.: Photoinduced optical transparency in dye-sensitized solar cells containing graphene nanoribbons. J. Phys. Chem. C **115**, 25125 (2011)
84. Gun, J., Kulkarni, S.A., Xiu, W., Batabyal, S.K., Sladkevich, S., Prikhodchenko, P.V., Gutkin, V., Lev, O.: Graphene oxide organogel electrolyte for quasi solid dye sensitized solar cells. Electrochem. Commun. **19**, 108 (2012)
85. Bai, H., Li, C., Wang, X., Shi, G.: On the gelation of graphene oxide. J. Phys. Chem. C **115**, 5545 (2011)

86. Neo, C.Y., Ouyang, J.: The production of organogels using graphene oxide as the gelator for use in high-performance quasi-solid state dye-sensitized solar cells. Carbon **54**, 48 (2013)
87. Xu, Y., Bai, H., Lu, G., Li, C., Shi, G.: Flexible graphene films via the filtration of water-soluble. J. Am. Chem. Soc. **130**, 5856 (2008)
88. Jang, H.S., Yun, J.M., Kim, D.Y., Park, D.W., Na, S.I., Kim, S.S.: Moderately reduced graphene oxide as transparent counter electrodes for dye-sensitized solar cells. Electrochim. Acta **81**, 301 (2012)
89. Wu, M.-S., Zheng, Y.-J.: Electrophoresis of randomly and vertically embedded graphene-nanosheets in activated carbon film as a counter electrode for dye-sensitized solar cells. Phys. Chem. Chem. Phys. **15**, 1782 (2013)
90. Zheng, H., Neo, C.Y., Mei, X., Qiu, J., Ouyang, J.: Reduced graphene oxide films fabricated by gel coating and their application as platinum-free counter electrodes of highly efficient iodide/triiodide dye-sensitized solar cells. J. Mater. Chem. **22**, 14465 (2012)
91. Dong, P., Zhu, Y., Zhang, J., Hao, F., Wu, J., Lei, S., Lin, H., Hauge, R.H., Tour, J.M., Lou, J.: Vertically aligned carbon nanotubes/graphene hybrid electrode as a TCO- and Pt-free flexible cathode for application in solar cells. J. Mater. Chem. A **2**, 20902 (2014)
92. Velten, J., Mozer, A.J., Li, D., Officer, D., Wallace, G., Baughman, R., Zakhidov, A.: Amplifying charge-transfer characteristics of graphene for triiodide reduction in dye-sensitized solar cells. Nanotechnology **23**, 085201 (2012)
93. Das, S., Sudhagar, P., Verma, V., Song, D., Ito, E., Lee, S.Y., Kang, Y.S., Choi, W.: Effect of HNO_3 functionalization on large scale graphene for enhanced tri-iodide reduction in dye-sensitized solar cells. Adv. Funct. Mater. **21**, 3729 (2011)
94. Das, S., Sudhagar, P., Ito, E., Lee, D.Y., Nagarajan, S., Lee, S.Y., Kang, Y.S., Choi, W.: Effect of HNO_3 functionalization on large scale graphene for enhanced tri-iodide reduction in dye-sensitized solar cells. J. Mater. Chem. **22**, 20490 (2012)
95. Kavan, L., Yum, J.H., Nazeeruddin, M.K., Grätzel, M.: Graphene nanoplatelet cathode for Co(III)/(II) mediated dye-sensitized solar sells. ACS Nano **5**, 9171 (2011)
96. Kavan, L., Yum, J.H., Grätzel, M.: Graphene nanoplatelets outperforming platinum as the electrocatalyst in co-bipyridine-mediated dye-sensitized solar cells. Nano Lett. **11**, 5501 (2011)
97. Schriver, M., Regan, W., Loster, M., Zettl, A.: Carbon nanostructure-aSi:H photovoltaic cells with high open-circuit voltage fabricated without dopants. Solid State Commun. **150**, 561 (2010)
98. Li, X., Li, C., Zhu, H., Wang, K., Wei, J., Li, X., Xu, E., Li, Z., Luo, S., Lei, Y., Wu, D.: Hybrid thin films of graphene nanowhiskers and amorphous carbon as transparent conductors. Chem. Commun. **46**, 3502 (2010)
99. Yin, L., Zhang, K., Luo, H., Cheng, G., Ma, X., Xiong, Z., Xiao, X.: Highly efficient graphene-based Cu(In, Ga)Se$_2$ solar cells with large active area. Nanoscale **6**, 10879 (2014)
100. Lu, Z., Guo, C.X., Bin Yang, H., Qiao, Y., Guo, J., Li, C.M.: One-step aqueous synthesis of graphene-CdTe quantum dot-composed nanosheet and its enhanced photoresponses. J. Colloid Interface Sci. **353**, 588 (2011)
101. Zhu, G., Xu, T., Lv, T., Pan, L., Zhao, Q., Sun, Z.: Graphene-incorporated nanocrystalline TiO_2 films for CdS quantum dot-sensitized solar cells. J. Electroanal. Chem. **650**, 248 (2011)
102. Dai, L.: Layered graphene/quantum dots : nanoassemblies for highly efficient solar cells. ChemSusChem **3**, 797 (2010)

Graphene and Its Modifications for Supercapacitor Applications

Mandira Majumder and Anukul K. Thakur

Abstract Supercapacitors, also termed as electrochemical capacitors or ultracapacitors store charge using high surface area conducting materials. However, their extensive use is limited by the low energy density delivered and relatively high effective series resistance. In order to improve the specific capacitance, energy density as well as the power density, combining materials with requisite properties resulting in hybrids seems to be an attractive way out. Carbon forms one of the most prime materials to be used as supercapacitor electrodes. Structurally modified graphene through chemical functionalization reveals numerous possibilities for attaining tunable structural and electrochemical properties. Till now several chemical and physical functionalization methods have been explored in order to augment the stabilization and result in modification of the graphene. This chapter is concerned detailing the variety of chemical modifications routes of graphene reported so far, their effect on the electrochemical properties of graphene and the applicability of the developed material as a supercapacitor electrode material.

Keywords Graphene · Covalent interactions · Non-covalent interactions · Supercapacitor

1 Introduction

Perpetually growing demand for energy with the continuous exhaustion of the available conventional sources obligate renovation in the upcoming energy supply and production-related strategies of various countries [1, 2]. Development of efficient, renewable, and clean energy sources together with compatible modes of storage as well as energy conversion becomes a primary aim for the technologists and

M. Majumder
Nanostructured Composite Materials Laboratory, Department of Physics, Indian Institute of Technology (Indian School of Mines) Dhanbad, Dhanbad 826004, India

A. K. Thakur (✉)
Department of Physics, Indian Institute of Science Education and Research Berhampur, Berhampur 760010, India
e-mail: anukulphyiitd@gmail.com

© Springer Nature Switzerland AG 2019
S. Sahoo et al. (eds.), *Surface Engineering of Graphene*, Carbon Nanostructures, https://doi.org/10.1007/978-3-030-30207-8_5

researchers all over the world [1, 3–5]. Various non-conventional renewable sources have been put to the rescue like wind energy and solar energy. However, solar energy can be harvested when exposed to the sun, while wind energy is not possible to be obtained at all times. In order to achieve the seamless power supply, it is necessary to integrate these non-conventional sources of energy along with compatible energy storage modes. The irregularity in the supply of power along with various other limitations results in an imposed restriction on the wide-range application of these renewable energy modes in day-to-day applications. Compared to the various energy storage modes that we have, supercapacitors form the most significant next class of energy storage systems. These can provide with very high power density as compared to the batteries together with higher energy density as compared to the conventional dielectric capacitors [1, 6, 7].

Supercapacitors (SCs), also known as electrochemical capacitors or ultracapacitors are advantageous to the batteries owing to their high power density (>1000 times), low cost, less heating, non-toxicity, wide range of operating temperatures, mobility, low maintenance, long cycle life (>100,000 cycles), etc. [8–11]. Supercapacitors are known to store charge non-faradically in the electric double layers (EDL) between the electrode active material and the electrolyte so formed as a consequence of the charge separation. Some materials like metal oxides and polymers are known to store charge through highly reversible faradic reactions with the electrolyte producing pseudocapacitance [12, 13]. Supercapacitors have been extensively used in various industrial equipment's, forklifts, load cranes, consumer electronics, and electric vehicles. Further, supercapacitors are also integrated into electronic memory back-ups, industrial energy, and power management systems [14–17]. Integration of supercapacitor device in Airbus A380 planes has confirmed their benign and consistent performance and is expected to spread their market more in the near future [18].

A suitable conducting electrode material providing a proficient electrode–electrolyte interface equipped with an adept transfer path for the electrons forms the most essential part of supercapacitor devices determining the efficiency of the device [19]. In addition, an ideal electrode is expected to exhibit high electrical conductivity, large specific surface area, thermal stability, chemical stability, mechanical stability, and hierarchical pore size distribution [19, 20]. Carbon materials form the premiere and the major class of EDL capacitive electrode materials. Various carbon-based materials, such as graphene-based materials including fullerenes, carbon nanotubes, sheet-like graphene, and graphite, have been studied widely as supercapacitive electrode material owed to their favorable electrical properties, mechanical properties, and unique structures [21–25]. Graphene usually consists of monolayers of sp^2 hybridized carbon atoms united in a continuous hexagonal lattice with the affirmation of its unique physical properties [24, 26]. However, the limited charge storage in case of pure graphene attributed to the involvement of only surface interaction with the electrolyte leads to limited specific capacitance restricting their practical applications to a large extent. Also, graphene suffers from very low out of plane conductivity and agglomeration which leads to the emergence of boosting their overall electrochemical performance by chemical modifications [27–30].

This chapter holds a brief discussion on the detailed study of the electrochemical performance exhibited by the graphene as previously reported, various electrode materials designed from the graphene through various chemical modifications and their electrochemical performance.

2 Fundamentals of Supercapacitor

The charge storage mechanism for the supercapacitors can broadly be divided into two categories: (I) Storing charge electrostatically (non-faradaic reactions) in the electric double layer (EDL) formed as a consequence of the charge separation between the electrode and the electrolyte see Fig. 1a, b. Storing charges through highly reversible faradaic reactions with the electrolyte at the electrode surface giving rise to the pseudocapacitance, see Fig. 1c, d [31]. EDL capacitance refers to the capacitance resulting from the potential-dependent charge storage within the two oppositely charged layers created at the electrode–electrolyte interfacial region [32]. The premier model of EDL is majorly accredited to Helmholtz [33], and hence, the double layers consisting of separated charges are also termed as Helmholtz double

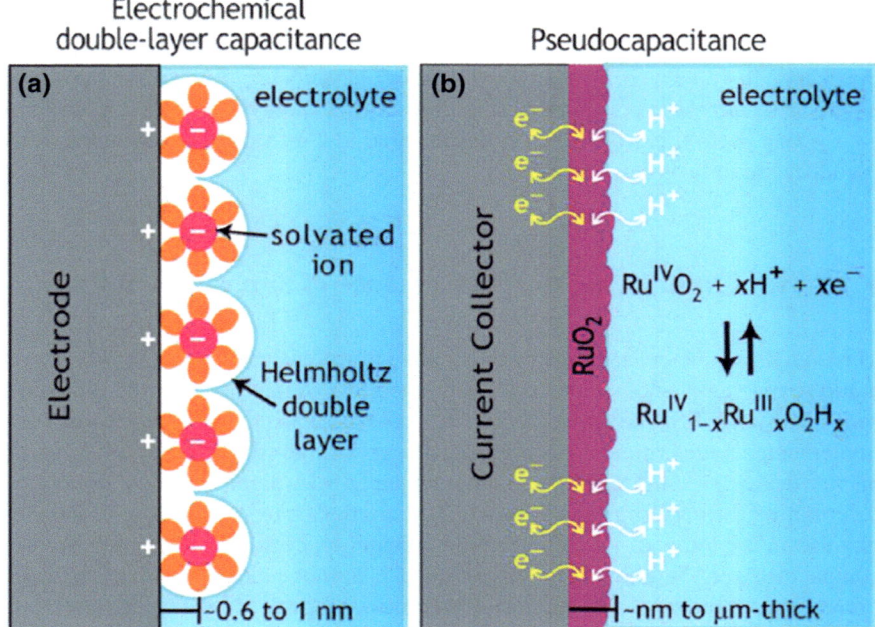

Fig. 1 Schematic of the charge storage mechanism for **a** EDL capacitance and **b** pseudocapacitance [31]

layers. In due time, the Gouy-Chapman-Stern model and the Gouy-Chapman model were introduced with some modifications to account for the description of the EDL structure more accurately [34, 35]. It was Becker who proposed first the mechanism of electricity storage and delivery for EDL in the year 1957, and hence, the term EDL capacitors were coined for the resulting supercapacitors storing charge in EDL [36, 37]. For EDL type of charge storage, the specific capacitance generated at each electrode follows property of a parallel-plate capacitor: $C = (\varepsilon_r \, \varepsilon_0/d)A$. Where ε_r is the relative permittivity, ε_0 is the permittivity of vacuum, A is the specific surface area of the electrode accessible to the electrolyte ions, and d is the effective thickness of the EDL [34, 35, 37]. Pseudocapacitance is generated as a result of the charge storage through highly reversible faradaic reactions between the electrode active material and the electrolyte. Attributed to the thermodynamic reasons, the faradaic charge transfer across the EDL leads to a unique potential-dependent charge buildup or discharge process so as the derivative dq/dV is equivalent to the pseudocapacitance produced. This type of faradaic charge transfer was familiarized by Trasatti et al. [38] and the pseudocapacitance is mostly attributed to the highly reversible redox reaction limited to the electrode active surface [39]. The mechanism for such charge storage can be explained as a result of the lowering of oxidation state leading to the reduction of the redox-active electrode materials in addition to the adsorption/intercalation of electrolytic cations from the at/near the electrode surfaces [40–42]. Pseudocapacitance is a result of the electrode–electrolyte interactions leading to charge acceptance and change in the voltage coming under the thermodynamic consideration [43]. The derivative of the accumulated charge with respect to the change in voltage corresponds to the pseudocapacitance. Since, EDL capacitors utilize an electrostatic non-faradic process of storing energy these electrodes exhibit a very long cycle life, higher power density, and low resistance as compared to the pseudocapacitors.

3 Factors Influencing the Performance of Supercapacitors

Three vital parameters are considered significant for evaluating the performance of a supercapacitor device: capacitance C_s, operating voltage V, and the equivalent series resistance ESR, as shown in Fig. 2. These three primary parameters are often implemented to determine the energy and power densities delivered by any supercapacitor device [37, 44]. Electrode materials play a very significant role in not only determining the performance of a supercapacitor device in terms of C_s, V, and ESR but also in the terms of its commercial applicability, cyclic stability, mobility, and compactness, etc. Besides searching for novel electrode materials with the requisite properties, cutting-edge manufacturing processes, innovative cell designs, and some additional factors become essential such as attaining high power density with substantially high energy density, high cyclic stability, rapid charge-discharge cycles, reduced self-discharging, safety, and cost-effectiveness [45, 46].

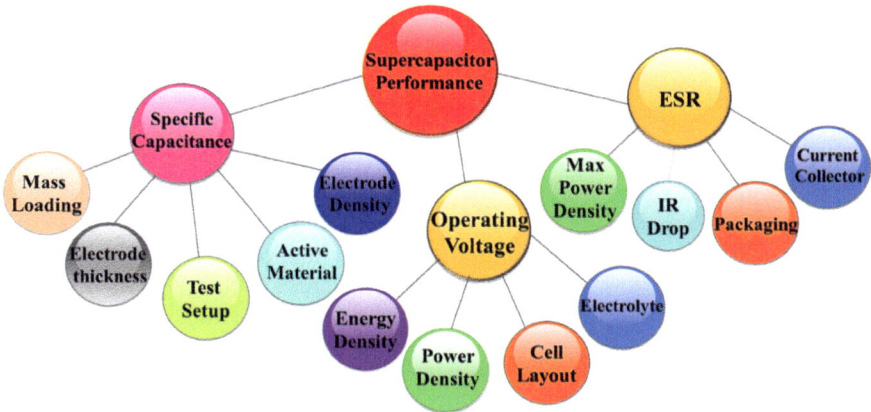

Fig. 2 Illustration of key performance metrics, test methods, major affecting factors for the evaluation of SCs

The energy density (E) and the power density (P) delivered by the supercapacitor are calculated using the following equations $E = 1/2\ CV^2$ and $P = V^2/4R_s$, where, $C(\text{F g}^{-1})$ is the total capacitance, V is the voltage, and R_s is the equivalent series resistance [37, 47–49]. A supercapacitor with high performance is a result of a combination of some indispensable factors such as high specific capacitance, large cell voltage, and reduced resistance. The assortment of suitable electrode materials and the electrolytic solution is the most significant measure to augment the performance of any supercapacitor device [50–52].

Electrode material for supercapacitor should exhibit a sufficiently high surface area and appropriate pore size. Further, the electrode material must be electrically conductive with a high robustness. Also, the density of the electrode material must be high enough to result into a high volumetric capacitance, energy and power density much necessary for fabrication of compact device [53–55]. Apart from the electrode material, another most important part of any supercapacitor device is the electrolyte used. The electrolyte is contained inside the separator and also inside the active material layers [54, 55]. An ideal electrolyte for a high-performance supercapacitor device must possess the following characteristics: a wide window of operating voltage, substantial electrochemical stability, sufficiently large ionic concentration, and small solvated ionic radius reduced resistivity, less viscosity, diminished volatility, non-toxicity, cost-effectiveness, and availability in a highly pure state [52, 56].

4 Modification of Graphene for the Preparation of Supercapacitor Electrodes

Graphene consists of sp^2 hybridized carbon atoms residing within a single plane consisting of two equivalent sub-lattices connected by σ bonds. Each carbon atom residing in the lattice has a π-orbital which is responsible for introducing delocalized electrons in the network [24, 57–60]. Attributed to this unique chemical structure graphene exhibits extraordinary physical and chemical properties such as high electron-hole symmetry, linear energy dispersion, and an internal degree of freedom [61–63]. In the field of supercapacitors graphene has attracted huge attention owed to its extraordinarily high specific surface area up to 2630 m^2 g^{-1}, highly flexible nature, good electrical conductivity, rich chemistry, and outstanding mechanical properties [61–65]. However, it is of utmost importance to understand the limitations that the graphene electrode materials face. Apart from the specific capacitance and the maximum energy density delivered by a graphene-based supercapacitor also depends on the density and the thickness of the graphene [66, 67]. The chemical modifications result in the improvement of the electronic properties, electrochemical properties, and the cyclic stability of the as obtained material. This chapter deals with an elaborate overview of the various kinds of graphene modifications, mainly dealing with their exclusive properties and various applications in the field of supercapacitor applications [68, 69]. The characteristics of the various materials synthesized by the graphene modification and investigated as supercapacitor electrodes have been discussed elaborately in the following sections.

5 Preparation of Graphene

As discussed earlier, graphene consists of sp^2-bonded carbon atoms residing in a single plane which is generally one-atom-thick. The planes are densely packed with carbons arranged in a honeycomb pattern, where the carbon–carbon bond exists [65, 70]. Several techniques that have been incorporated for the mass production of graphene includes both top-down and bottom-up methods [71–73]. These methods range from the mechanical exfoliation of highly pure graphite to the direct growth on the carbides or on some suitable metal substrates through various chemical routes, as outlined below:

The method of mechanical exfoliation produces sheets of monolayer graphene with varying thicknesses through a simple process of peeling [74]. Apart from these methods gradient ultrasonication and thermal treatment in a high-boiling point solvent with a high-boiling point such as N-methyl pyrrolidone are also well-known graphite synthesis methods with mechanical route [74, 75]. Commonly, these methods result in the production of defect-free monolayer graphene. However, synthesis route involving the use of a solvent with high-boiling point makes it difficult to deposit the resulting graphene sheets on the substrate. Surfactant-induced exfoliation of graphite in water has also been reported with the graphene sheets showing stability against the aggregation owed to the large potential barrier caused by the coulombic repulsion existing between the surfactant-induced sheets [74–77].

Chemical vapor deposition (CVD) is another well-known method implemented to produce mono- or a few-layered graphene using suitable metal surfaces at high temperatures [78–80]. During a typical CVD process, carbon is generally dissolved within a metal substrate followed by its precipitation on the substrate via cooling [81]. Apart from the thermal CVD, plasma-induced CVD also provides an alternate route for the synthesis of graphene at the relatively lower temperatures [82, 83]. In this process, the deposition instigates with the integration of a few carbon atoms in the Ni-substrate at a lower temperature, very similar to the process of carburization. The thickness of the material and crystalline order of the resultant graphene layers produced can be controlled through varying the rate of cooling and the concentration of the carbon dissolved with the Ni-substrate [83, 84].

Further, exfoliation of graphene through chemical route is another well-known practice used to produce graphene in large quantities [85]. This method comprises of magnifying the interlayer spacing of the graphite in order to result in a fading away interlayer Van der Waals force. This is attained by creating or intercalating various active functional groups in-between the layers of graphene so as to synthesize graphene-intercalated materials [86, 87]. Subsequently, the synthesized intercalated graphene-based compounds are then exfoliated to result into a mono- to a few-layered graphene through vigorous sonication, heating, or reduction. A very basic example of the chemical exfoliation constitutes the generation of the monolayer graphene oxide (GO) from multilayered graphite oxide via ultrasonication [88].

Another strategy is to use various techniques in order to synthesize microelectronics based on silicon have inspired the exploration of these techniques for the processing and the fabrication of graphene materials. According to this approach, the growth of graphene takes place on a single-crystal H-SiC at the silicon terminated face by using thermal desorption of silicon resulting in the production of a 2-3-layer-thick graphene sheet [89, 90]. However, it is unfortunate that this method is not implemented to synthesize monolayer graphene. Also, synthesis of the graphene domains with a uniform thickness throughout still remains a matter of challenge. As silicon automatically provides an insulating substrate, this method offers a direct route to the synthesis of a graphene wafer supported by silicon [91].

Apart from these methods, there are several other methods resulting in the preparation of graphene [92]. Some of the most significant and popular techniques for graphene modification has been discussed here. For the other methods, the chapter entitled "Graphene and its Derivatives for Battery Application" can be referred to.

6 Graphene Modification as Electrode Materials for Supercapacitor Application

The graphene or modifications graphene materials usually exploit properties such as the large surface area or high electrical conductivity. Generally speaking, graphene modification can be realized through either covalent bonding or non-covalent interactions which are shown in Fig. 3 [93–95]. The former is usually achieved via a few techniques, such as atom doping or reaction with the residual functional groups on graphene formed during the production or destruction of graphene's unsaturated structure. Chemical modifications of graphene would help to tackle the problems regarding its production, attain large surface area, increase storage capabilities, increase the stability, attain better mechanical properties, enhance the electrical conductivity, and improve handling and processing [96]. Reversible aggregation of graphene during the reduction process is a major source of concern. Aggregation makes the secondary Van der Waals forces come in effect. To overcome these strong Van der Waals forces, modification of graphene is necessary [97]. Figure 3 shows the schematic of classification of various interactions developed between the graphene sheets as a result of graphene modification through various chemical routes. As discussed earlier, graphene surfaces can be chemically modified using either covalent or non-covalent method [97, 98]. Lastly, the improvement in synthetic modifications on graphene is definitely here to stay and to bring promising opportunities to graphene in application of supercapacitor [96]. The various methods for the graphene modification are discussed in the upcoming section.

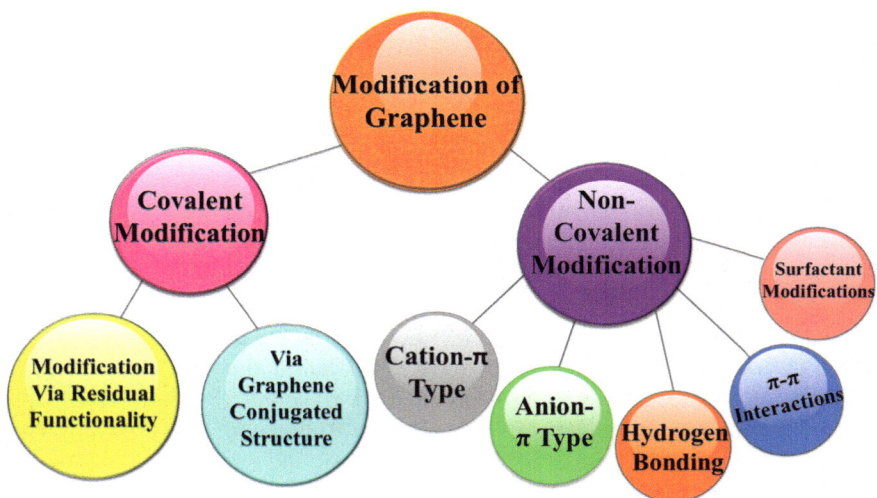

Fig. 3 Various methods for the graphene modification

7 Covalent Modification of Graphene

Attachment of different functionalities through covalent bonds on the graphene surfaces generally occurs by the existing oxygen linkages, termed as "oxygenated functional groups" or $\pi-\pi$ networks. Further, the existence of the carboxylic acid groups at the edges and functional groups like epoxy/hydroxyl above the surface of graphene oxide and graphene are also implemented to modify the surface functional properties of graphene [97, 99]. Modification of the graphene through covalent interaction is advantageous from the point of view that in case of the covalent modification stronger bonds are developed between the modifier and the graphene resulting in enhancement of the supercapacitor performance [100, 101]. The following sections briefly outline the major routes through which graphene can be modified covalently.

7.1 Covalent Attachment with Organic Functionalities

The formation of dispersed graphene sheets in organic solvents is considered to be a crucial step in order to synthesize graphene-based nanocomposites successfully. As a consequence, modification of graphene through covalent route with inducing various organic functional groups forms the key way to attain better dispersion. The reactions related to the organic covalent functionalization of graphene basically include two routes one is through the development of covalent bonds between the existing free radicals or dienophiles and C=C bonds of pure graphene and through the development of the covalent bonds between the organic functional groups and the oxygen groups of GO [97, 98]. Based on the previous theoretical interpretations and experimental observations with carbon nanotubes, fullerenes dienophiles, and the organic free radicals are considered to be the most striking organic species to react with sp^2 hybridized carbon atoms on graphene. A successful modification carried out chemically result in incorporation of redox-active sites on the surface, and also results in the enhancement of the conductivity of graphene. It has been known that created functional groups on the surface create additional pseudocapacitance to graphene. The enhanced wettability toward the electrolytes with incorporation of functional groups which are hydrophilic also significantly influences the characteristics of EDLC. Additional functional group content on the surface of graphene leads to dramatic boosting of the specific capacitance of the supercapacitor [102].

7.2 Modifications via Direct Doping of Atoms into the Graphene Lattice

The method of doping is a common approach to modify the electronic properties of the semiconductor materials. So far nitrogen or boron atoms have been used in case of the carbon materials to modify them into n-type or p-type materials, respectively [103, 104]. The n-type electronic doping of graphene can be carried at high annealing temperature of 1100 °C in ammonia [105]. Nitrogen and boron doping can greatly influence the charge storage capacity of graphene, owing to the modification of the electronic structure and surface energy [106]. N-doping improves the specific capacity of microporous graphene, which is attributed to the enhanced electrical conductivity and wettability [107, 108]. Density functional theory calculations showed that N-doping can increase the binding energy of electrolytic ions at the graphene surface, and accommodate more ions and thus have a higher charge storage capacity [109]. On the basis of these theoretical forecast, various wearable supercapacitors have been fabricated which consists of N-doped graphene (Fig. 4). According to a report, an N-doped graphene-based supercapacitor device exhibited much higher capacitance as compared to a device which is made of pure graphene tested in both organic electrolyte and KOH [110]. Besides nitrogen and boron, other

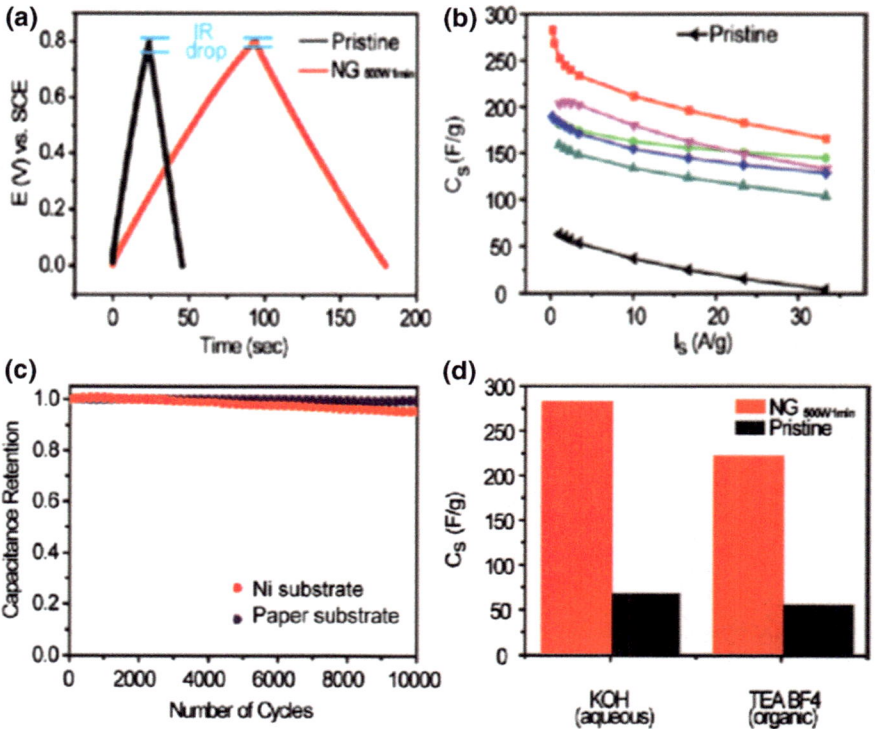

Fig. 4 **a** Charging-discharging curves, **b** gravimetric capacitances of nitrogen-doped graphene and pristine graphene measured at a different of current densities, **c** the cycling tests based on nickel-based and paper substrates up to 10,000 cycles, and **d** the specific capacitances measured in aqueous and organic electrolytes. Reprinted from Ref. [110], with permission from The American Chemical Society 2011

small molecules like poly (ethylene imine) have also been studied as a stable and complementary dopant. Apart from the doping of the graphene layers, the dopants itself can also be induced in between the consecutive graphene layers. Theoretical studies imply that the modification can be brought about by in the band structure of graphene by doping [111–113].

7.3 Modification Through the Free Radical Addition

Functionalization of carbon nanomaterials by the addition of free radical has been already achieved for graphite, fullerene, and carbon nanotubes. This approach has been effectively implemented for the graphene resulting in promotion of band gap and alteration of solubility. The functionalization of graphene via free radical addition approach has been performed by thermal as well as the photochemical treatments. Most common adducts which have been implemented so far are aryl diazonium salts, besides the other organic free radicals, as discussed below [95, 97, 98, 114].

7.3.1 Aryl Diazonium Salts

Mostly, aryl diazonium salts like diazonium tetrafluoroborate are used for synthesis by the free radical approach. Incorporation of the aryl group through developed covalent bond on the graphene carbon network consisting of sp^2 hybridized carbon atoms has been widely implemented [115]. When provided with acidic environment diazonium salt experiences a polar reaction attributed to it is ionic nature resulting in the formation of results in an aryl cation. On the other hand, a free radical step is followed when a neutral condition is maintained, as shown in the Scheme 1 [95]. Both these routes result in the expulsion of N_2 gas out of the diazonium salts. The reaction mechanism has been described as donating an electron by graphene to the diazonium

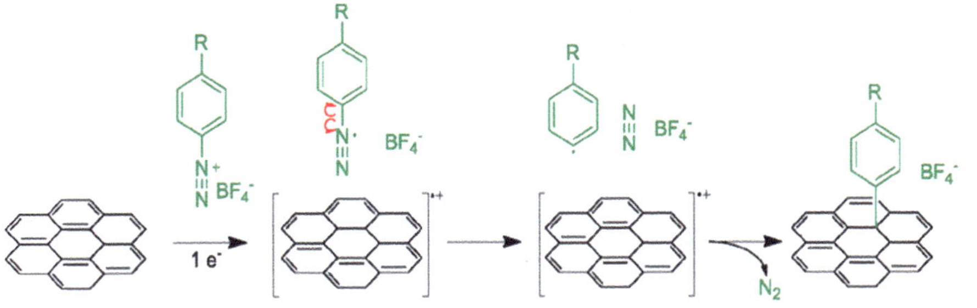

Scheme 1 Mechanism of the free radical addition for derivatives of phenyl onto graphene. Reprinted from Ref. [95], with permission from The Royal Society of Chemistry 2013

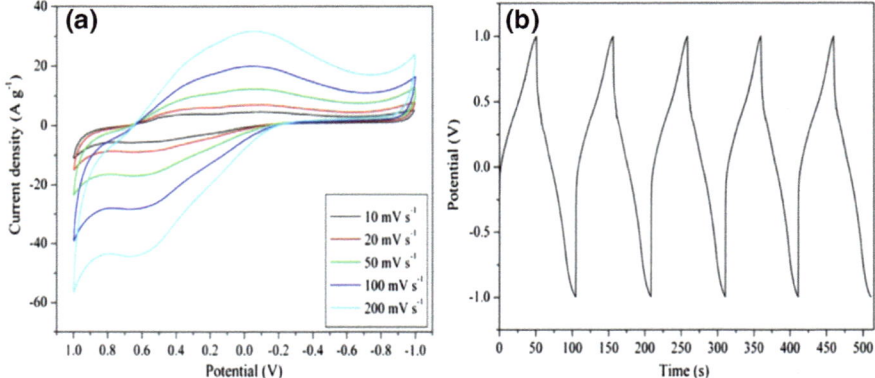

Fig. 5 **a** CV curve of ADS-G at various scan rates and **b** charge-discharge behavior of ADS-G-based supercapacitors performed at a constant current density of 4.5 A g^{-1}. Reprinted from Ref. [118], with permission from the Elsevier Ltd. 2012

salt resulting in radical aryl moiety. Further, it is believed that the functionalization reaction proceeds by a free radical mechanism [95, 116, 117].

Use of the aryldiazonium salts for carrying out surface modification by incorporating various functional molecules on the surface of graphene is a potential solution to weaken the Van der Waals force. However, the enhanced electrochemical properties as a result of graphene modification using aryl diazonium salts exhibit high specific capacitance values as compared to the pure graphene. According to a report functionalized graphene synthesized by aryl diazonium salts exhibits much improvement related to supercapacitive performance with enhanced specific capacitance as shown in Fig. 5 [118].

7.3.2 Peroxides

The incorporation of the aryl group into the sp^2 hybridized network of carbon atoms could also be accomplished by peroxides as well. However, only certain kind of peroxides including phenyl/phenyl incorporated with two or more than two nitro groups. The utility of these peroxides is to act as radical initiators when treated photo-chemically resulting in the liberation of the two molecules of CO_2 and in addition produce two radical species of phenyl (Scheme 2). Graphene is modified by the subsequent attachment of the phenyl radical species and the developed covalent linkage between them [95, 119, 120].

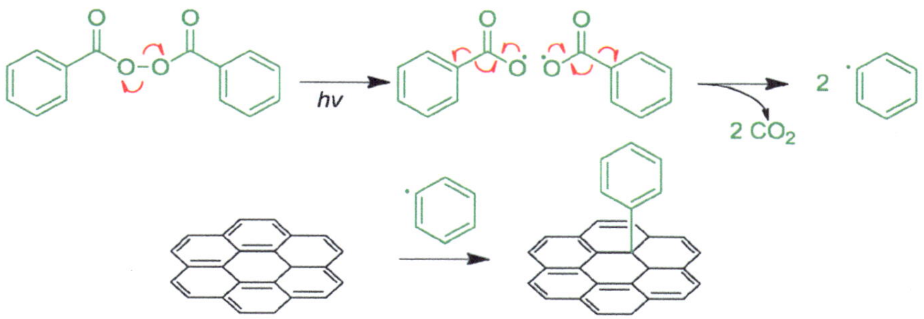

Scheme 2 Schematic representation of the mechanism of the creation of phenyl radical from benzoyl peroxide (top) and the addition of the phenyl moiety by a free radical mechanism on the graphene structure(bottom). Reprinted from Ref. [95], with permission from The Royal Society of Chemistry 2013

7.3.3 Bergman Cyclization

The Bergman cyclization takes place through the formation of a ring containing six members. The precursor includes a precursor which includes an enediyne group which gets cyclotrimerized under thermal treatment carried out at a high temperature by implementing the radical mechanism. The cyclization leads to the development of 1,4-benzenediyl bi-radical types. These highly reactive bi-radical species undergoes reaction with the graphene carbon atoms and hence establish covalent bonds on the one side of the existing bi-radical species. The other end of the radical has the possibility of getting involved either in a hydrogen bond or polymerization [121–123].

7.3.4 Kolbe Electrosynthesis

The Kolbe reaction occurs through electrolysis of carboxylate ions to lead to the generation of radical species by a succeeding decarboxylation step (Scheme 3). Most of the time only alkyl chain makes up the radical species [95, 124].

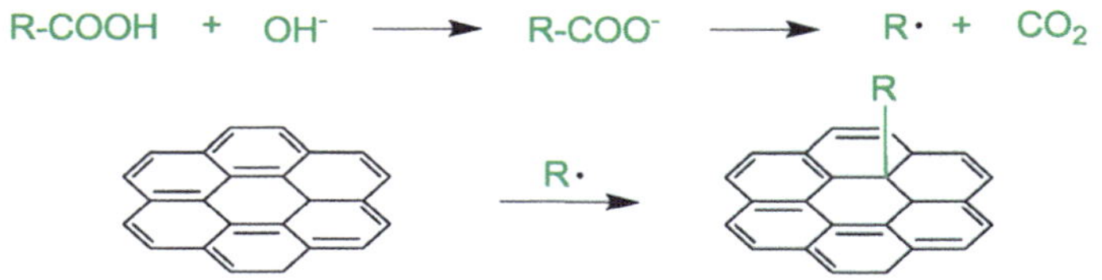

Scheme 3 Schematic representation of the Kolbe reaction toward the formation of radical species (top) and the free radical addition of radical species onto graphene (bottom). Reprinted from Ref. [95], with permission from The Royal Society of Chemistry 2013

7.4 Nucleophilic Addition

The Bingel reaction finds its origin from the chemistry of cyclopropanation of fullerene. This mechanism uses the halide derivative of diethyl malonate group and the reaction is carried out in the presence of a base like sodium hydride 1,8-diazabicyclo [22] or undec-7-ene. Its usefulness in graphene chemistry is appreciated owed to its ease of the reaction conditions. The halide-malonate is generated an in-situ form of a base and tetrahalomethane mixture. The base extracts a proton from the halide-malonate and provides an enolate which in turn nucleophilically bouts a C=C bond of the graphene framework. The resulting carbanion experiences a succeeding nucleophilic replacement [125–127].

8 Non-covalent Modification of Graphene

The modification route for graphene involving non-covalent approach seems to be more versatile and favorable attributed to minimum affect imparted to the graphene's natural structure. Various kinds of interactions such as nonpolar gas–π interaction, π–π interaction, H–π interaction, cation–π interaction, anion–π interaction, π–cation–π interaction, hydrogen bonding, and coordination bonds consists of the modification through the non-covalent route [97]. The interactions caused as a result of the non-covalent modifications are comparatively weak than the covalent interactions. However, multiple non-covalent interactions can always lead to better stability. In addition, they are easy to be acquired over the whole surface of graphene and these are also often reversible. These non-covalent synthesis approaches are appreciable when the synthesized graphene is to be equipped with both high electrical conductivity and also the high surface area. Covalent modification of the electrode materials based on graphene destroys the sp^2 carbon network by rehybridization of local sp^2-to-sp^3 bonds and diminishes the contribution from the fundamental double-layer capacitance of the electrode [96]. Another approach relates to the non-covalent modification of graphene-based electrodes, on the other hand, reduces the problem by utilization of the Van der Waals interactions to attach redox-active components onto the graphitic framework. This approach does not affect the uniform sp^2 carbon network and hence also does not disturb the power delivery or the double-layer capacitance for the graphene-based supercapacitor electrodes. The non-covalent modification also leads to the simplification of material synthesis method by reducing the number of synthesis steps and the reducing the number of reagents required. In the case of the non-covalent modification route, the group π–π interaction is the prime interaction is used. π–π interactions generally happen in-between two large aromatic rings with non-polarity and possessing overlapped p-orbitals. These interactions are as strong as the covalent attachment, and thereby provide much more stability as compared to the weaker electrostatic bonding, hydrogen bonding, and coordination bonding. In addition, modification by π–π stacking approach helps in retaining

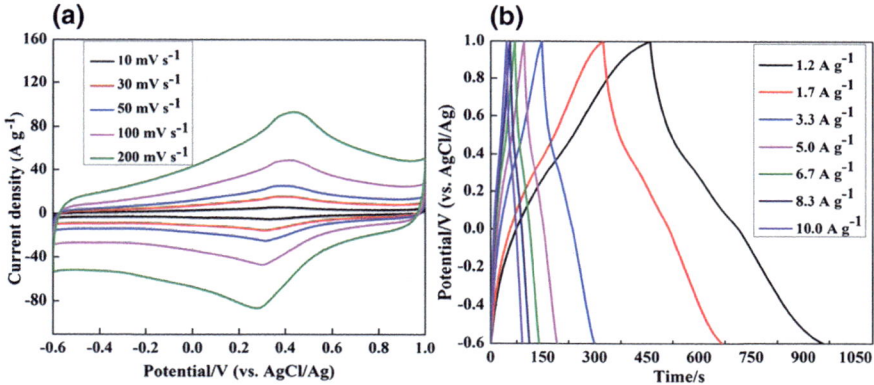

Fig. 6 **a** CV and **b** galvanometric charge-discharge of modified reduced graphene oxide. Reprinted from Ref. [131], with permission from The Royal Society of Chemistry 2015

conjugation of the graphene layers, and thus the electronic characteristics of the graphene remain unaltered, which is considered to be a great advantage [95–98]. According to a report use of hydrophobic Bu-hydroquinone led to the achievement of improved capacitance for the reduced graphene oxide electrodes and the electrode was observed to maintain 94% of the initial capacitance after 800 cycles [128]. Quinone derivatives of polyaromatic hydrocarbons, such as anthraquinone, are also more hydrophobic than hydroquinone/benzoquinone and have an extended p-electron system that can increase the strength of non-covalent attractions. In yet another report, the performance of the supercapacitor was evaluated for the material made of various polyaromatic-quinones adsorbed onto the onion-like carbons. These electrode materials were observed to retain excellent cycling stability, viz. 97% of the initial capacitance after 10,000 cycles [129, 130]. Yet in another report, sulfanilic acid azocromotrop was used for non-covalent modifications of the graphene oxide as a supercapacitor electrode material with the specific capacitance of 366 F g^{-1}, as shown in Fig. 6.

Apart from these, there are some more important interactions through which modification of graphene is resulted as discussed below.

8.1 Nonpolar Gas–π Interaction

In case of the π-interactions, when the counter molecule is a rare gas atom or a nonpolar molecule such as (gas dimers and hydrocarbons) the reaction is dominated with dispersion energies. On the other hand, both dispersion and the electrostatic energies direct the interaction for the case when the counter molecule is a Lewis acid or is a polar molecule [97, 132, 133].

8.2 H–π Interaction

Since H–π interaction is one of the hydrogen bond interactions so complexes with this kind of interaction are of much interest. The degree of polarizability availed by the π-electrons plays a significant role in prevailing the geometry and nature of the H–π interaction [97]. The amount of the dispersion energy inclines to be proportionate to the total number of electrons taking part in the interaction. For larger π-systems, like which are found in larger complexes containing several hydrogen bonds, a substantial contribution made by the dispersion energy results in the stabilization of the H–π complexes which are multidentate [134–136].

8.3 Hydrogen Bonding

Hydrogen bonding can be abundantly found in the energy storage materials. An individual hydrogen bond not sufficiently strong with a bond energy of 2–8 kcal mol^{-1}. The hydrogen bonding developed between the groups containing residual oxygen from the graphene sheets and the hydroxyl groups and resulting material with enhanced Young's modulus and tensile strength. For example, polymeric composites with graphene involving various epoxy, hydrophilic polymers, polyaniline, and poly (acrylonitrile) are observed to exhibit an outstanding enhancement in Young's modulus as a result of established hydrogen bonding interactions [96, 137, 138].

8.4 Cation–π Interaction

This is a special case of π–π interaction where a metal cation takes place of the counter molecule and both induction and electrostatic energies to dominate the developed interaction. In order to have a successful interaction between a positive organic cation and the negative π electron cloud, a precise estimation of the energy of interaction is required in turn necessitating consideration of the polarizability of the π system. The nature and property of the various interactions in cation-π interaction vary when transition metal complexes and π complexes are considered in places of alkali cations [97, 139–141].

8.5 π Cation–π Interaction

This is to be noted that the strength of the induction and the electrostatic energies case of the organic cation complexes of these π systems are much less as compared to that for π–alkali metal cation interactions. For the latter case, the dispersion energy becomes very significant. For the case, where an aromatic cation becomes the counter ion dispersion energy becomes very vital. This type of π cation–π interaction differs from the π–π interaction and metal cation–π interaction from the point of view of binding energies. The total binding energy for the π cation–π interaction is lesser than that involved for the cation–π interaction but largely stronger as compared to the π–π interaction [142–144].

8.6 Anion–π Interaction

The study on the lone pair–π and anion–π interactions are relatively new as compared to the π–π interaction, H–π interactions, and cation–π interaction. Recently, anion–π interactions have also been considered very interesting and have been extensively studied as electrode material various applications. The interaction energies for anion–π systems are comparable to those of the cation–π systems. The dispersion energies form a substantial part of the total energy in the anion–π complexes. The anion–π type interactions are also identified by a substantial augmentation in the amount of the energy of exchange-repulsion [145–148].

8.7 Coordination Bonds

Oxygen arbitrated coordination bonds, particularly which occurs for the transition metals, has been studied widely. It has been shown that the interaction between graphene and a metal can significantly be enhanced by the means of controlled modifications carried out on the macroscopic graphene-metal interface by the help of oxygen intercalation. It has been observed that when ruthenium residing beneath graphene was subjected to oxidization, the coupling between the metal and carbon was augmented [149, 150].

9 Graphene-Based Supercapacitors

Graphene can be obtained into various structures, viz. nanoparticles or dots, yarns or fibers, films, foams, and various composites. It was found that the rGO sheets could be inserted with various surfactants. The specific capacitance reported was 194 F g^{-1} for the graphene stabilized by TBAOH applied as a supercapacitor electrode at the current density of 1 A g^{-1}. This was partly attributed to the reduced degree of stacking in rGO enhanced wettability due to the surfactant intercalation. Also, the modification caused by the addition of surfactant leads the homogeneous dispersion of rGO into single or a few layers within aqueous solutions. These properties boost the chemical reactions between the rGO and also the secondary phases in the aqueous solvents.

Among the several surfactants studied, block copolymers poly-(ethylene oxide)-poly (phenylene)-poly (ethylene oxide) PEO106-PPO70-PEO106 (F127) were broadly investigated attributed to their well-known chemical properties [151]. According to a report GO was intercalated with triblock copolymer Pluronic F127 (F127) using the hydrothermal process, followed by the thermal annealing process to carry out the structural reconstruction. A specific surface area of 696 m^2 g^{-1} was reported in case of the surfactant-modified graphene, which is found to be three times larger than that of pure graphene. 210 F g^{-1} was the maximum specific capacitance measured at the scanning rate of 1 mV s^{-1} with 6 M KOH electrolyte, along with a superior cycling stability [152]. According to another report hollow graphene balls (GBs) synthesized by a template-directed method using Ni-nanoparticles as a template, yielded hollow GBs composed of multilayer graphene [153]. Another report accounts for a mass-producible mesoporous graphene nanoballs (MGB) via precursor-assisted method of chemical vapor deposition (CVD) for supercapacitor application The supercapacitor device so formed from the MGB as the electrode shows a specific capacitance of 206 F g^{-1} and greater than 96% retention of capacitance even after 10,000 cycles [154]. Recently, in another report graphene nanosheets were casted into micro/macro powdered particles. The supercapacitor electrode made from there exhibited a specific capacitance of 151 F g^{-1} at the scan rate of 10 mV s^{-1} [155]. Another report is about the study of NiO-graphene 3D nanocomposite exhibiting a high specific capacitance of about ~816 F g^{-1} at the scan rate of 5 mV s^{-1}, and also exhibiting better cycling performance without much degradation after 2000 cycles [156]. Also, nanocomposite materials designed by incorporating Co$_3$O$_4$ nanowires into the graphene foam synthesized via chemical vapor deposition by using nickel foam as substrates. This electrode material is reported with a specific capacitance of value ~1100 F g^{-1} rated at the current density of 10 A g^{-1} with an enhanced capacitance retention [157]. Graphene–MnO$_2$ 3D network engineered by depositing MnO$_2$ electrochemically with a mass loading of 9.8 mg cm^{-2} into 3D graphene with Ni-foam template [158]. This flexible nanocomposite electrode exhibited an aerial capacitance of value 1.42 F cm^{-2} measured at the rate of 2 mV s^{-1} at the desired cycling stability. This results also proved that implied that graphene networks can act as an excellent supporter for active materials. A prototype device with electrode material based on nitrogen and boron co-doped graphene aerogels synthesized by a hydrothermal reduction process shows a specific capacitance of ~62 F g^{-1}, together with a good rate capability and an improved energy density of ~8.65 W h kg^{-1} [159]. In a report, the specific capacitance measured for graphene hydrogel exhibiting well-defined and cross-linked 3D porous structure was obtained to be ~240 F g^{-1} at the current density of 1.2 A g^{-1} in 1 M aqueous solution of H$_2$SO$_4$ [160]. In another report, holey graphene hydrogel (HGH) implemented as a binder-free supercapacitor electrode exhibits a specific capacitance in the range between 283 and 234 F g^{-1} with cycling stability and excellent rate capability [161].

10 Conclusions and Future Design Concept

Supercapacitors forms a very important part of the modern charge storage devices family. These are very much effective for fast charge-discharge rates and long cyclic life. A subsistence amount of research is dedicated to design electrode material with enhance energy the density delivered along with high power density. Carbon materials are used preliminarily for the fabrication of electrode materials used in supercapacitors attributed to their long cycle life, light weight property, ease of synthesis, and eco-friendliness. Among the various forms of the carbon materials, graphene forms a very attractive form owed to its high surface area. Since its discovery in 2004, major progress has been realized by the implantation of graphene for various uses in electronics, novel materials, and chemistry. Graphene can be prepared via various synthesis techniques, among which mechanical exfoliation, thermal deposition, liquid-phase exfoliation of graphite, and oxidation of graphite are implemented the most. However, all of these are known to have some drawbacks though they prove advantageous for some particular kind of application such as a supercapacitor. Chemical modifications of graphene have several advantages as outlined: (i) This would reduce the major problems related to its commercial production, (ii) Also the modification makes the material more resistive to corrosion and structural and chemical alteration (iii) The modified graphene is easy to handle and process further. The composites involving modified graphene possess synergistic properties arising from both the modifiers and graphene. The modified graphene is possible to be tailored to result in materials with desired stability and solubility, along with tunable thermal, electric and mechanical properties, properties. These characteristics project them as very attractive materials for the production of electrode material for supercapacitors.

References

1. Yang, Z., Zhang, J., Kintner-Meyer, M.C.W., et al.: Electrochemical energy storage for green grid. Chem. Rev. **111**, 3577 (2011)
2. Choi, H.S., Park, C.R.: Theoretical guidelines to designing high performance energy storage device based on hybridization of lithium-ion battery and supercapacitor. J. Power Sources **259**, 1 (2014)
3. Simon, P., Gogotsi, Y.: Materials for electrochemical capacitors. Collect. Rev. Nat. J. **7**, 12647 (2010)
4. Cheng, F., Liang, J., Tao, Z., et al.: Functional materials for rechargeable batteries. Adv. Mater. **23**, 1695 (2011)
5. Marom, R., Francis, A.S., Leifer, N., et al.: A review of advanced and practical lithium battery materials. J. Mater. Chem. **21**, 9938 (2011)
6. Dubal, D.P., Ayyad, O., Ruiz, V., et al.: Hybrid energy storage: the merging of battery and supercapacitor chemistries. Chem. Soc. Rev. **44**, 1777 (2015)
7. Akram, U., Khalid, M., Shafiq, S., et al.: An innovative hybrid wind-solar and battery-supercapacitor microgrid system—Development and optimization. IEEE Access **5**, 25897 (2017)
8. Burke, A.: Ultracapacitors: why, how, and where is the technology. J. Power Sources **91**, 37 (2000)

9. Zhang, Y., Feng, H., Wu, X.: Progress of electrochemical capacitor electrode materials: A review. Int. J. Hyd. Energy **34**, 4889 (2009)
10. Thakur, A.K., Choudhary, R.B., Majumder, M., et al.: In-situ integration of waste coconut shell derived activated carbon/polypyrrole/rare earth metal oxide [Eu_2O_3]: a novel step towards ultrahigh volumetric capacitance. Electrochim. Acta **251**, 532 (2017)
11. Conway, B.E., Pell, J.: Double-layer and pseudocapacitance types of electrochemical capacitors and their applications to the development of hybrid devices. G. Pell J. Sol. State Electrochem. **7**, 637 (2003)
12. Thakur, A.K., Majumder, M., Choudhary, R.B., et al.: MoS_2 flakes integrated with boron and nitrogen-doped carbon: striking gravimetric and volumetric capacitive performance for supercapacitor applications. J. Power Sources **402**, 163 (2018)
13. Majumder, M., Choudhary, R.B., Koiry, S.P., et al.: Gravimetric and volumetric capacitive performance of polyindole/carbon black/MoS_2 hybrid electrode material for supercapacitor applications. Electrochim. Acta **248**, 98 (2017)
14. Conway, B.E., Murphy, O.J., Srinivasan, S. (eds.): Electrochemistry in Transition: From the 20th to the 21st Century. Springer Science & Business Media (2013)
15. Yu, A., Chabot, V., Zhang, J.: Electrochemical supercapacitors for energy storage and delivery: fundamentals and applications. J. Electrochem. Supercapacitors Energy Storage Deliv. Fundam. Appl. (2013)
16. Lu, M.: Supercapacitors: Materials, Systems, and Applications. Wiley, New York (2013)
17. Yassine, M., Drazen, F.: Performance of commercially available supercapacitors. Energies **10**, 1340 (2017)
18. Monthéard, R., Bafleur, M., Boitier, V., et al.: Coupling supercapacitors and aeroacoustic energy harvesting for autonomous wireless sensing in aeronautics applications. Energy Harvest. Syst. **3**, 265 (2016)
19. Chen, L., Ji, T., Mu, L., et al.: Cotton fabric derived hierarchically porous carbon and nitrogen doping for sustainable capacitor electrode. Carbon **111**, 839 (2017)
20. Wang, G., Zhang, L., Zhang, J.: A review of electrode materials for electrochemical supercapacitors. Chem. Soc. Rev. **41**, 797 (2012)
21. Zheng, Y., Yang, Y., Chen, S., et al.: Smart, stretchable and wearable supercapacitors: prospects and challenges. CrystEngComm **18**, 4218 (2016)
22. Bose, S., Kuila, T., Mishra, A.K., et al.: Carbon-based nanostructured materials and their composites as supercapacitor electrodes. J. Mater. Chem. **22**, 767 (2012)
23. Zhang, L., Zhou, L.R., Zhao, X.S.: Graphene-based materials as supercapacitor electrodes. J. Mater. Chem. **20**, 5983 (2010)
24. Brownson, D.A.C., Kampouris, D.K., Banks, C.E.: An overview of graphene in energy production and storage applications. J. Power Sources **196**, 4873 (2011)
25. Zhang, L.L., Zhao, X.S.: Carbon-based materials as supercapacitor electrodes. Chem. Soc. Rev. **38**, 2520 (2009)
26. Hou, J., Shao, Y., Ellis, M.W., et al.: Graphene-based electrochemical energy conversion and storage: fuel cells, supercapacitors and lithium ion batteries. Phys. Chem. Chem. Phys. **13**, 15384 (2011)
27. Seredych, M., Hulicova-Jurcakova, D., Lu, G.Q., et al.: Surface functional groups of carbons and the effects of their chemical character, density and accessibility to ions on electrochemical performance. Carbon **46**, 1475 (2008)
28. Li, H., Xi, H., Zhu, S., et al.: Preparation, structural characterization, and electrochemical properties of chemically modified mesoporous carbon. Micropor. Mesopor. Mat. **96**, 357 (2006)
29. Kim, S.S., Kadoma, Y., Ikuta, H.: Electrochemical performance of natural graphite by surface modification using aluminum. Electrochem. Solid-State Lett. **4**, 109 (2001)
30. Boukhvalov, D.W., Katsnelson, M.I.: Chemical functionalization of graphene with defects. Nano Lett. **8**, 4373 (2008)
31. Zhang, X., Cheng, X., Zhang, Q.J.: Where do batteries end and supercapacitors begin? Energy Chem. **25**, 967 (2016)

32. Roldan, S., Barreda, D., Granda, M., et al.: An approach to classification and capacitance expressions in electrochemical capacitors technology. Phys. Chem. Chem. Phys. **17**, 1084 (2015)
33. Helmholtz, H.V.: Ann. Phys. (Leipzig) **89**, 21 (1853)
34. Gouy, M.: Sur la constitution de la charge électrique à la surface d'un électrolyte. J. Phys. Theor. Appl. **9**, 457 (1910)
35. Chapman, D.: LI. A contribution to the theory of electrocapillarity. London, Edinburgh Dublin Philos. Mag. J. Sci. **25**, 475 (1913)
36. Becker, H.I.: General Electric. U.S. Patent 2 800 616 (1957)
37. Zhang, S., Pan.: Supercapacitors performance evaluation. Adv. Energy Mater. **5**, 1401401 (2015)
38. Angelinetta, C., Trasatti, S., Atanososka, L.D., et al.: Surface properties of $RuO_2 + IrO_2$ mixed oxide electrodes. J. Electroanal. Chem. Interf. Electrochem. **214**, 535 (1986)
39. Liu, T., Pell, W.G., Conway, B.E. Self-discharge and potential recovery phenomena at thermally and electrochemically prepared RuO_2 supercapacitor electrodes. Electrochim. Acta **42**, 3541 (1997)
40. Dupont, M.F., Donne, S.W.: Charge storage mechanisms in electrochemical capacitors: effects of electrode properties on performance. J. Power Sources **326**, 613 (2016)
41. Lin, Z., Taberna, P.L., Simon, P. Advanced analytical techniques to characterize materials for electrochemical capacitors. Curr. Opin. Electrochem. (2018)
42. Freeborn, T.J., Maundy, B., Elwakil, A.S.: Fractional-order models of supercapacitors, batteries and fuel cells: a survey. Mater. Renew. Sust. Energy **4**, 9 (2015)
43. Conway, B.E., Birss, V., Wojtowicz, J.: The role and utilization of pseudocapacitance for energy storage by supercapacitors. J. Power Sources **66**, 1 (1997)
44. Conway, B.E.: Electrochemical supercapacitors: scientific fundamentals and technological applications. Springer Science & Business Media (2013)
45. Stoller, M.D., Ruoff, R.S.: Best practice methods for determining an electrode material's performance for ultracapacitors. Energy Environ. Sci. **3**, 1294 (2010)
46. Korotcenkov, G.: Metal Oxides in Supercapacitors. Elsevier (2017)
47. Sevilla, M., Fuertes, A.B.: Direct synthesis of highly porous interconnected carbon nanosheets and their application as high-performance supercapacitors. ACS Nano **8**, 5069 (2014)
48. Wang, H., Xu, Z., Kohandehghan, A., et al.: Interconnected carbon nanosheets derived from hemp for ultrafast supercapacitors with high energy. ACS Nano **7**, 5131 (2013)
49. Qu, D., Shi, H.: Studies of activated carbons used in double-layer capacitors. J. Power Sources **74**, 99 (1998)
50. Frackowiak, E., Meller, M., Menzel, J., et al.: Redox-active electrolyte for supercapacitor application. Faraday Discuss. **172**, 179 (2014)
51. Chen, W., Rakhi, R.B., Alshareef, H.N., et al.: Capacitance enhancement of polyaniline coated curved-graphene supercapacitors in a redox-active electrolyte. Nanoscale **5**, 4134 (2013)
52. Tomiyasu, H., Shikata, H., Takao, K., et al.: An aqueous electrolyte of the widest potential window and its superior capability for capacitors. Sci. Rep. **7**, 45048 (2017)
53. Wang, Q., Yan, J., Fan, Z.: Carbon materials for high volumetric performance supercapacitors: design, progress, challenges and opportunities. Energy Environ. Sci. **9**, 729 (2016)
54. Zhang, C., Lv, W., Tao, Y., et al.: Towards superior volumetric performance: design and preparation of novel carbon materials for energy storage. Energy Environ. Sci. **8**, 1390 (2015)
55. Frackowiak, E.: Carbon materials for supercapacitor application. Phys. Chem. Chem. Phys. **9**, 1774 (2007)
56. Zhong, C., Deng, Y., Hu, W., et al.: A review of electrolyte materials and compositions for electrochemical supercapacitors. Chem. Soc. Rev. **44**, 7484 (2015)
57. Lim, E.L., Yap, C.C., Jumali, M.H.H.: A mini review: can graphene be a novel material for perovskite solar cell applications? Nano-Micro Lett. **10**, 27 (2018)
58. Dresselhaus, M.S., Dresselhaus, G.: Intercalation compounds of graphite. Dresselhaus Adv. Phys. **51**, 1 (2002)

59. Yu, Z., Tetard, L., Zhai, L., et al.: Supercapacitor electrode materials: nanostructures from 0 to 3 dimensions. Energy Environ. Sci. **8**, 702 (2015)
60. Stankovich, S., Dikin, D.A., Dommett, G.H.B., et al.: Graphene-based composite materials. Nature **442**, 282 (2006)
61. Gomez-Navarro, C., Weitz, R.T., Bittner, A.M., et al.: Electronic transport properties of individual chemically reduced graphene oxide sheets. Nano Lett. **7**, 3499 (2007)
62. Park, S., Ruoff, R.S.: Chemical methods for the production of graphenes. Nature Nanotechnol. **4**, 217 (2009)
63. Ng, S.W., Noor, N., Zheng, Z.: Graphene-based two-dimensional Janus materials. NPG Asia Mater. **1**, (2018)
64. Huang, Y., Liang, J., Chen, Y.: An overview of the applications of graphene-based materials in supercapacitors. Small **8**, 1805 (2012)
65. Novoselov, K.S., Fal, V.I., Colombo, L., et al.: A roadmap for graphene. Nature **490**, 192 (2012)
66. Ohta, T., Bostwick, A., Seyller, T.: Controlling the electronic structure of bilayer graphene. Science **313**, 951 (2006)
67. Loh, K.P., Bao, Q., Ang, P.K., et al.: The chemistry of graphene. J. Mater. Chem. **20**, 2277 (2010)
68. Yuan, W., Liu, A., Huang, L., et al.: High-performance NO_2 sensors based on chemically modified graphene. Adv. Mater. **25**, 766 (2013)
69. Kulkarni, H.B., Tambe, P., Joshi, G.M.: Influence of covalent and non-covalent modification of graphene on the mechanical, thermal and electrical properties of epoxy/graphene nanocomposites: a review. Compos. Interfaces **25**, 381 (2018)
70. Geim, A.K., Novoselov, K.S.: The rise of graphene. Nanosci. Technol. Collect. Rev. Nat. J. **11** (2010)
71. Chen, D., Feng, H., Li, J.: Graphene oxide: preparation, functionalization, and electrochemical applications. Chem. Rev. **112**, 6027 (2012)
72. Eswaraiah, V., Aravind, S.S.J., Ramaprabhu, S.: Top down method for synthesis of highly conducting graphene by exfoliation of graphite oxide using focused solar radiation. J. Mater. Chem. **21**, 6800 (2011)
73. Tour, J.M.: Top-down versus bottom-up fabrication of graphene-based electronics. Chem. Mater. **26**, 163 (2013)
74. Yi, M., Shen, Z.: A review on mechanical exfoliation for the scalable production of graphene. J. Mater. Chem. A **3**, 11700 (2015)
75. Dresselhaus, M.S., Araujo, P.T.: Perspectives on the nobel prize in physics for graphene, ACS Nano **4**, 6297 (2010)
76. Jayasena, B., Subbiah, S.: A novel mechanical cleavage method for synthesizing few-layer graphenes. Nanoscale Res. Lett. **6**, 95 (2011)
77. Chen, J., Duan, M., Chen, G.: Continuous mechanical exfoliation of graphene sheets via three-roll mill. J. Mater. Chem. **22**, 19625 (2012)
78. Li, X., Magnuson, C.W., Venugopal, A., et al.: Large-area graphene single crystals grown by low-pressure chemical vapor deposition of methane on copper. J. Am. Chem. Soc. **133**, 2816 (2011)
79. Wei, D., Liu, Y., Wang, Y., et al.: Synthesis of N-doped graphene by chemical vapor deposition and its electrical properties. Nano Lett. **9**, 1752 (2009)
80. Reina, A., Jia, X., Ho, J., et al.: Large area, few-layer graphene films on arbitrary substrates by chemical vapor deposition. Nano Lett. **9**, 30 (2008)
81. Mattevi, C., Kim, H., Chhowalla, M.: A review of chemical vapour deposition of graphene on copper. J. Mater. Chem. **21**, 3324 (2011)
82. Dey, A., Chroneos, A., Braithwaite, NStJ, et al.: Plasma engineering of graphene. Appl. Phys. Rev. **3**, 021301 (2016)
83. Cheng, L., Yun, K., Lucero, A., et al.: Low temperature synthesis of graphite on Ni films using inductively coupled plasma enhanced CVD. J. Mater. Chem. C **3**, 5192 (2015)

84. Zhang, X.Y., Ha Sun, S., Sun, X.J., et al.: Plasma-induced, nitrogen-doped graphene-based aerogels for high-performance supercapacitors. Light-Sci. Appl. **5**, 16130 (2016)
85. Hernandez, Y., Nicolosi, V., Lotya, M., et al.: High-yield production of graphene by liquid-phase exfoliation of graphite. Nat. Nanotechnol. **3**, 563 (2008)
86. Stankovich, S., Dikin, D.A., Piner, R.D., et al.: Synthesis of graphene-based nanosheets via chemical reduction of exfoliated graphite oxide. Carbon **45**, 1558 (2007)
87. Gilje, S., Song, H., Wang, M., Wang, K.L., Kaner, R.B.: A chemical route to graphene for device applications. Nano Lett. **7**, 3394 (2007)
88. Zhu, Y., Murali, S., Cai, W., et al.: Graphene and graphene oxide: synthesis, properties, and applications. Adv. Mater. **22**, 3906 (2010)
89. Gaskill, D.K., Jernigan, G., Campbell, P., et al.: Epitaxial graphene growth on SiC wafers. ECS Trans. **19**, 117 (2009)
90. Emtsev, K.V., Speck, F., Seyller, T., Ley, L., et al.: Interaction, growth, and ordering of epitaxial graphene on SiC {0001} surfaces: a comparative photoelectron spectroscopy study. Phys. Rev. B **77**, 155303 (2008)
91. Kang, K., Cho, Y., Yu, K.J.: Novel nano-materials and nano-fabrication techniques for flexible electronic systems. Micromachines **9**, 263 (2018)
92. Bhuyan, M.S.A., Uddin, M.N., Islam, M.M., et al.: Synthesis of graphene. Int. Nano Lett. **6**, 65 (2016)
93. Maiti, U.N., Lee, W.J., Lee, J.M., et al.: 25th anniversary article: chemically modified/doped carbon nanotubes & graphene for optimized nanostructures & nanodevices. Adv. Mater. **26**, 40 (2014)
94. Mishra, A.K., Ramaprabhu, S.: Functionalized graphene-based nanocomposites for supercapacitor application. J. Phys. Chem. C **115**, 14006 (2011)
95. Chua, C.K., Pumera, M.: Covalent chemistry on graphene. Chem. Soc. Rev. **42**, 3222 (2013)
96. Liu, J., Tang, J., Gooding, J.J.: Strategies for chemical modification of graphene and applications of chemically modified graphene. J. Mater. Chem. **22**, 12435 (2012)
97. Georgakilas, V., Otyepka, M., Bourlinos, A.B., et al.: Functionalization of graphene: covalent and non-covalent approaches, derivatives and applications. Chem. Rev. **112**, 6156 (2012)
98. Lonkar, S.P., Deshmukh, Y.S., Abdala, A.A.: Recent advances in chemical modifications of graphene. Nano Res. **8**, 1039 (2015)
99. Salavagione, H.J., Martínez, G., Ellis, G.: Recent advances in the covalent modification of graphene with polymers. Macromol. Rapid Comm. **32**, 1771 (2011)
100. Li, Z.F., Zhang, H., Liu, Q., et al.: Covalently-grafted polyaniline on graphene oxide sheets for high performance electrochemical supercapacitors. Carbon **71**, 257 (2014)
101. Xu, Y., Shi, G.: Assembly of chemically modified graphene: methods and applications. J. Mater. Chem. **21**, 3311 (2011)
102. Fang, Y., Luo, B., Jia, Y., et al.: Renewing functionalized graphene as electrodes for high-performance supercapacitors. Adv. Mater. **24**, 6348 (2012)
103. Panchakarla, L.S., Subrahmanyam, K.S., Saha, S.K.: Synthesis, structure, and properties of boron-and nitrogen-doped graphene. Adv. Mater. **21**, 4726 (2009)
104. Arnold, R., Melvin, B.: Method of doping epitaxially grown semiconductor material. U.S. Patent 3,361,600, Issued January 2 (1968)
105. Wang, X., Li, X., Zhang, L., et al.: N-doping of graphene through electrothermal reactions with ammonia. Science **324**, 768 (2009)
106. Reddy, A.L.M., Srivastava, A., Gowda, S.R.: Synthesis of nitrogen-doped graphene films for lithium battery application. ACS Nano **4**, 6337 (2010)
107. Wen, Z., Wang, X., Mao, S., et al.: Crumpled nitrogen-doped graphene nanosheets with ultrahigh pore volume for high-performance supercapacitor. Adv. Mater. **24**, 5610 (2012)
108. Ke, Q., John, W.: Graphene-based materials for supercapacitor electrodes–a review. Materiomics **2**, 37 (2016)
109. Singh, K.P., Bhattacharjya, D., Razmjooei, F., et al.: Effect of pristine graphene incorporation on charge storage mechanism of three-dimensional graphene oxide: superior energy and power density retention. Sci. Rep. **6**, 31555 (2016)

110. Jeong, H.M., Lee, J.W., Shin, W.H., et al.: Nitrogen-doped graphene for high-performance ultracapacitors and the importance of nitrogen-doped sites at basal planes. Nano Lett. **11**, 2472 (2011)
111. Giovannetti, G., Khomyakov, P.A., Brocks, G., et al.: Doping graphene with metal contacts. Phys. Rev. Lett. **101**, 026803 (2008)
112. Farmer, D.B., Golizadeh, M.R., Perebeinos, V., et al.: Chemical doping and electron—hole conduction asymmetry in graphene devices. Nano Lett. **9**, 388 (2008)
113. Liu, H., Kuila, T., Kim, N.H., et al.: In situ synthesis of the reduced graphene oxide–polyethyleneimine composite and its gas barrier properties. J. Mater. Chem. **A1**, 3739 (2013)
114. Zhang, L., Zhou, L., Yang, M., et al.: Photo-induced free radical modification of graphene. Small **9**, 1134 (2013)
115. Bahr, J.L., Yang, J., Kosynkin, D.V., et al.: Functionalization of carbon nanotubes by electro-chemical reduction of aryl diazonium salts: a bucky paper electrode. J. Am. Chem. Soc. **123**, 6536 (2001)
116. Bekyarova, E., Itkis, M.E., Ramesh, P., et al.: Chemical modification of epitaxial graphene: spontaneous grafting of aryl groups. J. Am. Chem. Soc. **131**, 1336 (2009)
117. Chehimi, M.M. (eds).: Aryl Diazonium Salts: New Coupling Agents in Polymer and Surface Science. Wiley, New York (2012)
118. Yu, D.S., Kuila, T., Kim, N.H., et al.: Effects of covalent surface modifications on the electrical and electrochemical properties of graphene using sodium 4-aminoazobenzene-4'-sulfonate. Carbon **54**, 310 (2013)
119. Hamilton, C.E., Lomeda, J.R., Sun, Z., et al.: Radical addition of perfluorinated alkyl iodides to multi-layered graphene and single-walled carbon nanotubes. Nano Res. **3**, 138 (2010)
120. Palanisamy, S., Chen, S.M., Sarawathi, R.: A novel nonenzymatic hydrogen peroxide sensor based on reduced graphene oxide/ZnO composite modified electrode. Sens. Actuator B-Chem. **166**, 372 (2012)
121. Chen, S., Hu, A.: Recent advances of the Bergman cyclization in polymer science. Sci. China Chem. **58**, 1710 (2015)
122. Xiao, Y., Hu, A.: Bergman cyclization in polymer chemistry and material science. Macromol. Rapid Comm. **32**, 1688 (2011)
123. Layek, R.K., Nandi, A.K.: A review on synthesis and properties of polymer functionalized graphene. Polymer **54**, 5087 (2013)
124. Dickert, F.L., Alkire, C.R., Kolb, M.D., Lipkowski, J., Ross, P.N., Richard, C., Alkire, Dieter M. Kolb, Jacek Lipkowski, Philipp, N. Ross. (eds).: Chemically modified electrodes. Anal. Bioanal. Chem. **398**, 579 (2010)
125. Kuila, T., Bose, S., Mishra, A.K., et al.: Chemical functionalization of graphene and its applications. Prog. Mater Sci. **57**, 1061 (2012)
126. Hsiao, M.C., Liao, S.H., Yen, M.Y., et al.: Preparation of covalently functionalized graphene using residual oxygen-containing functional groups. ACS Appl. Mater. Interfaces **2**, 3092 (2010)
127. Shen, J., Shi, M., Ma, H., Yan, B., Li, N., Hu, Y., Ye, M.: Synthesis of hydrophilic and organophilic chemically modified graphene oxide sheets. J. Colloid Interf. Sci. **352**, 366 (2010)
128. Wang, H.W., Wu, H.Y., Chang, Y.Q., et al.: Tert-butylhydroquinone-decorated graphene nanosheets and their enhanced capacitive behaviors. Chinese Sci. Bull. **56**, 2092 (2011)
129. Anjos, D.M., McDonough, J.K., Perre, E., et al.: Pseudocapacitance and performance stability of quinone-coated carbon onions. Nano Energy **2**, 702 (2013)
130. Anjos, D.M., Kolesnikov, A.I., Wu, Z., et al.: Inelastic neutron scattering, Raman and DFT investigations of the adsorption of phenanthrenequinone on onion-like carbon. Carbon **52**, 150 (2013)
131. Jana, M., Saha, S., Khanra, P., et al.: Non-covalent functionalization of reduced graphene oxide using sulfanilic acid azocromotrop and its application as a supercapacitor electrode material. J. Mater. Chem. A **3**, 7323 (2015)

132. Shang, Q.Y., Bernstein, E.R.: Energetics, dynamics, and reactions of Rydberg state molecules in van der Waals clusters. Chem. Rev. **94**, 2015 (1994)
133. Tarakeshwar, P., Kim, K.S., Kraka, E., et al.: Structure and stability of fluorine-substituted benzene-argon complexes: The decisive role of exchange-repulsion and dispersion interactions. J. Chem. Phys. **115**, 6018 (2001)
134. Burley, S.K., Petsko, G.A.: Aromatic-aromatic interaction: a mechanism of protein structure stabilization. Science **229**, 23 (1985)
135. Kwon, J.Y., Singh, N.J., Kim, H.N., et al.: Fluorescent GTP-sensing in aqueous solution of physiological pH. J. Am. Chem. Soc. **126**, 8892 (2004)
136. Tarakeshwar, P., Choi, H.S., Kim, K.S.: Olefinic vs aromatic $\pi-h$ interaction: a theoretical investigation of the nature of interaction of first-row hydrides with ethene and benzene. J. Am. Chem. Soc. **123**, 3323 (2001)
137. Medhekar, N.V., Ramasubramaniam, A., Ruoff, R.S., et al.: Hydrogen bond networks in graphene oxide composite paper: structure and mechanical properties. ACS Nano **4**, 2300 (2010)
138. Liang, J., Huang, Y., Zhang, L., et al.: Molecular-level dispersion of graphene into poly [vinyl alcohol] and effective reinforcement of their nanocomposites. Adv. Funct. Mater. **19**, 2297 (2009)
139. Du, Q.S., Wang, Q.Y., Du, L.Q., et al.: Theoretical study on the polar hydrogen-π [Hp-π] interactions between protein side chains. Chem. Cent. J. **7**, 92 (2013)
140. Kim, D., Hu, S., Tarakeshwar, P., et al.: Cation$-\pi$ interactions: a theoretical investigation of the interaction of metallic and organic cations with alkenes, arenes, and heteroarenes. J. Phys. Chem. A **107**, 1228 (2003)
141. Wang, W., Hobza, P.: Chemphyschem: theoretical study on the complexes of benzene with isoelectronic nitrogen-containing heterocycles. Eur. J. Chem. Phys. Phys. Chem. **9**, 1003 (2008)
142. Dougherty, D.A., Stauffer, D.A.: Acetylcholine binding by a synthetic receptor: implications for biological recognition. Science **250**, 1558 (1990)
143. Singh, N.J., Shin, D., Lee, H.M., et al.: Structural basis of triclosan resistance. J. Struct. Biol. **174**, 173 (2011)
144. Ma, J.C., Dougherty, D.A.: The cation-π interaction. Chem. Rev. **97**, 1303 (1997)
145. Quiñonero, D., Garau, C., Rotger, C., et al.: Anion–π interactions: do they exist? Angew. Chem. **114**, 3539 (2002)
146. Guha, S., Saha, S.: Fluoride ion sensing by an anion$-\pi$ interaction. J. Am. Chem. Soc. **132**, 17674 (2010)
147. Schottel, B.L., Chifotides, H.T., Dunbar, K.R.: Anion-π interactions. Chem. Soc. Rev. **37**, 68 (2008)
148. Mooibroek, T.J., Black, C.A., Gamez, P., et al.: What's new in the realm of anion$-\pi$ binding interactions? putting the anion-π interaction in perspective. Cryst. Growth Des. **8**, 1082 (2008)
149. Liu, J., Yang, W., Zareie, H.M., et al.: pH-detachable polymer brushes formed using titanium$-$diol coordination chemistry and living radical polymerization [RAFT]. Macromolecules **42**, 2931 (2009)
150. Sutter, P., Sadowski, J.T., Sutter, E.A.: Chemistry under cover: tuning metal$-$graphene interaction by reactive intercalation. J. Am. Chem. Soc. **132**, 8175 (2010)
151. Liang, C., Li, Z., Dai, S.: Mesoporous carbon materials: synthesis and modification. Angew Chem. Int. Ed. **47**, 3696 (2008)
152. Ke, Q., Liu, Y., Liu, H., et al.: Surfactant-modified chemically reduced graphene oxide for electrochemical supercapacitors. RSC Adv. **4**, 26398 (2014)
153. Yoon, M., Choi, W.M., Baik, H., et al.: Synthesis of multilayer graphene balls by carbon segregation from nickel nanoparticles. ACS Nano **6**, 6803 (2012)
154. Lee, J.S., Kim, S.I., Yoon, J.C.: Chemical vapor deposition of mesoporous graphene nanoballs for supercapacitor. ACS Nano **7**, 6047 (2013)
155. Park, S.H., Kim, H.K., Yoon, S.B., et al.: Spray-assisted deep-frying process for the in situ spherical assembly of graphene for energy-storage devices. Chem. Mater. **27**, 457 (2015)

156. Cao, X., Shi, Y., Shi, W., et al.: Preparation of novel 3D graphene networks for supercapacitor applications. Small **7**, 316 (2011)
157. Dong, X.C., Xu, H., Wang, X.W., et al.: 3D graphene–cobalt oxide electrode for high-performance supercapacitor and enzymeless glucose detection. ACS Nano **6**, 3206 (2012)
158. He, Y., Chen, W., Li, X., et al.: Freestanding three-dimensional graphene/MnO_2 composite networks as ultralight and flexible supercapacitor electrodes. ACS Nano **7**, 174 (2012)
159. Cong, H.P., Ren, X.C., Wang, P.: Macroscopic multifunctional graphene-based hydrogels and aerogels by a metal ion induced self-assembly process. ACS Nano **6**, 2693 (2012)
160. Xu, Y., Sheng, K., Li, C., et al.: Self-assembled graphene hydrogel via a one-step hydrothermal process. ACS Nano **4**, 4324 (2010)
161. Xu, Y., Chen, C., Zhao, Z., et al.: Solution processable holey graphene oxide and its derived macrostructures for high-performance supercapacitors. Nano Lett. **15**, 4605 (2015)

Functionalization of Graphene—A Critical Overview of its Improved Physical, Chemical and Electrochemical Properties

Ramesh Kumar Singh, Naresh Nalajala, Tathagata Kar and Alex Schechter

Abstract Graphene, the 2D allotrope of carbon, is reported to be functionalized with a plethora of organic and inorganic species. This functionalization imparts significant improvement in the physical, chemical and electrochemical properties of graphene. The covalent and non-covalent functionalization of graphene with electron-rich organic moieties and heteroatoms is focused on different sections of this chapter. The focus is laid on the improvement in physical, chemical and electrochemical properties of graphene achieved through this functionalization. The enhancement in electrocatalytic activity of non-metal-doped graphene towards the oxygen reduction reaction, methanol oxidation reaction and photocatalysis is covered. Towards the end, the potential uses of functionalized graphene for selected applications like biosensors, fuel cells and dye-sensitized solar cells are also discussed.

Keywords Graphene · Functionalization · Doping · Electrochemistry · Applications

1 Introduction

Graphene a single atomic layer of sp^2-hybridized carbon atoms placed in the hexagonal structure has received tremendous attention because of its outstanding physical, chemical and electrochemical properties [1–3]. This newly emerged allotrope of carbon with unique properties like high electrical conductivity, sheet-like structure and high surface area [4] offers a wide range of applicability in the field of material science and electrochemistry. Due to these intrinsic properties, it has been used both as a catalyst and as a support material in electrochemical applications such as fuel

R. K. Singh (✉) · A. Schechter
Department of Chemical Sciences, Ariel University, Ariel, Israel
e-mail: rameshkumariitb@gmail.com

N. Nalajala
National Chemical Laboratory [NCL], Pune, India
e-mail: nareshmse@gmail.com

T. Kar
Department of Materials Science and Engineering, Tel Aviv University, Tel Aviv, Israel

© Springer Nature Switzerland AG 2019
S. Sahoo et al. (eds.), *Surface Engineering of Graphene*, Carbon Nanostructures,
https://doi.org/10.1007/978-3-030-30207-8_6

cell [5], supercapacitors [4, 6] and Li-ion batteries [7–9]. As a catalyst support, it has been shown to promote electro-reduction of molecular oxygen when decorated with precious/non-precious metals, and alloys [10]. Graphene has also been employed as an electrode in electrochemical double-layer capacitors which require high surface area [11–13].

Graphene is synthesized by a variety of synthetic routes such as chemical [14, 15], hydrothermal [16], chemical vapour deposition (CVD) [17, 18] and epitaxial growth [19]. Chemical and hydrothermal methods are applied for the bulk synthesis of graphene for electrocatalysis, while the CVD growth techniques are followed for obtaining monolayer graphene structures. For catalytic purposes, graphene is derived from its parent non-stoichiometric material, graphene oxide (GO), the main difference being the presence of the insulating oxides in the later. The oxides when removed from GO, it forms the highly conducting sp^2-hybridized carbon structure, reduced graphene oxide (rGO). However, the rGO is not completely devoid of all the oxygen. There is some reminiscent oxide content in the rGO which is a blessing in disguise because these oxide groups prevent the carbon layers from restacking and add to the faradaic behaviour when tested in an aqueous electrolyte.

In recent years, the functionalization of graphene has received significant attention due to the interaction of specific nitrogenated [20, 21] (diazonium and nitrene) and oxygenated [22] functional groups on graphene, which further enhances its application in the electrocatalysis [23] and other applications [19] compared to that of bare graphene. In this context, the classic and recent articles are reviewed which deal with graphene functionalization and it is categorized into two different classes: (i) covalent [24] and (ii) non-covalent [25, 26]. In general, covalent functionalization is performed on sp^2-hybridized carbon which changes the physical structure and electronic properties. In contrast, the non-covalent functionalization is achieved by van der Waals and dispersive force interactions which do not change the electronic properties of the graphene but can modify the morphology while simultaneously introducing new chemical groups on the structure [27]. Several successful attempts were made by the different research groups to dope graphene with organic and inorganic compounds, various transition metals [28], nitrogen [29, 30] and sulphur [31]. These added dopants are electron-rich and enhance the electronic and catalytic property of bare graphene. In addition of these dopants to graphene, the activation barrier for the electron transfer decreases, thereby making the doped graphene a very efficient electron donor. For example, when graphene is doped electron-rich nitrogen, the 'N-doped graphene' becomes an extremely suitable material for catalysing sluggish oxygen reduction reaction (ORR) in aqueous electrolytes [5]. Different organic and inorganic dopants have also been used to improve the dispersion of hydrophobic graphene making it suitable for catalytic and device-related applications. At the same time, nitrogen [5], sulphur [31], boron [32, 33], halogen [34–36], [36–38] and precious (Pt, Pd) [39–41] and non-precious (Co, Fe) [28] metal-doped graphenes are extensively reported for electrocatalytic applications.

The influence of functionalization on the physical, chemical and electrochemical properties of graphene/GO and the applications of functionalized graphene-based materials is discussed in the different sections of this chapter. In brief, Sect. 2 discusses the fundamental aspects of graphene functionalization; Sects. 3 and 4 demonstrate the influence on the physical and chemical properties, respectively, of functionalized graphene; Sect. 5 deals with metal heteroatoms doping and its implication in electrochemical reactions such as oxygen reduction, methanol oxidation, formic acid oxidation, hydrogen evolution and photocatalysis. Section 6 deals with the application of the functionalized graphene to electrochemical devices such as biosensors, fuel cells and dye-sensitized solar cells. The last section includes the concluding remarks in the chapter along with the future outlook.

2 Fundamental Aspects of Graphene Functionalization

Graphene has been chemically functionalized for different applications either by covalent or non-covalent bonding. Graphene is hydrophobic in nature; i.e. it has a poor dispersion in polar/non-polar solvents. The main aim of functionalization of graphene is to improve its dispersion and to add required chemical and physical properties such as higher conductivity or good interaction with other molecules [42].

The covalent functionalization of graphene is mainly achieved either by establishing a covalent bond between the functional groups of the native organic moieties on oxidized graphene surface or by making the bond with the sp^2 C=C structure of graphene basal plane. There are several reports on the addition of free radicals, viz. nitrophenyls [43], diazonium salts [20, 24], hydroxylated aryl groups and benzoyl peroxide [44] to the C=C structure of graphene. Apart from these, the functionalization of graphene with dienophiles (azomethine ylide [45], tetraphenylporphyrin (TPP) [46], phenyl [47]/alkyl azides [48], nitrenes [21], arynes [49]) is reported for improving the graphene dispersion and making it feasible for several applications.

The driving forces for the non-covalent functionalization of the graphene or GO are mainly π–π interactions [50], van der Waals forces, ionic interactions and hydrogen bonding [27]. Graphene is reported to be stabilized in an aqueous suspension by functionalization using pyrenebutyrate [51], pyrenebutanoic acid succinimidyl ester (PYR-NHS) [52], 1-pyrenecarboxylic acid (PCA) [22], DNA [26], polyaniline (PANI) [53], sulphonated polyaniline (SPANI) [25]. Chemically reduced exfoliated graphite oxide (EGO) has been stabilized using sodium lignosulphonate (SLS) [54], pyrenebutanoic acid succinimidyl ester (PYR-NHS) [55], pyrene and perylenediimide [56].

For deeper insight about the covalent and non-covalent functionalization, the readers are directed to the review article by Georgakilas et al. [27, 57].

3   Effect of Graphene Functionalization on its Physical Properties

In this section, physical properties with respect to different functionalizations (covalent/non-covalent) with different methods are presented. By attaching different functional groups to basal plane and by tuning the charge carrier concentration and its type (via surface doping with various metal, non-metal or organic molecules), an improvement in the physical properties (electronic, optical, magnetic and mechanical) is observed compared to those of pristine graphene. A brief overview of the literature on these physical properties of functionalized graphene is discussed below.

3.1   Electronic Properties

Elias et al. studied the hydrogenation of a single atomic plane of graphite for the electronic properties [58]. The results are supported with transmission electron microscopy (TEM) and Raman spectroscopy along with electronic properties of prepared materials. The material exhibited an electrical insulting behaviour by an increase of the resistivity of the order of 2 with decreasing temperature from 300 to 4 K compared to that of without hydrogenation of graphene. Therefore, it is demonstrated that after treatment with atomic hydrogen, graphene can become an insulator; usually graphene exhibits semimetal properties. Moreover, very good reversibility of graphene also observed that is metallic in nature by annealing the hydrogenated devices at 450 °C in argon atmosphere [58]. In another study, Schniepp et al. demonstrated thermal exfoliation method for producing functionalized single GO sheets which are highly conducting compared to graphitic oxide that is an insulator [59]. The band gap engineering of graphene is explored by periodic and partially suspended graphene sheets over a Si substrate (periodically modulated graphene, PMG) which is prepared using lithography technique by Lee et al. [60]. The electronic properties of prepared devices over nanopatterned surface suggest that graphene may possess of semiconductor behaviour from semi-metallic nature because of periodic change in coordination chemistry and repeatable partial contacts with the substrate. Moreover, the induced strain at periodic nanotrenches also may provide the modification of electronic band structure. The semiconductor behaviour of the PMG is demonstrated by measuring the resistivity of graphene sheets with decreasing temperature from 300 K (3 kΩ/sq) to 100 K (200 kΩ/sq); the resistivity increases by two orders of magnitude [60]. Schiros et al. carried a remarkable investigation on DFT and core-level spectroscopy of different nitrogen bond types in graphene such as graphitic, pyrilic and nitrilic [61]. They found that the type of nitrogen bond with its neighbours (here carbon) has a profound impact on its carrier concentration and thus provided new guidelines for the design of next generation of graphene-based electronic devices [61]. Macedo et al. investigated the effect of an electrochemically induced covalent bond of 4-carboxyphenyl (4-CP) units over graphene for its electrical properties using combination of infrared spectroscopy, electrochemical analysis and computational studies [62]. The spectroscopy results (near-field optical microscopy and atomic spectroscopy) revealed that plasmon–phonon coupling upon functionalization strongly affects the physical and optical properties. Moreover, the computational studies suggest that electron transfer became weaker because of the retention of charges at sp^3-hybridized carbon atoms which is the result of covalent anchoring of 4-CP units over graphene [62].

Cervantes-Sodi et al. carried a remarkable investigation over chemically modified graphene nanoribbons for electronic properties using density functional theory (DFT) calculations [63]. They observed very interesting variations in DOS with respect to different functionalizations such as edge, bulk substitutions and chemisorption. The results suggest that chemical modifications can lead to semiconductor–metal transitions, lifting of spin degeneracy and increase in band gap and thus concluded that chemical modification can significantly alter and be useful in the design of advanced GNR-based electronic devices. Recently, Torre et al. described a method to tune the electronic properties of N-doped graphene by weak non-covalent functionalization with iron(II) phthalocyanine (FePc) molecule [64]. The observations are complemented with each other by scanning tunnelling microscopy (STM), atomic force microscopy (AFM), DFT studies and AFM simulations. The results suggest that the electronic properties of FePc molecule can be altered locally by reordering of selected d-orbitals of Fe with p_z-orbitals of single graphitic nitrogen defect because of weak non-covalent interaction between graphene and molecule. Furthermore, the high-resolution AFM images could distinguish different spin states of the molecule over the surface using non-magnetic CO functionalized tip. On the other hand, it is reported that non-covalent functionalization exhibits superior electronic properties compared to covalent functionalization that has few-layer graphene with low defect content and average crystallite size. The reason for enhanced electronic properties (donor density) with non-covalent functionalized rGO is attributed to the availability of π electron from sp^2 network of graphene and surface functionality. Moreover, Saha et al. observed superior electrochemical properties at the electrode–electrolyte interface because of facile charge transfer with non-covalent functionalized rGO [65]. Zhou et al. obtained half loaded hydrogen–graphene by applying external electric field from fully hydrogenated graphene because unpaired electrons in the unsaturated carbon sites give rise to magnetic moments [66]. Moreover, the fine-tuning of electronic properties is established with half hydrogenated graphene with a substitution of F atoms. Therefore, it is demonstrated that depending on the type and coverage of atomic species used for surface modification one can derive semiconductor from metallic, magnetic from non-magnetic and indirect band gap from direct band gap of materials. An insightful study by Dedkovet et al. provided evidence for the spin filtering property with graphene/Ni(111) interface including the fine-tuning of electronic properties by adsorption of graphene over the ferromagnetic substrate (Ni(111)), and the results are supported with various imaging and spectroscopy techniques [67]. The reason for the aforementioned observations arises because of considerable hybridization of graphene π and Ni 3d valence band states that are in fact associated with partial charge transfer of spin-polarized electrons from Ni to carbon atoms and thus contribute to effective magnetic moment in the graphene layer.

3.2 Optical Properties

Johari et al. carried out ab initio DFT calculations to investigate the optical properties of GO with a different coverage and compositions of major functional groups like epoxy, hydroxyl and carbonyls [68]. The calculations from the electron energy loss spectroscopy (EELS) results suggest that the addition of carbonyl group can significantly decrease the optical gap and open the band gap because of increase in width of the hole that created by carbonyl. Hence, the authors suggested that the rGO with the coverage of carbonyl groups could be considered for the design of electronic devices with improved optoelectronic properties. Xu et al. observed enhanced optical limiting properties of graphene hybrid materials which are modified covalently by porphyrins [69]. The authors synthesized proposed graphene hybrid material (TPP-NHCO-SPF graphene) by following standard chemistry of amine-functionalized porphyrin (TPP-NH$_2$) and GO in N,N-dimethylformamide. In this case, the amine-functionalized porphyrin (TPP-NH$_2$) and GO are covalently bonded together via an amide group. The transmittance results show that the largest dip with TPP-NHCO-SPF graphene compared to other materials demonstrates enhanced optical limiting properties. The reason is provided that TPP-NHCO-SPF graphene may have an efficient electron transfer from TPP-NH$_2$ to graphene moiety after photoexcitation. In another study, Du et al. designed zinc-based porphyrin–graphene composites from the covalently bonded polymer-based phenyl sulphone, (p-amino)-phenylhydroquinone and asymmetrical dinaphthylporphyrin with graphene oxide sheet [70]. The polymer-based graphene hybrid materials show stronger optical repose and larger nonlinear extinction coefficient than its individual components at the same linear transmittance response (Fig. 1a, b). Wang et al. designed graphene–porphyrin nanohybrids (rGO-TPP1 and rGO-TPP2) by covalent functionalization of rGO with diazonium, and this interaction causes to enhance optical limiting properties (Fig. 1c, d).

The enhanced optical performance of the hybrid material is attributed to the cumulative effect of nonlinear scattering, reverse saturable absorption and efficient photoinduced electron transfer from donor polymer moieties of polymer backbone to acceptor GO moieties [71]. In support of the increase in optical limiting properties with covalent bonding of polypherene with rGO, Li et al. established tetracarboxylic Zn(II) phthalocyanine–amino-functionalized GO (ZnPcC$_4$–NGO) [72]. Covalent functionalization of zinc phthalocyanine with three graphene materials (pure graphene, GO and rGO) is carried out for the investigation of nonlinear and optical limiting properties by Zhao et al. [73]. The authors found that among these three graphene materials, rGO-ZnPC exhibited better optical limiting effect, larger nonlinear attenuation coefficient (β) and high linear transmittance and the reason is attributed to the strong covalent bond nature between graphene and phthalocyanine. It is established that the covalent functionalization of porphyrin and fullerene with

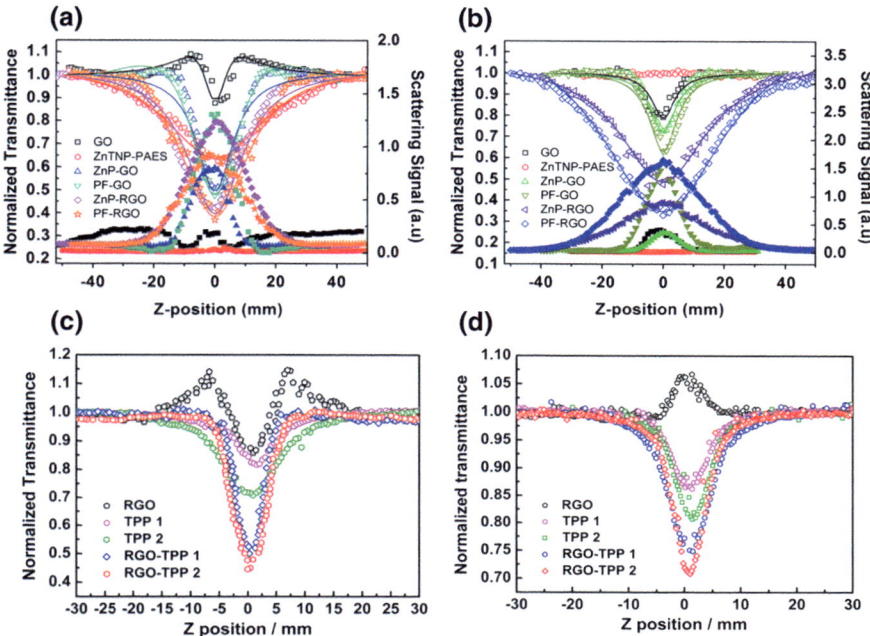

Fig. 1 Comparison of the open aperture Z scan results (transmittance) of graphene–porphyrin-based hybrid materials (**a**) and **b** zinc-based from Ref. [70] **c** with 21 ps and **d** with 4 ns diazonium salts from Ref. [71] with its individual components and standard C60. These figures are reproduced with the permission from RSC and Nature publishers

graphene also provided enhanced nonlinear optical properties because of efficient electron transfer between photoexcited porphyrin and fullerene to graphene [74]. The covalent functionalization of three conjugated polymers with same reaction sites but different monomers with graphene exhibited excellent optical limiting properties [75]. On the other hand, the non-covalent functionalization of porphyrins with graphene also found to be with enhanced optical properties. In this context, Orellana et al. carried DFT calculations over non-covalently functionalized tetraphenylpor-phyrin (TPP) molecules with graphene and found that G-TPP has excellent stability properties while preserving the absorption properties by employing n-type doping mechanism; in fact, it is important for light-harvesting applications [76].

3.3 Magnetic Properties

Liu et al. demonstrated by first principle calculations a novel material graphitic C_2O with embedment of macrocycle molecule (crown ether) in the graphene which in fact both have a similar geometry [77]. The authors found that the resultant material can effectively bind metal atoms and perhaps provide a new method to further alter graphene properties. In this context, the authors observed enhanced superconducting properties with crown graphene material that bind with alkali metals. The reason attributed to the favourable/stronger electron–phonon interaction that is the origin for superconducting transition in the hybrid crown graphene materials. In the case of silicone/graphene hybrid system, it is observed that there is no magnetism for single carbon vacancy due to the absence of unpaired electrons, on the other hand highest magnetic moment (order of 4.0 Bohr magneton) for a single silicon atom vacancy because of the presence of unpaired electrons which in fact provide mid-gap states [78]. Ray et al. demonstrated plasma functionalization as a facile method to improve the magnetic properties of graphene materials (for instance, few-layer graphene (FLG)) for the design of advanced spintronic devices [79]. In this study, the authors observed superior magnetic properties such as saturation magnetization (Ms) after hydrogen-doping (graphone, 13.94×10^{-4} emu/g) and nitrogen-doping (N-FLG, 118.62×10^{-4} emu g^{-1}) compared to pristine FLG (3.4794×10^{-4} emu/g). The corresponding hysteresis loops those are observed for the above-mentioned materials are shown in Fig. 2. The maximum ferromagnetic behaviour is observed with N-doped FLG (Fig. 2c) and with siliphene, and mixed features of dia and ferromagnetic behaviour are observed (Fig. 2d) with least saturation magnetization (0.11×10^{-4} emu/g). The reason for higher Ms value for the hydrogenated FLG compared to pristine graphene is due to the formation of sp^3 hybridization, thus providing favourable ferromagnetism. Besides that, with nitrogen-doped FLG, the

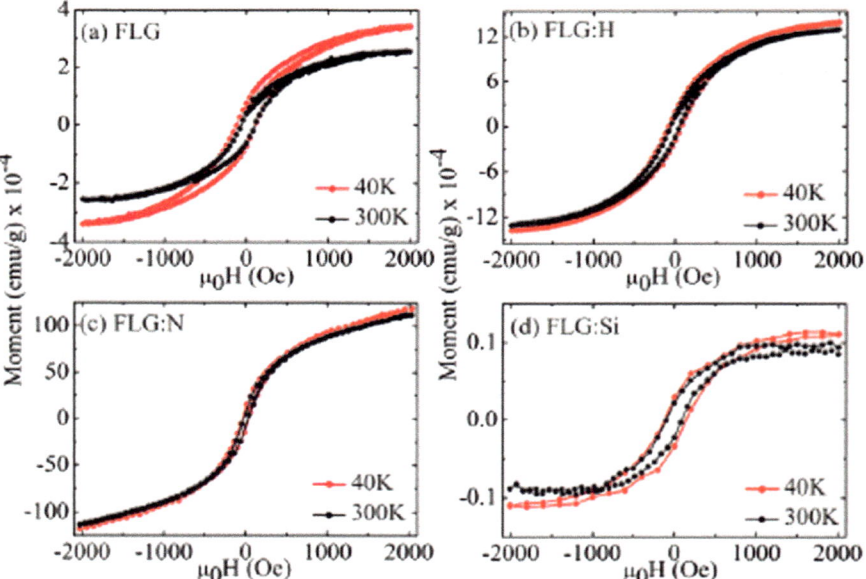

Fig. 2 Magnetic hysteresis loop of various functionalized graphenes: **a** pristine few-layer graphene (FLG), **b** hydrogenated FLG/graphone, **c** nitrogen-doped FLG, **d** silicon-modified FLG/siliphene. This figure is reproduced with permission from Ref. [79]; RSC publishers

extra π electron with nitrogen enables electron-rich structure, hence providing strong magnetic coupling between magnetic moments.

In another study, a similar group established plasma-enhanced CVD methods to prepare hydrogenated graphene/graphone over graphene for the investigation of magnetic properties in spintronic devices. It is found that hydrogenation of graphene can provide free spins by the conversion of sp^2- to sp^3-hybridized structure, and the possibility of having unpaired electrons from defect sites could cause observed ferromagnetic behaviour [80]. Liu et al. demonstrated that N-doping with GO could be an effective way of making intrinsic non-magnetic graphene as highly magnetized ferromagnetic materials and thus paved the way for potential applications such as in spintronic devices [81]. Yazyev et al. carried first principle calculations to investigate the magnetic properties of graphene when two types of defects are presented on the surface: one is defect due to chemisorption of hydrogen, and the other is vacancy defect [82]. The calculated magnetic moments are found to be different for different types and concentration of defects, it is 1 μ_B for hydrogen chemisorbed defect, and it is 1.12–1.53 μ_B for vacancy defect. An insightful study by Herrero et al. states that one can tune the magnetism of selected graphene regions by precise control at the atomic scale of chemisorbed hydrogen and the results are complemented with each other by first principle investigation and STM studies [83]. The reason is attributed to the prevailing of long-range coupling of spin-polarized state essentially localized on the carbon sub-lattice opposite to the one where the hydrogen atom is chemisorbed.

3.4 Mechanical Properties

Park et al. reported the enhanced mechanical properties such as mechanical stiffness (10–200%) and fracture strength (50%) of paper like GO after modification with less than 1 wt% of Mg^{2+}, Ca^{2+} [84]. It is found that the presence of oxygen groups on basal planes and carboxylate groups at edges of the sheets can strongly bond the Mg^{2+}, Ca^{2+} and thus contribute to excellent mechanical properties. Goncalves et al. employed atom transfer radical polymerization (ATRP) to modify the GO surface with poly(methyl methacrylate) (PMMA) chains which are readily soluble in organic solvents and therefore used as fillers to investigate the mechanical properties of composite materials (graphene-modified PMMA, GPMMA) [85]. In this context, the observed load vs. displacement nano-indentation curves (Fig. 3a) of different wt% of fillers into the matrix suggests that 1% (w/w) GO and GPMMA have shown significant reinforcement effect because of surface modification and well adhesion with the polymer matrix. All the hybrid materials exhibited enhanced stiffness properties compared to virgin PMMA material such as maximum depth of 380 nm, 290 nm and 350 nm for 0.5, 1 and 3% (w/w of GO and GPMMA), respectively. On the other hand, the hybrid materials exhibit much better Young's modulus such as 21.8 MPa with 1 (w/w) GPMMA compared to 18.7 MPa with PMMA (Fig. 3b). Moreover, the hybrid materials exhibited both elastic–plastic regions and reasonable ductility. Song et al. deployed a method of non-covalent functionalization of pyrene block

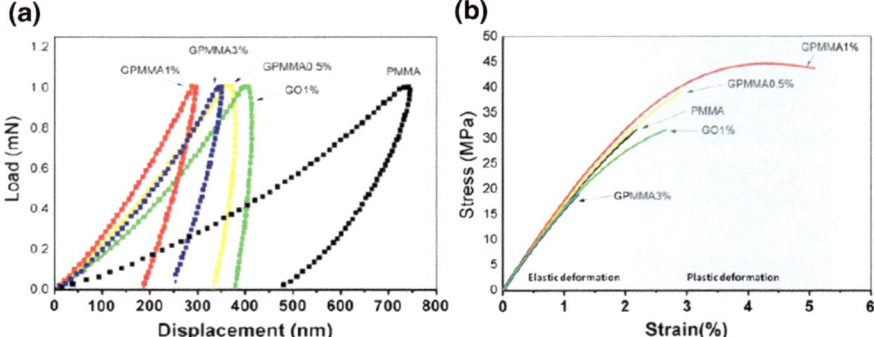

Fig. 3 Load versus displacement nano-indentation curves (**a**) and stress versus strain curves of PMMA films with and without graphene fillers. Reproduced with permission from Ref. [85]; RSC publisher

copolymers (poly(methyl methacrylate)-block-polydimethylsiloxane, Py-PMMA-b-PDMS) with GO for enhanced mechanical properties [86]. The authors observed a 23, 54, 117 and 218% increase in tensile strength, Young's Modulus, elongation at break and toughness, respectively, with 0.05 wt% of GO @ Py-PMMA-b-PDMS, and it is attributed to cumulative effects of reinforcing and toughness of functionalized GO particles.

Dikin et al. discovered a paper like GO materials which outperforms many other similar materials in terms of stiffness and strength [87]. The enhanced mechanical properties of the paper like graphene are attributed to the interlocking-tile arrangement of the nanoscale GO sheets. Suk et al. reported Young's modulus of graphene sheets of different thicknesses (one layer, two layers and three layer) and found that monolayer graphene has low effective Young's modulus compared to that of pristine graphene [88]. Very recently, Ruoff group established a folding strategy of graphene into polymer materials for the improved mechanical properties compared to pristine-stacked polymer and reinforced graphene polymer films [89]. The authors observed that 73.5, 73.2 and 59.1% increase in Young's modulus, strength and toughness modulus, respectively, with 0.085% of graphene-folded polycarbonate compared to the pristine-stacked polymer. Therefore, it is established that the folding approach can provide enhanced properties which are having positive implications in the design of advanced materials for different applications but not limited including energy conversion, storage and optical sensors.

4 Effect of Graphene Functionalization on its Chemical Properties

4.1 CO Oxidation

For decades, it is of crucial importance to convert the main emission of combustion of hydrocarbons that is environmentally harmful CO to less harmful CO_2. In this context, numerous efforts are made to oxidize CO to CO_2 at ambient conditions and are well documented in the literature. In this section, studies that are relevant to modified graphene for CO oxidation will be addressed. Tang et al. reported on DFT studies based on Pt modification (as a single atom) over pristine and defective (vacancy defect) graphene for CO oxidation [90]. It is found that the deposition of Pt over defective graphene exhibited a very high activity (at low temperature) towards the target reaction compared to Pt on pristine graphene. The following reasons are attributed to the observed activity difference: (1) the defective graphene can stabilize Pt adatom and hence provide more positive charge at this active site. (2) This situation provided weakens the CO adsorption while increasing the O_2 adsorption, which leads to enhanced CO oxidation and limits CO poisoning of the Pt active site. (3) Predicted low energy barriers (<0.6 V) with Pt-modified defective graphene are compared to pristine counterparts. In search of low cost and green catalyst to oxidize CO at low temperature, Fe modification with GO is reported in the literature [91]. The strong hybridization between Fe (via 3d states) and activated O_2 from the adsorbed O_2 (via 2p states) leads to carbonate like intermediate state with low energy barrier (0.6 eV) via Eley–Rideal mechanism. The formed carbonate like intermediate interacts with another CO molecule and produces two CO_2 molecules. In another study by Guo et al., conducted DFT calculations on graphene deposited on impurity free and Zn atom doped Fe/Ni(111) substrates, for probing the catalytic activity of graphene towards CO oxidation [92]. It is observed that activation barriers of two-step mechanism of CO oxidation are found to be less (<0.5 eV) in the case of graphene-supported Zn-doped Fe/Ni(111) substrate and thus provided enhanced activity compared to impurity-free counterparts. Moreover, the authors emphasized that CO poisoning is not an issue in the proposed system unlike other established systems (graphene-supported/covered catalysts) because graphene itself is an active site for the desired catalysis reaction and it is well known that CO does not bind to graphene. The embedded metal atoms in the graphene are proposed as an active catalyst for CO oxidation at room temperature. In general, it is known that CO oxidizes to CO_2 at room temperature at an activation barrier of less than 0.5 eV. In this context, Liu et al. established DFT calculations over Au-embedded graphene system as an active catalyst for oxidation of CO to CO_2 at room temperature. It is found that the calculated activation barrier is small as that of 0.31 eV and thus provided enhanced activity compared to pristine counterparts [93]. Even they generalized the method for other transition metal such as titanium-embedded systems as well for the CO oxidation at room temperature. The reason attributed to the enhanced activity is because of partially localized d-orbital states in the vicinity of Fermi level due to the interaction between metal (Au and Ti) and graphene. Another interesting study by Zhou et al. reported first principles approach on catalytic activity (CO oxidation) of $Au_{(8)}$ and $Pt_{(4)}$ clusters on single carbon vacancy (defective) graphene [94]. They were found that there is a great decrease in the reaction barriers of the aforementioned clusters over defective graphene compared to pristine (defect-free) graphene; the values estimated to be 3 eV (0.5 eV) and 0.2 eV (0.13 eV) for $Au_{(8)}$ ($Pt_{(4)}$) over defect-free and defective graphene, respectively. This study sets up the guidelines for the design of advanced graphene-based materials for various applications since the formation of defect over graphene is inevitable during the synthesis. The addition of Co_3O_4/graphene over polymeric materials [polybutylene succinate (PBS) and polylactide (PLA)] found to be active towards the CO oxidation compared to their virgin counterparts. The plausible mechanism is explained to be the CO molecule that is absorbed by the Co^{3+} in Co_3O_4 followed by adsorbed CO oxidation by consuming the surface oxygen that might be coordinated with three Co^{3+} cations [95]. Towards the search for cost-effective catalysts without comprising the catalytic activity, Wang et al. established graphene decorated with CuO nanorods [96]. The proposed system is synthesized using a hydrothermal method and further subjected to CO oxidation. The activity results suggest that, among the other catalysts studied, the composition of CuO nanorods with 10 wt% graphene exhibits better activity; 100% CO oxidation to CO_2 at 165 °C. Further, they established the stability of the best possible active catalyst (10%-rGO-CuO) and found that at initial 6 h of reaction, there is a decrease in activity and then it is stabilized throughout the reaction time (42 h). The reasons ascribed to decrease in activity at initial hours are unstable surface compositions, and the interaction between reactants and catalyst surface has not yet reached the redox equilibrium. In order to overcome the obstacles with graphene materials such as aggregation and stacking via π–π interactions that cause less utilization of graphene, graphene aerogels of the interconnected 3D network are being proposed. In this context, Li et al. prepared Ru-modified graphene aerogels for CO oxidation under different experimental conditions, for instance, freeze-dried aerogel material as starting material subjected to treatment at 150 °C for 6 h in dry air (air-150-6), 180 °C for 6 h in H_2 flow (H_2-180-6) and 180 °C for 24 h in dry air (air-180-24), respectively [97]. The CO oxidation activities and stability of highly active material suggest that mildly oxidized sample (air-150-6) showed the 100% conversion of CO including stability (70 h) at as low as temperature 25 °C. The high activity and stability of these materials are ascribed to channels enabling the easy access and diffusion of reactants, intermediates and products including access of active sites. Thus far, in this section, several contributions from the literature have been reviewed in the context of functionalization of graphene towards CO oxidation.

4.2 Fischer–Tropsch (F-T) Synthesis

It is of the urgent need to convert the environmentally harmful gas CO to useful chemicals such as light olefins (C_2–C_4 olefins) and LPG (C_2–C_4 paraffins). Therefore, F-T synthesis is considered as an important heterogeneous reaction and thus received great interest from the research community to mitigate the major global warming issue by combining CO with H_2 and produce the aforementioned useful chemicals. To pave the way for cost-effective catalysts with high activity and product selectivity, several catalyst systems have been established and reported in the literature [98, 99]. Recently, graphene-based materials have received considerable attention as the catalyst for F-T synthesis because of its unique physical, chemical and structural properties. Therefore, with this objective, in this section various graphene-based materials those were synthesized and used for F-T synthesis are reviewed. The consideration of alkali promoters during the synthesis of F-T catalysts, in particular Fe-based materials, is a well-known approach to achieve the high activity, stability and selectivity of product. To prove this concept, Cheng et al. carried an experimental study with a systematic increase in potassium contents from 0 to 2 wt% in their GO-supported Fe-based F-T catalyst and checked the activity for both CO conversion and selectivity to lower olefin content [100]. They found that the high activity and stability towards %CO conversion with functionalized graphene (nitrogen functional groups) supported cobalt compared to that with pristine graphene supported cobalt catalysts. On the other hand, the trend in the increase in lower olefin selectivity with an increase in K content is observed. The reasons are attributed to the increase in surface area, surface basicity and adsorption capacities of H_2 and CO, Hagg's carbide size accordingly with an increase in K content. Recently, the carbon-supported Fe-based particles being considered as active catalysts with very good activities towards the production of C_{5+} hydrocarbons with low methane and CO_2 by-products. Zhao et al. established GO-supported Fe-based nanoparticles as an active F-T synthesis with CO conversions of ~37% (with 47 wt% of C_{5+} hydrocarbon), and it is high compared to that of activated carbon-supported Fe nanoparticles that have only ~14% (~39 wt%) [101]. It is found that thermal treatment at different temperatures under reducing atmosphere (H_2) can decrease the number of surface oxygen and sulphur species which lead to the change of surface properties of active Fe phase and thus exhibit unique catalytic activities. Therefore, the tuning of a number of surface species of the catalyst has the profound impact on interaction with support (GO) and the active catalyst (Fe) and thus provides the new guidelines towards the design of better F-T catalysts. Very recently, Abbas et al. reported an innovative and eco-friendly rapid sonochemical method for preparation of Fe_2O_3 nanocubes over GO sheets as F-T catalysts [102]. In this investigation, they compared the activities towards CO conversions of prepared catalysts using sonochemical and hydrothermal methods. Among the catalysts prepared using the sonochemical method, graphene-supported Fe_3O_4 nanocubes showed better activity and selectivity of C_{5+} hydrocarbons compared to their virgin counterparts (without support). The reasons for high activity and selectivity are ascribed to high surface area, the extent of reducibility and presence of

Hagg's carbide caused by the incorporation of graphene as support. Moreover, they emphasized that the sonochemical method of preparation can produce stable catalysts compared to a hydrothermal method by showing TEM images of spent catalysts with the retention of Fe_3O_4 shape after the reaction. Sun et al. proposed a one-pot synthesis of rGO-supported iron oxide nanoparticles (Fe-rGO) as active catalysts for F-T synthesis [103]. They found that Fe-rGO exhibits higher FTS activity and better selectivity for C_{5+} and C_5–C_{11} hydrocarbons compared to pre-reduced graphene-supported and activated carbon-supported iron oxide nanoparticles. The reason for high activity with Fe-rGO materials is mentioned to be the presence of high populated defect sites over rGO which in fact provide highly dispersed iron oxide nanoparticles and strong metal–support interaction. One important issue with graphene material as a support is that agglomeration and sintering of metal nanoparticles because of its inert and hydrophobic nature. Therefore, it is very important to provide strong metal–support interactions to obtain better catalytic activities. It is reported that the presence of dopants such as nitrogen or its functional groups can provide strong metal–support (graphene) and thus one can tune the activity of the catalysts. The reason attributed to this subtle interaction is the metal can accommodate and bind at nitrogen species instead of the carbon which is known to be inert in nature. In this context, the proof of concept was experimentally proved and reported in the literature by Taghavi et al., with 15 wt% cobalt over nitrogen-functionalized graphene sheets and used as F-T catalysts [104]. They found that the high activity and stability towards %CO conversion with cobalt supported functionalized graphene compared to that with pristine graphene supported cobalt catalysts. The reasons for functionalized graphene being good support for cobalt catalyst are providing strong metal–support interaction and thus enable a decrease of sintering of nanoparticles. In similar lines, Hajjar et al. reported that Co/GO and Co/nanoporous GO are better F-T catalysts compared to that of alumina-supported Co nanoparticles [105].

4.3   Other Chemical Properties

The detection of peroxide is of paramount importance since it behaves as a signalling molecule in the regulation of various biological signal transduction processes. In this context, Yadav et al. established a very good sensing ability of graphene towards H_2O_2 detection [106]. It is mentioned that the graphene exhibits very good sensing ability of 355 $\mu A\ mM^{-1}\ cm^{-2}$ with response time less than 5 s at 25 °C because of the presence of edge defects. It is important to design recyclable heterogeneous catalysts with features of high stability, activity, low cost and easily scalable. In order to establish this, Rostamnia et al. designed GO-supported Pd nanoparticles (SE_{P123}-GO/Pd) for the synthesis of amides in the presence of different surfactants; in fact, the use of amide functionalities is exceptional in many of the materials such as polymers, organic compounds, natural and pharmaceutical compounds [107]. The authors found that the prepared material is able to oxidize aliphatic and aromatic alcohols for the production of aldehydes with excellent yields. Moreover, the produced aldehydes could be further used to oxidize with various amines including secondary aliphatic amine. Rahimi et al. proved that the graphene efficiently works towards the accumulation of electron–hole pairs with desired recombination time to produce active radical species such as OH* and O_2* species on sulphur-doped graphene-supported Fe_2O_3 nanoparticles and is successfully used for degradation of various dyes and oxidation of alcohols [108]. The particle aggregation and leaching of particles from the support are serious obstacles in providing very stable and recycling catalysts in the regime of any catalytic processes. Therefore, it is of high interest in the design of graphene-based supported after surface modification (hydroxyl, carbonyl, and epoxy can acts nucleation sites) which can provide the proper adhesion of nanoparticles with the support. In this context, an insightful study has been carried out by Zahed et al., the deposition of silver nanoparticles (~20–25 nm) over different supports such as reduced GO (Ag NPs/rGO), partially reduced GO (Ag NPs/GO) and thiolated partially reduced GO (Ag NPs/GOSH) [109]. From the activity and stability studies of oxidation of benzyl alcohol using molecular oxygen as an oxidant, they found that completely reduced graphene is not a better support for Ag nanoparticles and it indicated the need for functionalization of support to bind the Ag nanoparticles. On the other hand, thiolated functionalized partially reduced graphene support provided excellent activity and stability because uniform distribution and stabilization of Ag NPs lead to prevent the agglomeration and leaching of Ag NPs. The enhanced performance with Ag NPs/GOSH is attributed to the strong adhesion of Ag NPs with GOSH. Another important study by Song et al. reported a co-reduction strategy to produce glutathione-protected gold clusters over graphene support (Au @ HSG/rGO) followed by modification with poly(2-(dimethylamino) ethyl acrylate (PDMAEA) through π–π interactions [110]. The purpose of PDMAEA is to provide proper dispersion of Au @ HSG on graphene and also to promote temperature-controlled activity. For the reduction of 4-nitrophenol, Au @ HSG/rGO exhibits 20 times higher activity compared to Au/rGO. It is proposed that the pyrene-functionalized polymer PDMAEA can act as polymer shelter to minimize the particle aggregation and thus provide high dispersion and stability of Au @ HSG/rGO.

5 Effect of Graphene Functionalization on its Electrochemical Properties

In this section, the influence of graphene functionalization on various electrochemical reactions such as oxygen reduction, methanol oxidation, formic acid oxidation and ethanol oxidation is presented. By surface modification of graphene, the reaction rates of these reactions enhanced significantly due to the change in the chemical property of carbon adjacent to the dopant in the graphene structure.

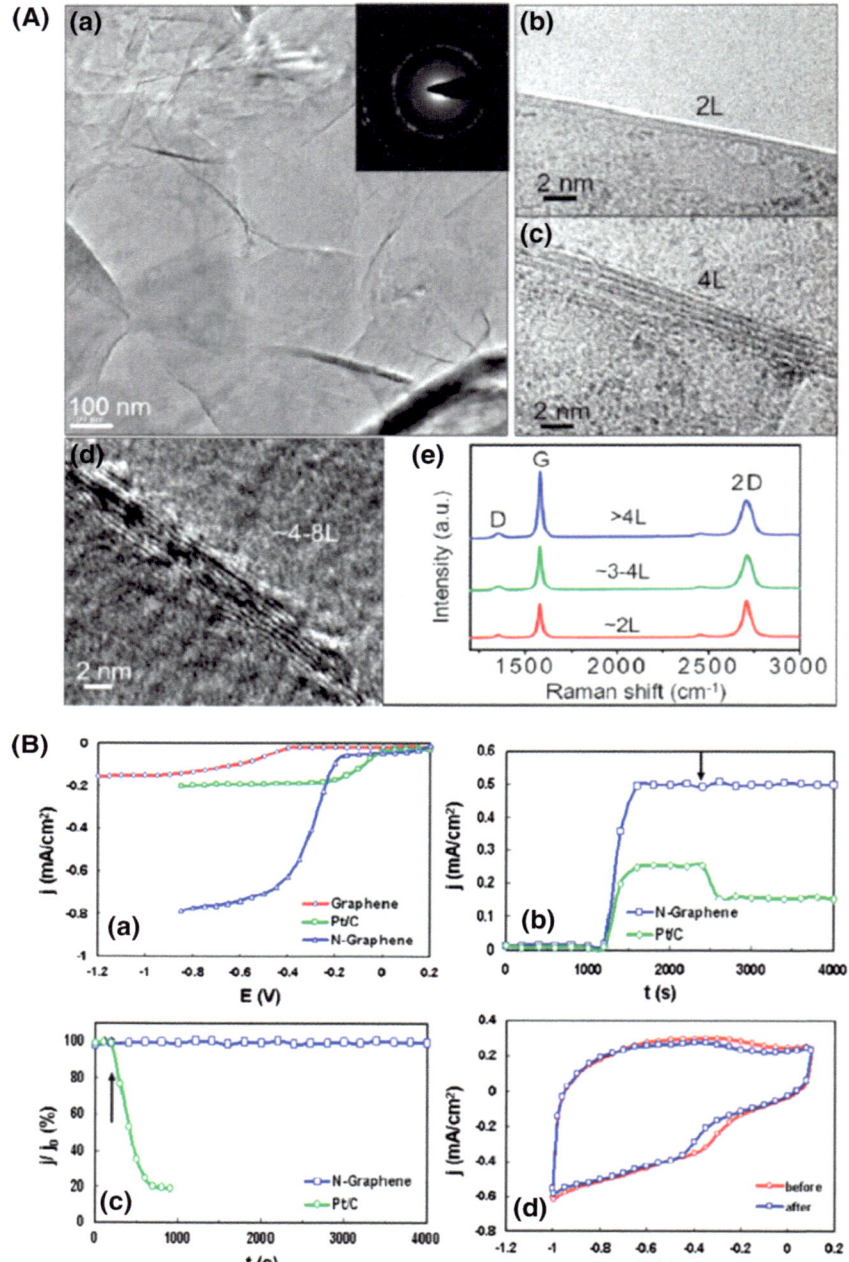

◄**Fig. 4** **Panel A**: TEM images and Raman spectra of N-doped graphene and **Panel B**: ORR voltammograms of graphene, Pt/C and N-graphene at 0.10 V s^{-1} in air-saturated 0.1 M KOH at 1000 rpm (**a**), the chronoamperometric response of Pt/C and N-graphene at −0.4 V (**b**). The arrow shows the addition of 2 M methanol in air-saturated 0.1 M KOH. The chronoamperometric response of these electrodes to CO (**c**) and CV before and after 200,000 cycles in air-saturated 0.1 M KOH solution at 0.1 V s^{-1}. Reprinted with permission from Ref. [5]. Copyright (2010) American Chemical Society

5.1 Electrochemical Reactions

5.1.1 Oxygen Reduction Reaction

Oxygen reduction reaction (ORR) is the classical electrochemical reaction in electrochemical energy conversion and storage devices such as fuel cells [111–118] and metal–air batteries [119–123]. Heteroatom (N [5, 23, 124–126], P [33], S [127], B [32, 33], Cl [42], Br [42], I [42])-doped graphene has been extensively studied for stable ORR in replacement with Pt in alkaline fuel cells. Among them, N-doping is even of higher importance due to improved activity and stability.

In this regard, Gong et al. have carried out pioneering work on ORR using vertically aligned nitrogen-containing carbon nanotube structure in alkaline electrolyte and performed DFT calculations to understand the reason for activity improvement [124]. This catalyst outperforms the state-of-the-art Pt/C in terms of the half-wave potential of ORR. The catalyst shows a very low yield of the peroxide which generates parallel to ORR and follows 4e^{-} reduction to water. The enhanced activity was attributed to the positive charge created on the neighbouring carbon atom by nitrogen, and it is further supported with the density functional calculation. This work is one of the significant contributions in the field and therefore received high attention among the researchers. Later, the same groups [128–130] have worked on the nitrogen-doping on various carbon structures which shows the improved ORR performance in line with outstanding contribution by Gong et al. In another work, Schechter and co-workers have reported nitrogen-doped vertically aligned multiwalled carbon nanotubes (VA-MWCNTs) using plasma treatment for ORR in 0.1 M KOH solution. After plasma treatment, the carbon nanotubes exhibited 15 times higher surface area and thus the higher ORR activity. This catalyst can be used as an effective metal-free ORR catalyst in the alkaline fuel cells [131].

Essentially, both carbon nanotubes and graphene have similar structures in terms of the lattice orientation. However, due to better electronic, optical and mechanical properties of graphene it is exploited where carbon nanotubes have been used previously in order to facilitate the charge transfer kinetics. The N-doped graphene has been demonstrated as a metal-free ORR catalyst by the Qu et al. [5]. It was synthesized by CVD from methane in the presence of ammonia. The reason for selecting graphene also as carbon support is mainly due to its outstanding chemical properties, which may help improve the interaction with oxygen. The high purity of synthesized N-graphene was verified by XPS, Raman, TEM and AFM studies (Fig. 4a). This study claimed that the N-graphene catalysts were synthesized for the first time that catalyses the O_2 reduction via a $4e^-$ direct pathway (confirmed by the Koutecky-Levich equation). These metal-free electrocatalysts possess remarkable catalytic activity for ORR and resistance to the CO poisoning, methanol tolerance and low crossover, thus, can be demonstrated in the fuel cell applications (Fig. 4b). Later, Zhang et al. discussed the mechanism of ORR on N-doped graphene by DFT simulations [126]. They carried out the simulation on N-graphene and demonstrated that ORR on N-graphene follows the direct $4e^-$ pathway.

The electrocatalytic site of single N-graphene was identified and claimed to be either due to the high positive spin density or due to the high positive charge density. The major conclusion from the work was to explain the high activity of N-doped graphene. Most of the work on the doping of carbon-based material is carried out either by high electronegative element such as nitrogen or by low electronegative elements such as P and B. However, reports on the electronegativity of the doped atom, such as sulphur or selenium, similar to the carbon are rare. In this regard, the work carried out by Yang et al. shows yet another class of graphene material doped with sulphur for efficient ORR [127]. The S-doping in the graphene structure was characterized by TEM, STEM, XPS and Raman spectroscopy techniques. The proposed active site has high positive spin created by the dopant, and it is supported with DFT simulations by Zhang et al. [126]. This catalyst shows higher activity than the Pt/C, higher resistance to poisoning and long-term stability and fits as an outstanding material for the alkaline fuel cell application. For more details on the heteroatom doping of the carbon structure, the readers are directed to follow the review article by Yan et al. [132]. Another study by Jahan et al. demonstrated the enhanced activity of the graphene–porphyrin metal–organic framework (MOF) for ORR in direct methanol fuel cell application (Fig. 5) [133].

Their study reveals that the functionalization of rGO sheets with pyridine changes the degree of the crystallization, porosity and the charge-transfer rate of iron-porphyrin that facilitate the direct $4e^-$ ORR pathways. Recently, Schechter et al. have found direct evidence for the defect origin on graphitic surfaces with various structural and chemical structures (nitrogen-doped and iron-modified). The authors used a novel scanning electrochemical microscopy (SECM) technique combined with simultaneous topography measurements using atomic force microscopy (AFM) to visualize the effect of surface structure and nitrogen-doping on ORR on a nanometric scale [134].

In the quest of developing low-cost ORR catalyst, an excellent work by Liang et al. demonstrated that the Co_3O_4 nanocrystals on graphene are a superior ORR catalyst in the alkaline medium compared to that of Co_3O_4 alone [33]. The activity of the catalyst was further improved by the N-doping of Co_3O_4/graphene composite. The stability of the catalyst is even better than of Pt in an alkaline electrolyte with a Tafel slope of ~42 mV decade^{-1} exhibiting $4e^-$ ORR pathway. Another group reported the N-doping on the carbon nanostructure as support for Pt-based catalysts in the fuel cell [135]. A critical review on the nitrogen-doped carbon support for energy conversion/storage application is done by Wood et al., and the reader is directed to read this paper for more detail [136]. In brief, the catalyst–support interaction plays a key role in the long-term stability, which is highly desirable for the fuel cell

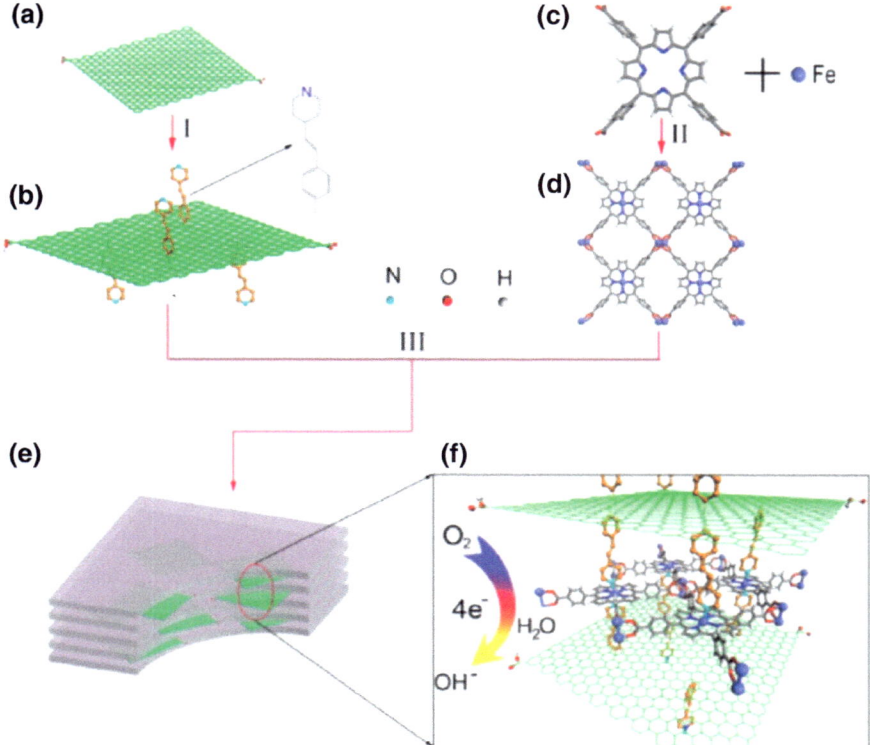

Fig. 5 Schematic of structure of **a** reduced graphene oxide (rGO), **b** G-dye, **c** TCPP, **d** (FeP)$_n$, **e** (G-dye-FeP)$_n$ MOF and **f** enlarged view of graphene layer inside the framework of (G-dye-FeP)$_n$ MOF. Reprinted with permission from the Ref. [133]

application. N-doping of graphene-based support imparts (i) change in electronic property by creating net positive change on carbon structure, (ii) better dispersion and (iii) the stability through increased catalyst-supported chemical bonding also knows as 'tethering effect' [135]. Effect of nitrogen-doping on carbon nanostructure for ORR is recently demonstrated by Neergat and co-workers where the enhanced ORR activity of N-doped carbon/graphene is correlated with the effective density of states, carrier concentration and flat band potential [137].

5.1.2 Electrochemical Fuel Oxidation

Along with long-term stability, the activity of Pd-based catalysts is comparable to that of Pt towards the electrochemical oxidation of methanol, ethanol and formic acid with lower CO-poisoning effect. Inspired from the high stability and active surface area of functionalized graphene and resistance to CO poisoning, Pd is combined with graphene to increase the performance as demonstrated in Fig. 6 for formic acid [41], methanol [40] and ethanol [39] electrochemical oxidation.

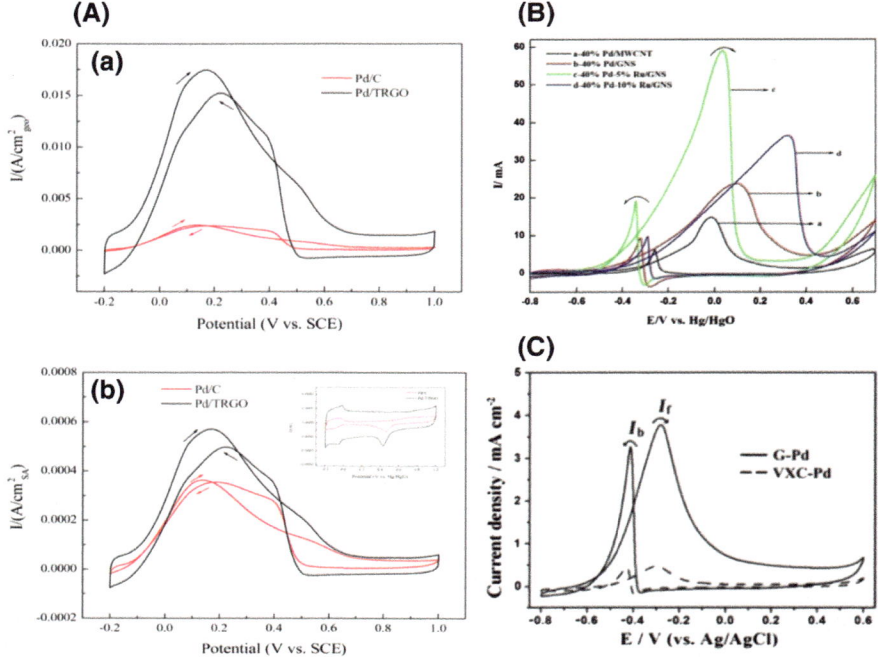

Fig. 6 **A** CVs of formic acid oxidation in 0.5 M H_2SO_4 and 0.5 M HCOOH: **a** normalized with geometric surface area and **B** normalized with Pd electrochemical surface area (CV in 0.5 M H_2SO_4 is shown in the inset to (b)) (reproduced with permission from Ref. [41]). **C** CVs of methanol oxidation on 1 M KOH and 1 M MeOH at a scan rate of 50 mV s^{-1} (reproduced with permission from Ref. 40). **c** CVs of ethanol oxidation of Pd/graphene and Pd/Vulcan XC in 1 M ethanol and 1 M NaOH (reproduced with permission from Ref. [39])

5.1.3   Hydrogen Evolution Reaction

Hydrogen evolution reaction (HER) is very important for the development of the electrolysers and conversion of H_2 fuel. In this regard, extensive research has been devoted to the development of highly efficient HER catalyst. Li et al. synthesized MoS_2 on rGO sheets by a single-step solvothermal method with enhanced HER activity (Fig. 7a, b) [138]. The superior HER activity of the composite material with a Tafel slope of ~41 mV decade^{-1} suggests the Volmer–Heyrovsky mechanism with hydrogen desorption as a rate-determining step (Fig. 7c, d). This catalyst outperforms most of the MoS_2 materials as previously reported in terms of activity; the reason was attributed to the abundant active edge sites of MoS_2 and electronic coupling with the graphene network.

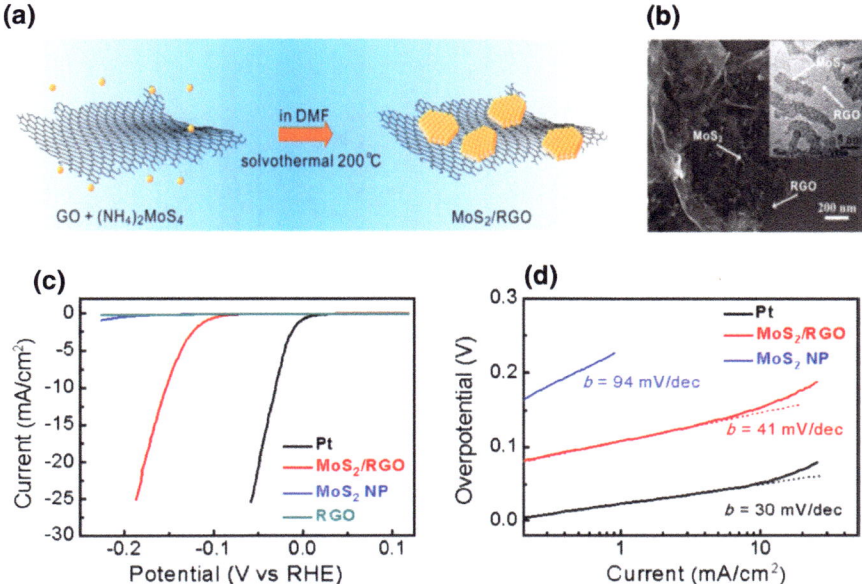

Fig. 7 **a** Synthesis of MoS$_2$/rGO composites, **b** TEM images of the rGO/MoS$_2$, **c** electrochemical performance of HER on MoS$_2$/rGO in comparison with Pt, rGO and MoS$_2$, and **d** corresponding Tafel plots. Reproduced with permission from Ref. [138]

5.1.4 Photocatalysis

Graphene, owing to its high surface area (2630 m^2 g^{-1}), electrical conductivity and optical transmittance of 97.7%, has attracted attention towards the photocatalytic application. Therefore, several graphene-based photocatalysts are synthesized recently for energy conversion and storage. Among other graphene–metal oxide-based photocatalysts, the TiO$_2$–graphene composites are mostly investigated for their exceptional photocatalytic activity [139]. In a composite, the interface between TiO$_2$ and graphene plays a role in transferring the electron because of the difference in their energy level. The Schottky barrier in the TiO$_2$–graphene heterojunction prevents the electron–hole pair recombination, thus making it suitable for effective photocatalysis. Zhang et al. reported the hydrothermal synthesis of TiO$_2$–graphene composite and its excellent photocatalytic behaviour towards degradation of methylene blue compared to bare TiO$_2$ and TiO$_2$-CNT composite as shown in Fig. 8a [140]. At the same time, Zhow et al. also solvothermally synthesized a TiO$_2$–graphene composite for the photodegradation of methylene blue under simulated sunlight irradiation [141]. Gao et al. reported the hydrothermal synthesis of Bi$_2$WO$_6$–graphene composite for the photocatalysis of Rhodamine B under visible sunlight (λ > 420 nm) [142]. A ZnO–graphene composite was reported by Xu et al. for the photocatalysis of methylene blue degradation under the irradiation of UV light [143]. On the other hand, Li et al. reported the synthesis and photocatalytic activity of a CdS–graphene

(a)   **(b)**

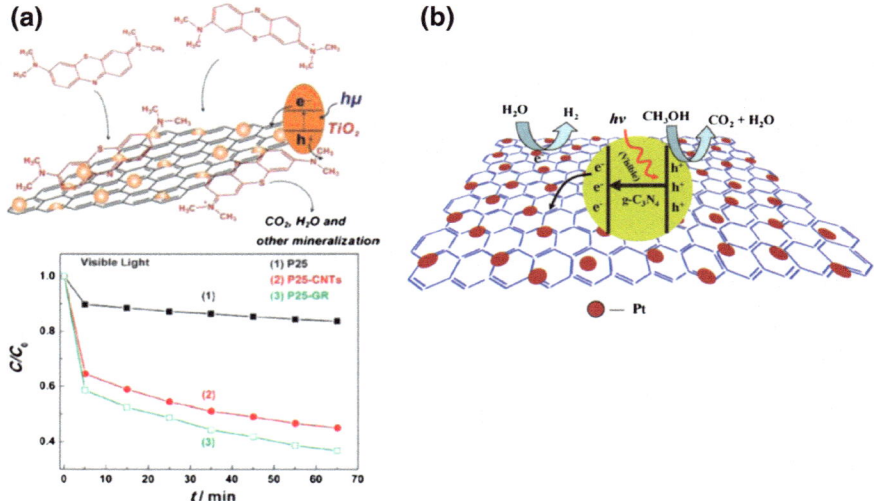

Fig. 8 **a** Schematic of the photocatalytic activity of a TiO$_2$–graphene composite compared with the bare TiO$_2$ and TiO$_2$-CNT counterparts. Reproduced with permission from Ref. [140]. **b** Schematic of the photocatalytic activity of C$_3$N$_4$–graphene composite for the photocatalytic evolution hydrogen from water. Reproduced with permission from Ref. [145]

composite for the photoinduced production of hydrogen [144]. The photocatalytic hydrogen generation at a rate of 451 μmol h^{-1} g^{-1} under the irradiation of visible light on a graphene–C$_3$N$_4$ composite is reported by Xiang et al. (Fig. 8b) [145]. Nitrogen-doped graphene/CdS nanocomposites are reported by Jia et al. for the efficient evolution of hydrogen using solar energy [146].

6 Device Fabrication with Functionalized Graphene

Several devices like biosensors, fuel cells and dye-sensitized solar cells (DSSCs) are fabricated with functionalized/doped graphene and, an improvement in the performance is observed as compared to those obtained with bare graphene. A brief overview of the importance and performance of these devices is given below.

6.1 Biosensors

Graphene-based biosensors are of significant importance because of their intrinsic property of sensing biomolecules present in trace quantity. Zhou et al. demonstrated a graphene-based field-effect transistor (FET) modified with antibodies targeting carcinoembryonic antigen (anti-CEA) [147]. The modified graphene structure could electrically sense the detection of the antigen with 100 pg/mL detection limit during the reaction of anti-CE and CEA protein [147]. Li et al. reported the detection of the poisonous Pb^{+2} ions in aqueous media on graphene modified with single-stranded DNA with a detection limit of 160 ng/L [148]. A pentacene-based FET supported on GO has been demonstrated by Lee et al. to detect tumour cells and artificial DNA [149]. Reduced graphene oxide (rGO)-supported Pt nanoparticles are reported to detect brain natriuretic peptide (BNP) in blood samples with the detection limit as low as 100 fM [150]. The work by Lei et al. is of extreme importance in the field of medical technology because BNP is widely recognized as biomarkers for the diagnosis of heart failure [150]. Ng et al. reported the use of Ni-doped graphene for the detection of 3-nitro-L-tyrosine (3-NT) which is considered to be a biomarker of neurodegenerative diseases [151]. Choi et al. reported the use of Nafion/rGO composite for the detection of trace quantity (1.37×10^{-7} M) organophosphate with the response time of less than 3 s; the schematic of the process is shown in Fig. 9 [152].

A MoS_2 graphene-based biosensor is reported for the detection of single-stranded DNA (ssDNA) [153].

A polydopamine/graphene-based and glucose oxidase/polydopamine/graphene biosensor have been demonstrated by Kanyong et al. for the quantitative detection of hydrogen peroxide and glucose, respectively [154].

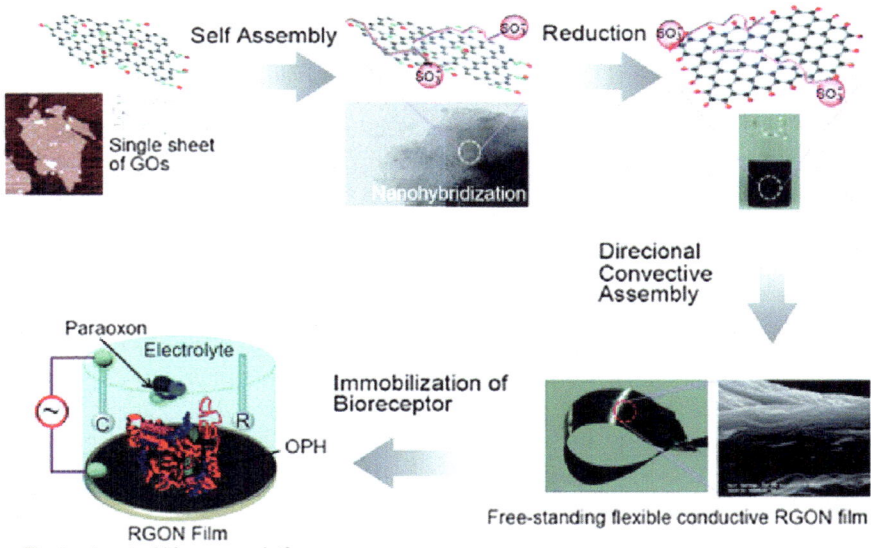

Fig. 9 Schematic representation of an electrochemical biosensor for the detection of organophosphates on a Nafion/graphene (rGON) composite. Reprinted with permission from Ref. [152]. Copyright (2010) American Chemical Society

6.2 Fuel Cells

Due to the enhanced performance of the doped graphene for the ORR, it starts to feature in the low-temperature fuel cell mainly anion-exchange membrane fuel cells (AEMFCs) as a catalyst, and even graphene shows appreciable ORR activity in alkaline media. The AEMFCs are still in the irinitial stage of the development; the major challenges include the stability and poor conductivity of the anion-exchange membrane [29–31, 155, 156]. Palaniselvam et al. reported N-graphene for alkaline fuel cell reaching to the peak power density of ~27 m W cm^{-2} [30]. Recently, Peng et al. carried out outstanding work on N-doped carbon-CoO$_x$ nanohybrids—a precious metal-free cathode with a power density of ~1 W cm^{-2} and 100 h stability in anion-exchange membrane fuel cells [157]. The stability of the cell was successfully demonstrated for 100 h at 600 mA cm^{-2} under H$_2$-air operation. With further development, functionalized graphene can be used as highly efficient and durable ORR catalysts in the near future.

6.3 Dye-Sensitized Solar Cells

A dye-sensitized solar cell (DSSC) is a low-cost solar cell belonging to the category of thin-film solar cells based on the semiconductor formed between photosensitized anode and an electrolyte. In this regard, Bi et al. performed a novel study on quasi-core–shell structure of N-doped graphene (NDG)/CoS as a counter electrode (CE) for DSSCs. A short-circuit current of 20.3 mA cm^{-2} with 10.7% energy conversion efficiency is demonstrated by this cell (Fig. 10a) [158]. The high activity and conductivity of NDG/CoS for DSSCs are attributed to the interaction between the cores of CoS particle coated with an ultrathin layer of NDG that acts as a conductive path. Moreover, this strategy opens a window to replace Pt in DSSC with less expensive and practically viable material [27, 158]. Later, Ma et al. used porous activated graphene nanoplatelets (a-GNPs) incorporated into photoanodes of TiO$_2$ to enhance the efficiency of DSSC [159]. The incorporation of a-GNP (0.02 wt%) leads to an increase in the short-circuit current and conversion efficiency of 37 and 28%, respectively (Fig. 10b). However, excessive increase in GNP leads to dye absorption resulting in a decrease in cell efficiency. They also reported that a-GNP efficiently conducts the electrolyte ions and electrons. In this way, electron transfer and charge separation are accelerated, thus lowering electron recombination.

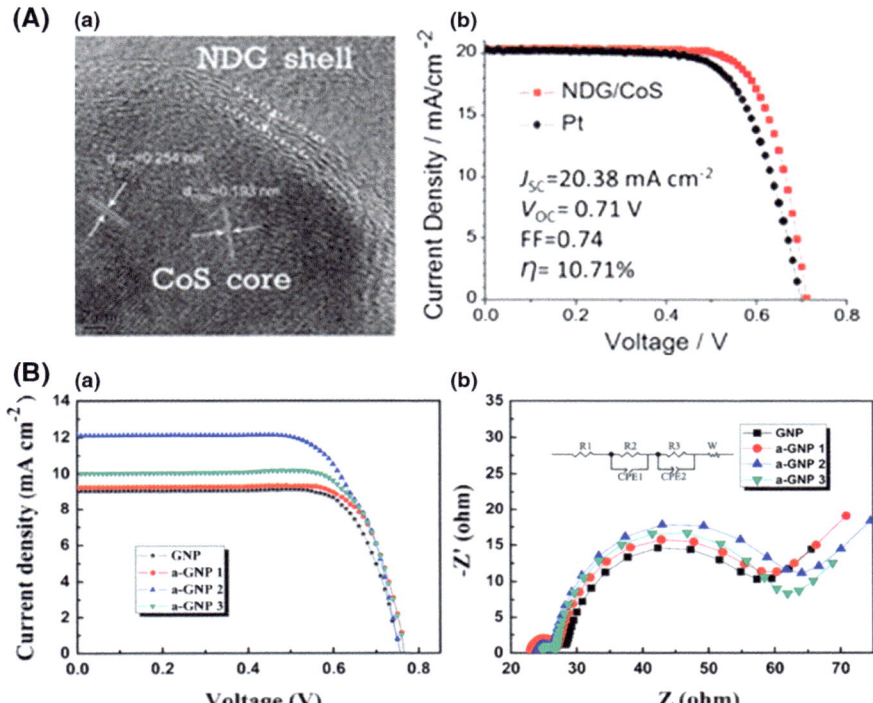

Fig. 10 **A** TEM image of N-doped graphene (NDG)/CoS (a) and IV characteristics of DSSCs using N-doped graphene/CoS (b) (reproduced with permission from the Ref. [158]). **B** IV characterization and impedance response of DSSC with various a-GNP samples (reproduced with permission from Ref. [159])

7 Conclusions and Future Outlook

Based on the recent literature, this chapter summarizes the graphene functionalization, viz. covalent and non-covalent approaches and their improved physical (electronic, mechanical, optical and magnetic), chemical and electrochemical properties. Moreover, functionalized graphene is explored thoroughly for various electrochemical reactions such as ORR, methanol oxidation, hydrogen evolution, photocatalysis. Functionalized graphene electrocatalysts are reported to be stable and active in the alkaline medium in comparison with the conventional Pt-based catalyst. At the same time, they are tolerant towards methanol and CO which enables their long-term operation in the electrochemical devices. The reason for the enhancement of the activity is associated with the improved active charge density of atomic state and change of the electronic configuration of adjacent carbon. The use of functionalized graphene in various physical, chemical and electrochemical devices can be realized as a low-cost alternative to the conventional precious materials and can steam the research towards a future achievement of desirable performance in a vast number of applications.

References

1. Geim, A.K., Novoselov, K.S.: The rise of graphene. Nat. Mater. **6**, 183 (2007)
2. Novoselov, K.S., Geim, A.K., Morozov, S.V., Jiang, D., Zhang, Y., Dubonos, S.V., Grigorieva, I.V., Firsov, A.A.: Electric field effect in atomically thin carbon films. Science **306**(5596), 666 LP–669 LP (2004). https://doi.org/10.1126/science.1102896
3. Kuila, T., Bose, S., Mishra, A.K., Khanra, P., Kim, N.H., Lee, J.H.: Chemical functionalization of graphene and its applications. Prog. Mater Sci. **57**(7), 1061–1105 (2012). https://doi.org/10.1016/j.pmatsci.2012.03.002
4. Stoller, M.D., Park, S., Zhu, Y., An, J., Ruoff, R.S.: Graphene-based ultracapacitors. Nano Lett. **8**(10), 3498–3502 (2008). https://doi.org/10.1021/nl802558y
5. Qu, L., Liu, Y., Baek, J.-B., Dai, L.: Nitrogen-doped graphene as efficient metal-free electrocatalyst for oxygen reduction in fuel cells. ACS Nano **4**(3), 1321–1326 (2010). https://doi.org/10.1021/nn901850u
6. Wang, H., Hao, Q., Yang, X., Lu, L., Wang, X.: Graphene oxide doped polyaniline for supercapacitors. Electrochem. commun. **11**(6), 1158–1161 (2009). https://doi.org/10.1016/j.elecom.2009.03.036
7. Yoo, E., Kim, J., Hosono, E., Zhou, H., Kudo, T., Honma, I.: Large reversible Li storage of graphene nanosheet families for use in rechargeable lithium ion batteries. Nano Lett. **8**(8), 2277–2282 (2008). https://doi.org/10.1021/nl800957b
8. Lian, P., Zhu, X., Liang, S., Li, Z., Yang, W., Wang, H.: Large reversible capacity of high quality graphene sheets as an anode material for lithium-ion batteries. Electrochim. Acta **55**(12), 3909–3914 (2010). https://doi.org/10.1016/j.electacta.2010.02.025
9. Paek, S.-M., Yoo, E., Honma, I.: Enhanced cyclic performance and lithium storage capacity of SnO_2/graphene nanoporous electrodes with three-dimensionally delaminated flexible structure. Nano Lett. **9**(1), 72–75 (2009). https://doi.org/10.1021/nl802484w
10. Guo, S., Sun, S.: Fept nanoparticles assembled on graphene as enhanced catalyst for oxygen reduction reaction. J. Am. Chem. Soc. **134**(5), 2492–2495 (2012). https://doi.org/10.1021/ja2104334
11. Vivekchand, S.R.C., Rout, C.S., Subrahmanyam, K.S., Govindaraj, A., Rao, C.N.R.: Graphene-based electrochemical supercapacitors. J. Chem. Sci. **120**(1), 9–13 (2008). https://doi.org/10.1007/s12039-008-0002-7
12. Liu, C., Yu, Z., Neff, D., Zhamu, A., Jang, B.Z.: Graphene-based supercapacitor with an ultrahigh energy density. Nano Lett. **10**(12), 4863–4868 (2010). https://doi.org/10.1021/nl102661q
13. Wang, Y., Shi, Z., Huang, Y.; Ma, Y., Wang, C., Chen, M., Chen, Y. : Supercapacitor devices based on graphene materials. J. Phys. Chem. C **113**(30), 13103–13107 (2009). https://doi.org/10.1021/jp902214f
14. Stankovich, S., Dikin, D.A., Piner, R.D., Kohlhaas, K.A., Kleinhammes, A., Jia, Y., Wu, Y., Nguyen, S.T., Ruoff, R.S.: Synthesis of graphene-based nanosheets via chemical reduction of exfoliated graphite oxide. Carbon **45**(7), 1558–1565 (2007). https://doi.org/10.1016/j.carbon.2007.02.034
15. Marcano, D.C., Kosynkin, D.V, Berlin, J. M., Sinitskii, A., Sun, Z., Slesarev, A., Alemany, L.B., Lu, W., Tour, J.M.: Improved synthesis of graphene oxide. ACS Nano **4**(8), 4806–4814 (2010). https://doi.org/10.1021/nn1006368
16. Xu, Y., Sheng, K., Li, C., Shi, G.: Self-assembled graphene hydrogel via a one-step hydrothermal process. ACS Nano **4**(7), 4324–4330 (2010). https://doi.org/10.1021/nn101187z
17. Muñoz, R., Gómez-Aleixandre, C.: Review of cvd synthesis of graphene. Chem. Vap. Depos. **19**(10–12), 297–322 (2013). https://doi.org/10.1002/cvde.201300051
18. Guermoune, A., Chari, T., Popescu, F., Sabri, S.S., Guillemette, J., Skulason, H.S., Szkopek, T.: Chemical vapor deposition synthesis of graphene on copper with methanol, ethanol, and propanol precursors. Carbon N. Y. **49**(13), 4204–4210 (2011). https://doi.org/10.1016/j.carbon.2011.05.054

19. Yang, W., Chen, G., Shi, Z., Liu, C.C., Zhang, L., Xie, G., Cheng, M., Wang, D., Yang, R., Shi, D., et al.: Epitaxial growth of single-domain graphene on hexagonal boron nitride. Nat. Mater. **12**(9), 792–797 (2013). https://doi.org/10.1038/nmat3695

20. Sinitskii, A., Dimiev, A., Corley, D.A., Fursina, A.A., Kosynkin, D.V, Tour, J.M.: Kinetics of diazonium functionalization of chemically converted graphene nanoribbons. ACS Nano **4**(4), 1949–1954 (2010). https://doi.org/10.1021/nn901899j

21. Strom, T.A., Dillon, E.P., Hamilton, C.E., Barron, A.R.: Nitrene addition to exfoliated graphene: a one-step route to highly functionalized graphene. Chem. Commun. **46**(23), 4097–4099 (2010). https://doi.org/10.1039/C001488E

22. An, X., Butler, T.W., Washington, M., Nayak, S.K., Kar, S.: Optical and sensing properties of 1-pyrenecarboxylic acid-functionalized graphene films laminated on polydimethylsiloxane membranes. ACS Nano **5**(2), 1003–1011 (2011). https://doi.org/10.1021/nn102415c

23. Imran Jafri, R., Rajalakshmi, N., Ramaprabhu, S.: Nitrogen doped graphene nanoplatelets as catalyst support for oxygen reduction reaction in proton exchange membrane fuel cell. J. Mater. Chem. **20**(34), 7114–7117 (2010). https://doi.org/10.1039/C0JM00467G

24. Fang, M., Wang, K., Lu, H., Yang, Y., Nutt, S.: Covalent polymer functionalization of graphene nanosheets and mechanical properties of composites. J. Mater. Chem. **19**(38), 7098–7105 (2009). https://doi.org/10.1039/B908220D

25. Bai, H., Xu, Y., Zhao, L., Li, C., Shi, G.: Non-covalent functionalization of graphene sheets by sulfonated polyaniline. Chem. Commun. (13), 1667–1669 (2009). https://doi.org/10.1039/B821805F

26. Liu, J., Li, Y., Li, Y., Li, J., Deng, Z.: Noncovalent DNA decorations of graphene oxide and reduced graphene oxide toward water-soluble metal–carbon hybrid nanostructures via self-assembly. J. Mater. Chem. **20**(5), 900–906 (2010). https://doi.org/10.1039/B917752C

27. Georgakilas, V., Tiwari, J.N., Kemp, K.C., Perman, J.A., Bourlinos, A.B., Kim, K.S., Zboril, R.: Noncovalent functionalization of graphene and graphene oxide for energy materials, biosensing, catalytic, and biomedical applications. Chem. Rev. **116**(9), 5464–5519 (2016). https://doi.org/10.1021/acs.chemrev.5b00620

28. Toh, R.J., Poh, H.L., Sofer, Z., Pumera, M.: Transition metal (Mn, Fe, Co, Ni)-doped graphene hybrids for electrocatalysis. Chem. Asian J. **8**(6), 1295–1300 (2013). https://doi.org/10.1002/asia.201300068

29. Li, J., Zhang, Y., Zhang, X., Han, J., Wang, Y., Gu, L., Zhang, Z., Wang, X., Jian, J., Xu, P., et al.: Direct transformation from graphitic C_3N_4 to nitrogen-doped graphene: an efficient metal-free electrocatalyst for oxygen reduction reaction. ACS Appl. Mater. Interfaces **7**(35), 19626–19634 (2015). https://doi.org/10.1021/acsami.5b03845

30. Palaniselvam, T., Valappil, M.O., Illathvalappil, R., Kurungot, S.: Nanoporous graphene by quantum dots removal from graphene and its conversion to a potential oxygen reduction electrocatalyst via nitrogen doping. Energy Environ. Sci. **7**(3), 1059–1067 (2014). https://doi.org/10.1039/C3EE43648A

31. Klingele, M., Pham, C., Vuyyuru, K. R., Britton, B., Holdcroft, S., Fischer, A., Thiele, S.: Sulfur doped reduced graphene oxide as metal-free catalyst for the oxygen reduction reaction in anion and proton exchange fuel cells. Electrochem. Commun. **77**, 71–75 (2017). https://doi.org/10.1016/j.elecom.2017.02.015

32. Zheng, Y., Jiao, Y., Ge, L., Jaroniec, M., Qiao, S.Z.: Two-step boron and nitrogen doping in graphene for enhanced synergistic catalysis. Angew. Chem. Int. Ed. **125**(11), 3192–3198 (2013). https://doi.org/10.1002/ange.201209548

33. Sheng, Z.-H., Gao, H.-L., Bao, W.-J., Wang, F.-B., Xia, X.-H.: Synthesis of boron doped graphene for oxygen reduction reaction in fuel cells. J. Mater. Chem. **22**(2), 390–395 (2012). https://doi.org/10.1039/C1JM14694G

34. Li, J., Li, X., Xiong, D., Hao, Y., Kou, H., Liu, W., Li, D., Niu, Z.: Novel iodine-doped reduced graphene oxide anode for sodium ion batteries. RSC Adv. **7**(87), 55060–55066 (2017). https://doi.org/10.1039/c7ra09349g

35. Zhan, Y., Huang, J., Lin, Z., Yu, X., Zeng, D., Zhang, X., Xie, F., Zhang, W., Chen, J., Meng, H.: Iodine/Nitrogen co-doped graphene as metal free catalyst for oxygen reduction reaction. Carbon N. Y. **95**, 930–939 (2015). https://doi.org/10.1016/j.carbon.2015.09.024

36. Huang, H., Ming, K., Fang, Y., Zhao, H., Wang, X., Chen, J., Guo, J., Zhang, J.: Fluorine-doped graphene with an outstanding electrocatalytic performance for efficient oxygen reduction reaction in alkaline solution. R. Soc. Open Sci. **5**(10), 180925 (2018). https://doi.org/10.1098/rsos.180925

37. Ion-Ebrasu, D., Varlam, M., Balan, D., Enachescu, M., Raceanu, M., Carcadea, E., Marinoiu, A., Stefanescu, I.: Iodine-doped graphene for enhanced electrocatalytic oxygen reduction reaction in proton exchange membrane fuel cell applications. J. Electrochem. Energy Convers. Storage **14**(3), 031001 (2017). https://doi.org/10.1115/1.4036684

38. Park, M., Jeon, I.Y., Ryu, J., Jang, H., Back, J.B., Cho, J.: Edge-halogenated graphene nanoplatelets with F, Cl, or Br as electrocatalysts for all-vanadium redox flow batteries. Nano Energy **26**, 233–240 (2016). https://doi.org/10.1016/j.nanoen.2016.05.027

39. Gao, L., Yue, W., Tao, S., Fan, L.: Novel strategy for preparation of graphene-pd, pt composite, and its enhanced electrocatalytic activity for alcohol oxidation. Langmuir **29**(3), 957–964 (2013). https://doi.org/10.1021/la303663x

40. Awasthi, R., Singh, R.N.: Graphene-supported Pd–Ru nanoparticles with superior methanol electrooxidation activity. Carbon N. Y. **51**, 282–289 (2013). https://doi.org/10.1016/j.carbon.2012.08.055

41. Wang, Y., Liu, H., Wang, L., Wang, H., Du, X., Wang, F., Qi, T., Lee, J.-M., Wang, X.: Pd catalyst supported on a chitosan-functionalized large-area 3d reduced graphene oxide for formic acid electrooxidation reaction. J. Mater. Chem. **A1**(23), 6839–6848 (2013). https://doi.org/10.1039/C3TA10214A

42. Jeon, I.Y., Choi, H.J., Choi, M., Seo, J.M., Jung, S.M., Kim, M.J., Zhang, S., Zhang, L., Xia, Z., Dai, L., et al.: Facile, scalable synthesis of edge-halogenated graphene nanoplatelets as efficient metal-free eletrocatalysts for oxygen reduction reaction. Sci. Rep. **3**, 1–7 (2013). https://doi.org/10.1038/srep01810

43. Niyogi, S., Bekyarova, E., Itkis, M.E., Zhang, H., Shepperd, K., Hicks, J., Sprinkle, M., Berger, C., Lau, C.N., deHeer, W.A., et al.: Spectroscopy of covalently functionalized graphene. Nano Lett. **10**(10), 4061–4066 (2010). https://doi.org/10.1021/nl1021128

44. Liu, H., Ryu, S., Chen, Z., Steigerwald, M.L., Nuckolls, C., Brus, L.E.: Photochemical reactivity of graphene. J. Am. Chem. Soc. **131**(47), 17099–17101 (2009). https://doi.org/10.1021/ja9043906

45. Georgakilas, V., Bourlinos, A.B., Zboril, R., Steriotis, T.A., Dallas, P., Stubos, A.K., Trapalis, C.: Organic functionalisation of graphenes. Chem. Commun. **46**(10), 1766–1768 (2010). https://doi.org/10.1039/B922081J

46. Zhang, X., Hou, L., Cnossen, A., Coleman, A.C., Ivashenko, O., Rudolf, P., van Wees, B.J., Browne, W.R., Feringa, B.L.: One-pot functionalization of graphene with porphyrin through cycloaddition reactions. Chem. Eur. J. **17**(32), 8957–8964 (2011). https://doi.org/10.1002/chem.201100980

47. Liu, L.-H., Lerner, M.M., Yan, M.: Derivitization of pristine graphene with well-defined chemical functionalities. Nano Lett. **10**(9), 3754–3756 (2010). https://doi.org/10.1021/nl1024744

48. Vadukumpully, S., Gupta, J., Zhang, Y., Xu, G.Q., Valiyaveettil, S.: Functionalization of surfactant wrapped graphene nanosheets with alkylazides for enhanced dispersibility. Nanoscale **3**(1), 303–308 (2011). https://doi.org/10.1039/C0NR00547A

49. Zhong, X., Jin, J., Li, S., Niu, Z., Hu, W., Li, R., Ma, J.: Aryne cycloaddition: highly efficient chemical modification of graphene. Chem. Commun. **46**(39), 7340–7342 (2010). https://doi.org/10.1039/C0CC02389B

50. Riley, K.E., Pitoňák, M., Jurečka, P., Hobza, P.: Stabilization and structure calculations for noncovalent interactions in extended molecular systems based on wave function and density functional theories. Chem. Rev. **110**(9), 5023–5063 (2010). https://doi.org/10.1021/cr1000173

51. Xu, Y., Bai, H., Lu, G., Li, C., Shi, G.: Flexible graphene films via the filtration of watersoluble noncovalent functionalized graphene sheets. J. Am. Chem. Soc. **130**(18), 5856–5857 (2008). https://doi.org/10.1021/ja800745y

52. Wang, Y., Chen, X., Zhong, Y., Zhu, F., Loh, K.P.: Large area, continuous, few-layered graphene as anodes in organic photovoltaic devices. Appl. Phys. Lett. **95**(6), 1–4 (2009). https://doi.org/10.1063/1.3204698

53. Zhang, K., Zhang, L.L., Zhao, X.S., Wu, J.: Graphene/Polyaniline nanofiber composites as supercapacitor electrodes. Chem. Mater. **22**(4), 1392–1401 (2010). https://doi.org/10.1021/cm902876u

54. Yang, Q., Pan, X., Huang, F., Li, K.: Fabrication of high-concentration and stable aqueous suspensions of graphene nanosheets by noncovalent functionalization with lignin and cellulose derivatives. J. Phys. Chem. C **114**(9), 3811–3816 (2010). https://doi.org/10.1021/jp910232x

55. Kodali, V.K., Scrimgeour, J., Kim, S., Hankinson, J.H., Carroll, K.M., de Heer, W.A., Berger, C., Curtis, J.E.: Nonperturbative chemical modification of graphene for protein micropatterning. Langmuir **27**(3), 863–865 (2011). https://doi.org/10.1021/la1033178

56. Su, Q., Pang, S., Alijani, V., Li, C., Feng, X., Müllen, K.: Composites of graphene with large aromatic molecules. Adv. Mater. **21**(31), 3191–3195 (2009). https://doi.org/10.1002/adma.200803808

57. Georgakilas, V., Otyepka, M., Bourlinos, A.B., Chandra, V., Kim, N., Kemp, K.C., Hobza, P., Zboril, R., Kim, K.S.: Functionalization of graphene: covalent and non-covalent approaches, derivatives and applications. Chem. Rev. **112**(11), 6156–6214 (2012). https://doi.org/10.1021/cr3000412

58. Elias, D.C., Nair, R.R., Mohiuddin, T.M.G., Morozov, S.V., Blake, P., Halsall, M.P., Ferrari, A.C., Boukhvalov, D.W., Katsnelson, M.I., Geim, A.K., et al.: Control of graphene's properties by reversible hydrogenation: evidence for graphane. Science (80-) **323**(5914), 610 LP–613 LP (2009). https://doi.org/10.1126/science.1167130

59. Schniepp, H.C., Li, J.L., McAllister, M.J., Sai, H., Herrera-Alonson, M., Adamson, D.H., Prud'homme, R.K., Car, R., Seville, D.A., Aksay, I.A.: Functionalized single graphene sheets derived from splitting graphite oxide. J. Phys. Chem. B **110**(17), 8535–8539 (2006). https://doi.org/10.1021/jp060936f

60. Lee, J.K., Yamazaki, S., Yun, H., Park, J., Kennedy, G.P., Kim, G.T., Pietzsch, O., Wiesendanger, R., Lee, S., Hong, S., et al.: Modification of electrical properties of graphene by substrate-induced nanomodulation. Nano Lett. **13**(8), 3494–3500 (2013). https://doi.org/10.1021/nl400827p

61. Schiros, T., Nordlund, D., Pálová, L., Prezzi, D., Zhao, L., Kim, K. S., Wurstbauer, U., Gutiérrez, C., Delongchamp, D., Jaye, C., et al.: Connecting dopant bond type with electronic structure in N-doped graphene. Nano Lett. **12**(8), 4025–4031 (2012). https://doi.org/10.1021/nl301409h

62. Macedo, L.J.A., Lima, F.C.D.A., Amorim, R.G., Freitas, R.O., Yadav, A., Iost, R.M., Balasubramanian, K., Crespilho, F.N.: Interplay of non-uniform charge distribution on the electrochemical modification of graphene. Nanoscale **10**(31), 15048–15057 (2018). https://doi.org/10.1039/c8nr03893g

63. Cervantes-Sodi, F., Csányi, G., Piscanec, S., Ferrari, A.C.: Electronic properties of chemically modified graphene ribbons. Phys. Status Solidi Basic Res. **245**(10), 2068–2071 (2008). https://doi.org/10.1002/pssb.200879640

64. de la Torre, B., Švec, M., Hapala, P., Redondo, J., Krejčí, O., Lo, R., Manna, D., Sarmah, A., Nachtigallová, D., Tuček, J., et al.: Non-covalent control of spin-state in metal-organic complex by positioning on N-doped graphene. Nat. Commun. **9**(1), 1–9 (2018). https://doi.org/10.1038/s41467-018-05163-y

65. Saha, S., Samanta, P., Chandra Murmu, N., Kuila, T.: Investigation of the surface plasmon polariton and electrochemical properties of covalent and non-covalent functionalized reduced graphene oxide. Phys. Chem. Chem. Phys. **19**(42), 28588–28595 (2017). https://doi.org/10.1039/c7cp05923j

66. Zhou, J., Wu, M.M., Zhou, X., Sun, Q.: Tuning electronic and magnetic properties of graphene by surface modification. Appl. Phys. Lett. **95**(10) (2009). https://doi.org/10.1063/1.3225154

67. Dedkov, Y.S., Fonin, M.: Electronic and magnetic properties of the graphene-ferromagnet interface. New J. Phys. **12**, 125004 (2010). https://doi.org/10.1088/1367-2630/12/12/125004

68. Johari, P., Shenoy, V.B.: Modulating optical properties of graphene oxide: role of prominent functional groups. ACS Nano **5**(9), 7640–7647 (2011). https://doi.org/10.1021/nn202732t

69. Xu, Y., Liu, Z., Zhang, X., Wang, Y., Tian, J., Huang, Y., Ma, Y., Zhang, X., Chen, Y.: A graphene hybrid material covalently functionalized with porphyrin: synthesis and optical limiting property. Adv. Mater. **21**(12), 1275–1279 (2009). https://doi.org/10.1002/adma. 200801617

70. Du, Y., Dong, N., Zhang, M., Zhu, K., Na, R., Zhang, S., Sun, N., Wang, G., Wang, J.: Covalent functionalization of graphene oxide with porphyrin and porphyrin incorporated polymers for optical limiting. Phys. Chem. Chem. Phys. **19**(3), 2252–2260 (2017). https://doi.org/10.1039/c6cp05920a

71. Wang, A., Yu, W., Huang, Z., Zhou, F., Song, J., Song, Y., Long, L., Cifuentes, M.P., Humphrey, M.G., Zhang, L., et al.: Covalent functionalization of reduced graphene oxide with porphyrin by means of diazonium chemistry for nonlinear optical performance. Sci. Rep. **6**, 1–12 (2016). https://doi.org/10.1038/srep23325

72. Li, Z., He, C., Wang, Z., Gao, Y., Dong, Y., Zhao, C., Chen, Z., Wu, Y., Song, W.: Ethylenediamine-modified graphene oxide covalently functionalized with a tetracarboxylic Zn(ii) phthalocyanine hybrid for enhanced nonlinear optical properties. Photochem. Photobiol. Sci. **15**(7), 910–919 (2016). https://doi.org/10.1039/c6pp00063k

73. Zhao, X., Yan, X. Q., Ma, Q., Yao, J., Zhang, X. L., Liu, Z. B., Tian, J. G.: Nonlinear optical and optical limiting properties of graphene hybrids covalently functionalized by phthalocyanine. Chem. Phys. Lett. **577**, 62–67 (2013). https://doi.org/10.1016/j.cplett.2013.04.023

74. Liu, Z., Xu, Y., Zhang, X., Zhang, X., Chen, Y., Tian, J.: Porphyrin and fullerene covalently functionalized graphene hybrid materials with large nonlinear optical properties. J. Phys. Chem. B (ACS Publ.) 9681–9686 (2009)

75. Xu, X., Li, P., Zhang, L., Liu, X., Zhang, H.L., Shi, Q., He, B., Zhang, W., Qu, Z., Liu, P.: Covalent functionalization of graphene by nucleophilic addition reaction: synthesis and optical-limiting properties. Chem. Asian J. **12**(19), 2583–2590 (2017). https://doi.org/10. 1002/asia.201700899

76. Orellana, W., Correa, J.D.: Noncovalent functionalization of carbon nanotubes and graphene with tetraphenylporphyrins: stability and optical properties from ab initio calculations. J. Mater. Sci. **50**(2), 898–905 (2014). https://doi.org/10.1007/s10853-014-8650-0

77. Liu, W., Liu, J.Y., Miao, M.S.: Macrocycles inserted in graphene: from coordination chemistry on graphene to graphitic carbon oxide. Nanoscale **8**(41), 17976–17983 (2016). https://doi.org/10.1039/c6nr04178g

78. Chowdhury, S., Jana, D.: Electronic and magnetic properties of modified silicene/graphene hybrid: ab initio study. Mater. Chem. Phys. **183**, 580–587 (2016). https://doi.org/10.1016/j.matchemphys.2016.09.018

79. Ray, S.C., Soin, N., Pong, W. F., Roy, S.S., Strydom, A.M., McLaughlin, J.A., Papakonstantinou, P.: Plasma modification of the electronic and magnetic properties of vertically aligned Bi-/Tri-layered graphene nanoflakes. RSC Adv. **6**(75), 70913–70924 (2016). https://doi.org/10.1039/c6ra14457h

80. Ray, S.C., Soin, N., Makgato, T., Chuang, C.H., Pong, W.F., Roy, S.S., Ghosh, S.K., Strydom, A.M., McLaughlin, J.A.: Graphene supported graphone/graphane bilayer nanostructure material for spintronics. Sci. Rep. **4** (2014). https://doi.org/10.1038/srep03862

81. Liu, Y., Tang, N., Wan, X., Feng, Q., Li, M., Xu, Q., Liu, F., Du, Y.: Realization of ferromagnetic graphene oxide with high magnetization by doping graphene oxide with nitrogen. Sci. Rep. **3** (2013). https://doi.org/10.1038/srep02566

82. Yazyev, O.V., Helm, L.: Defect-induced magnetism in graphene. Phys. Rev. B Condens. Matter Mater. Phys. **75**(12), 1–5 (2007). https://doi.org/10.1103/PhysRevB.75.125408

83. Gonzalez-Herrero, H., Gomez-Rodriguez, J.M., Mallet, P., Moaied, M., Palacios, J.J., Salgado, C., Ugeda, M.M., Veuillen, J.-Y., Yndurain, F., Brihuega, I.: Supplementary materials for atomic-scale control of graphene magnetism by using hydrogen atoms. Science (80-) **352**(6284), 437–441 (2016). https://doi.org/10.1126/science.aad8038

84. Park, R.S., Lee, S., Bozoklu, K.-S., Cai, G., Nguyen, W., Ruoff, S.T.: Graphene oxide papers. ACS Nano **2**(3), 572–578 (2008). https://doi.org/10.1021/nn700349a

85. Gonalves, G., Marques, P.A.A.P., Barros-Timmons, A., Bdkin, I., Singh, M.K., Emami, N., Grácio, J.: Graphene oxide modified with pmma via atrp as a reinforcement filler. J. Mater. Chem. **20**(44), 9927–9934 (2010). https://doi.org/10.1039/c0jm01674h

86. Song, S., Wan, C., Zhang, Y.: Non-covalent functionalization of Graphene oxide by Pyrene-block copolymers for enhancing physical properties of Poly(Methyl Methacrylate). RSC Adv. **5**(97), 79947–79955 (2015). https://doi.org/10.1039/c5ra14967c

87. Dikin, D.A., Stankovich, S., Zimney, E.J., Piner, R.D., Dommett, G.H.B., Evmenenko, G., Nguyen, S.T., Ruoff, R.S.: Preparation and characterization of graphene oxide paper. Nature **448**(7152), 457–460 (2007). https://doi.org/10.1038/nature06016

88. Suk, J. W., Piner, R. D., An, J., Ruoff, R. S.: Mechanical properties of monolayer graphene oxide. ACS Nano. 4, 6557−6564 (2010)

89. Wang, B., Li, Z., Wang, C., Signetti, S., Cunning, B.V., Wu, X., Huang, Y., Jiang, Y., Shi, H., Ryu, S., et al.: Folding large graphene-on-polymer films yields laminated composites with enhanced mechanical performance. Adv. Mater. **30**(35), 1–10 (2018). https://doi.org/10.1002/adma.201707449

90. Tang, Y., Yang, Z., Dai, X.: A theoretical simulation on the catalytic oxidation of co on pt/graphene. Phys. Chem. Chem. Phys. **14**(48), 16566–16572 (2012). https://doi.org/10.1039/c2cp41441d

91. Li, F., Zhao, J., Chen, Z.: Fe-anchored graphene oxide: a low-cost and easily accessible catalyst for low-temperature CO oxidation. J. Phys. Chem. C **116**(3), 2507–2514 (2012). https://doi.org/10.1021/jp209572d

92. Guo, N., Xi, Y., Liu, S., Zhang, C.: Greatly enhancing catalytic activity of graphene by doping the underlying metal substrate. Sci. Rep. **5**, 1–7 (2015). https://doi.org/10.1038/srep12058

93. Lu, Y.-H., Zhou, M., Zhang, C., Feng, Y.-P.: Metal-embedded graphene: a possible catalyst with high activity. J. Phys. Chem. C **113**(47), 20156–20160 (2009). https://doi.org/10.1021/jp908829m

94. Zhou, M., Zhang, A., Dai, Z., Zhang, C., Feng, Y.P.: Greatly enhanced adsorption and catalytic activity of au and pt clusters on defective graphene. J. Chem. Phys. **132**(19), 7–10 (2010). https://doi.org/10.1063/1.3427246

95. Wang, X., Song, L., Yang, H., Xing, W., Lu, H., Hu, Y.: Cobalt Oxide/Graphene composite for highly efficient CO oxidation and its application in reducing the fire hazards of aliphatic polyesters. J. Mater. Chem. **22**(8), 3426–3431 (2012). https://doi.org/10.1039/c2jm15637g

96. Wang, Y., Wen, Z., Zhang, H., Cao, G., Sun, Q., Cao, J.: CuO Nanorods-decorated reduced graphene oxide nanocatalysts for catalytic oxidation of Co. Catalysts **6**(12), 214 (2016). https://doi.org/10.3390/catal6120214

97. Li, W., Zhang, H., Wang, J., Qiao, W., Ling, L., Long, D.: Flexible Ru/Graphene aerogel with switchable surface chemistry: highly efficient catalyst for room-temperature CO oxidation. Adv. Mater. Interfaces **3**(10), 1–8 (2016). https://doi.org/10.1002/admi.201500711

98. Mahmoudi, H., Mahmoudi, M., Doustdar, O., Jahangiri, H., Tsolakis, A., Gu, S., LechWyszynski, M.: A review of Fischer Tropsch synthesis process, mechanism, surface chemistry and catalyst formulation. Biofuels Eng. **2**(1), 11–31 (2017). https://doi.org/10.1515/bfuel-2017-0002

99. Jahangiri, H., Bennett, J., Mahjoubi, P., Wilson, K., Gu, S.: A review of advanced catalyst development for fischer-tropsch synthesis of hydrocarbons from biomass derived syn-gas. Catal. Sci. Technol. **4**(8), 2210–2229 (2014). https://doi.org/10.1039/c4cy00327f

100. Cheng, Y., Lin, J., Xu, K., Wang, H., Yao, X., Pei, Y., Yan, S., Qiao, M., Zong, B.: Fischer-tropsch synthesis to lower olefins over potassium-promoted reduced graphene oxide supported iron catalysts. ACS Catal. **6**(1), 389–399 (2016). https://doi.org/10.1021/acscatal.5b02024

101. Zhao, H., Zhu, Q., Gao, Y., Zhai, P., Ma, D.: Iron oxide nanoparticles supported on pyrolytic graphene oxide as model catalysts for fischer tropsch synthesis. Appl. Catal. A Gen. **456**, 233–239 (2013). https://doi.org/10.1016/j.apcata.2013.03.006

102. Abbas, M., Zhang, J., Lin, K., Chen, J.: Fe_3O_4 nanocubes assembled on RGO nanosheets: ultrasound induced in-situ and eco-friendly synthesis, characterization and their excellent catalytic performance for the production of liquid fuel in fischer-tropsch synthesis. Ultrason. Sonochem. **42**, 271–282 (2018). https://doi.org/10.1016/j.ultsonch.2017.11.031

103. Sun, B., Jiang, Z., Fang, D., Xu, K., Pei, Y., Yan, S., Qiao, M., Fan, K., Zong, B.: One-pot approach to a highly robust iron oxide/reduced graphene oxide nanocatalyst for fischer-tropsch synthesis. ChemCatChem **5**(3), 714–719 (2013). https://doi.org/10.1002/cctc.201200653
104. Taghavi, S., Asghari, A., Tavasoli, A.: Enhancement of performance and stability of graphene nano sheets supported cobalt catalyst in fischer–tropsch synthesis using graphene functionalization. Chem. Eng. Res. Des. **119**, 198–208 (2017). https://doi.org/10.1016/j.cherd.2017. 01.021
105. Hajjar, Z., Doroudian Rad, M., Soltanali, S.: Novel CO/Graphene oxide and CO/nanoporous graphene catalysts for fischer–tropsch reaction. Res. Chem. Intermed. **43**(3), 1341–1353 (2017). https://doi.org/10.1007/s11164-016-2701-x
106. Yadav, M.D., Dasgupta, K., Kushwaha, A., Srivastava, A.P., Patwardhan, A.W., Srivastava, D., Joshi, J.B.: Few layered graphene by floating catalyst chemical vapour deposition and its extraordinary H_2O_2 sensing property. Mater. Lett. **199**, 180–183 (2017). https://doi.org/10. 1016/j.matlet.2017.04.085
107. Rostamnia, S., Doustkhah, E., Golchin-Hosseini, H., Zeynizadeh, B., Xin, H., Luque, R.: Efficient tandem aqueous room temperature oxidative amidations catalysed by supported Pd nanoparticles on graphene oxide. Catal. Sci. Technol. **6**(12), 4124–4133 (2016). https://doi. org/10.1039/c5cy01596k
108. Rahimi, R., Moshari, M., Rabbani, M., Azad, A.: Photooxidation of benzyl alcohols and photodegradation of cationic dyes by Fe_3O_4@sulfur/reduced graphene oxide as catalyst. RSC Adv. **6**(47), 41156–41164 (2016). https://doi.org/10.1039/c6ra00137h
109. Zahed, B., Hosseini-Monfared, H.: A comparative study of silver-graphene oxide nanocomposites as a recyclable catalyst for the aerobic oxidation of benzyl alcohol: support effect. Appl. Surf. Sci. **328**, 536–547 (2015). https://doi.org/10.1016/j.apsusc.2014.12.078
110. Song, Z., Li, W., Niu, F., Xu, Y., Niu, L., Yang, W., Wang, Y., Liu, J.: A novel method to decorate au clusters onto graphene via a mild co-reduction process for ultrahigh catalytic activity. J. Mater. Chem. A **5**(1), 230–239 (2017). https://doi.org/10.1039/c6ta08284j
111. Chung, H.T., Cullen, D. A., Higgins, D., Sneed, B.T., Holby, E.F. More, K.L., Zelenay, P.: Direct atomic-level insight into the active sites of a high-performance pgm-free ORR catalyst. Science (80-) **357**(6350), 479–484 (2017). https://doi.org/10.1126/science.aan2255
112. Gasteiger, H.A., Markovi, N.M.: Chemistry: just a dream--or future reality? Science (80-) **324**(5923), 48–49 (2009). https://doi.org/10.1126/science.1172083
113. Jaouen, F., Proietti, E., Lefèvre, M., Chenitz, R., Dodelet, J.-P., Wu, G., Chung, H.T., Johnston, C.M., Zelenay, P.: Recent advances in non-precious metal catalysis for oxygen-reduction reaction in polymer electrolyte fuel cells. Energy Environ. Sci. **4**(1), 114 (2011). https://doi. org/10.1039/c0ee00011f
114. Gasteiger, H.A., Kocha, S.S., Sompalli, B., Wagner, F.T.: Activity benchmarks and requirements for Pt, Pt-alloy, and Non-Pt oxygen reduction catalysts for PEMFCs. Appl. Catal. B Environ. **56**(1-2 Special issue), 9–35 (2005). https://doi.org/10.1016/j.apcatb.2004.06.021
115. Wang, J.X., Inada, H.; Wu, L.; Zhu, Y.; Choi, Y.; Liu, P.; Zhou, W. P., Adzic, R.R.: Oxygen reduction on well-defined core-shell nanocatalysts: particle size, facet, and pt shell thickness effects. J. Am. Chem. Soc. **131**(47), 17299–17302 (2009). https://doi.org/10.1021/ja9067645
116. Stamenkovic, V.R., Fowler, B., Mun, B.S., Wang, G., Ross, P.N., Lucas, C.A., Markovic, N.M.:Improved oxygen reduction activity on $Pt_3Ni(111)$ via increased surface site availability. Science (80-) **315**(5811), 493–497 (2007). https://doi.org/10.1126/science.1135941
117. Li, Y., Zhou, W., Wang, H., Xie, L., Liang, Y., Wei, F., Idrobo, J.C., Pennycook, S.J., Dai, H.: An oxygen reduction electrocatalyst based on carbon nanotubeĝ€ graphene complexes. Nat. Nanotechnol. **7**(6), 394–400 (2012). https://doi.org/10.1038/nnano.2012.72
118. Greeley, J., Stephens, I.E.L., Bondarenko, A.S., Johansson, T.P., Hansen, H.A., Jaramillo, T.F., Rossmeisl, J., Chorkendorff, I., Nørskov, J.K.: Alloys of platinum and early transition metals as oxygen reduction electrocatalysts. Nat. Chem. **1**(7), 552–556 (2009). https://doi. org/10.1038/nchem.367
119. Xu, M., Ivey, D.G., Xie, Z., Qu, W.: Rechargeable Zn-air batteries: progress in electrolyte development and cell configuration advancement. J. Power Sources **283**, 358–371 (2015). https://doi.org/10.1016/j.jpowsour.2015.02.114

120. Blurton, K.F., Sammells, A.F.: Metal/air batteries: their status and potential - a review. J. Power Sources **4**(4), 263–279 (1979). https://doi.org/10.1016/0378-7753(79)80001-4

121. Kraytsberg, A., Ein-Eli, Y.: Review on Li-air batteries - opportunities, limitations and perspective. J. Power Sources **196**(3), 886–893 (2011). https://doi.org/10.1016/j.jpowsour.2010.09.031

122. Lee, J.S., Kim, S.T., Cao, R., Choi, N.S., Liu, M.; Lee, K.T., Cho, J.: Adv. Energy Mater. **1**(1), 34–50 (2011). https://doi.org/10.1002/aenm.201000010

123. Gelman, D., Shvartsev, B., Ein-Eli, Y.: Aluminum-air battery based on an ionic liquid electrolyte. J. Mater. Chem. A **2**(47), 20237–20242 (2014). https://doi.org/10.1002/aenm.201000010

124. Gong, K., Du, F., Xia, Z., Durstock, M., Dai, L.: Nitrogen-doped carbon nanotube arrays with high electrocatalytic activity for oxygen reduction. Science (80-) **323**(5915), 760 LP–764 LP (2009). https://doi.org/10.1126/science.1168049

125. Liang, J., Jiao, Y., Jaroniec, M., Qiao, S.Z.: Sulfur and nitrogen dual-doped mesoporous graphene electrocatalyst for oxygen reduction with synergistically enhanced performance. Angew. Chemie. Int. Ed. **51**(46), 11496–11500 (2012). https://doi.org/10.1002/anie.201206720

126. Zhang, L., Xia, Z.: Mechanisms of oxygen reduction reaction on nitrogen-doped graphene for fuel cells. J. Phys. Chem. C **115**(22), 11170–11176 (2011). https://doi.org/10.1021/jp201991j

127. Yang, Z., Yao, Z., Li, G., Fang, G., Nie, H., Liu, Z., Zhou, X., Chen, X., Huang, S.: Sulfurdoped graphene as an efficient metal-free cathode catalyst for oxygen reduction. ACS Nano **6**(1), 205–211 (2012). https://doi.org/10.1021/nn203393d

128. Xiong, W., Du, F., Liu, Y., Perez, A., Supp, M., Ramakrishnan, T.S., Dai, L., Jiang, L.: 3-D carbon nanotube structures used as high performance catalyst for oxygen reduction reaction. J. Am. Chem. Soc. **132**(45), 15839–15841 (2010). https://doi.org/10.1021/ja104425h

129. Wang, S., Yu, D., Dai, L.: Polyelectrolyte functionalized carbon nanotubes as efficient metalfree electrocatalysts for oxygen reduction. J. Am. Chem. Soc. **133**(14), 5182–5185 (2011). https://doi.org/10.1021/ja1112904

130. Yu, D., Nagelli, E., Du, F., Dai, L.: Metal-free carbon nanomaterials become more active than metal catalysts and last longer. J. Phys. Chem. Lett. **1**(14), 2165–2173 (2010). https://doi.org/10.1021/jz100533t

131. Subramanian, P., Cohen, A., Teblum, E., Nessim, G.D., Bormasheko, E., Schechter, A.: Electrocatalytic activity of nitrogen plasma treated vertically aligned carbon nanotube carpets towards oxygen reduction reaction. Electrochem. Commun. **49**, 42–46 (2014). https://doi.org/10.1016/j.elecom.2014.10.005

132. Yan, X., Jia, Y., Yao, X.: Defects on carbons for electrocatalytic oxygen reduction. Chem. Soc. Rev. **47**, 7628–7658 (2018). https://doi.org/10.1039/C7CS00690J

133. Jahan, M., Bao, Q., Loh, K.P.: Electrocatalytically active graphene-porphyrin MOF composite for oxygen reduction reaction. J. Am. Chem. Soc. **134**(15), 6707–6713 (2012). https://doi.org/10.1021/ja211433h

134. Kolagatla, S., Subramanian, P., Schechter, A.: Nanoscale mapping of catalytic hotspots on Fe, N-Modified HOPG by scanning electrochemical microscopy-atomic force microscopy. Nanoscale **10**(15), 6962–6970 (2018). https://doi.org/10.1039/C8NR00849C

135. Zhou, Y., Neyerlin, K., Olson, T.S., Pylypenko, S., Bult, J., Dinh, H.N., Gennett, T., Shao, Z., O'Hayre, R.: Enhancement of Pt and Pt-Alloy fuel cell catalyst activity and durability via nitrogen-modified carbon supports. Energy Environ. Sci. **3**(10), 1437–1446 (2010). https://doi.org/10.1039/C003710A

136. Wood, K.N., O'Hayre, R., Pylypenko, S.: Recent progress on nitrogen/carbon structures designed for use in energy and sustainability applications. Energy Environ. Sci. **7**(4), 1212–1249 (2014). https://doi.org/10.1039/C3EE44078H

137. Bera, B., Chakraborty, A., Kar, T., Leuaa, P., Neergat, M.: Density of states, carrier concentration, and flat band potential derived from electrochemical impedance measurements of N-Doped carbon and their influence on electrocatalysis of oxygen reduction reaction. J. Phys. Chem. C **121**(38), 20850–20856 (2017). https://doi.org/10.1021/acs.jpcc.7b06735

138. Li, Y., Wang, H., Xie, L., Liang, Y., Hong, G., Dai, H.: MoS_2 nanoparticles grown on graphene: an advanced catalyst for the hydrogen evolution reaction. J. Am. Chem. Soc. **133**(19), 7296–7299 (2011). https://doi.org/10.1021/ja201269b

139. Liang, Y., Wang, H., Casalongue, H.S., Chen, Z., Dai, H.: TiO_2 Nanocrystals grown on graphene as advanced photocatalytic hybrid materials. Nano Res. **3**(10), 701–705 (2010). https://doi.org/10.1007/s12274-010-0033-5

140. Zhang, H., Lv, X., Li, Y., Wang, Y., Li, J.: P25-graphene composite as a high performance photocatalyst. ACS Nano **4**(1), 380–386 (2010). https://doi.org/10.1021/nn901221k

141. Zhou, K., Zhu, Y., Yang, X., Jiang, X., Li, C.: Preparation of graphene-TiO_2 composites with enhanced photocatalytic activity. New J. Chem. **35**(2), 353–359 (2011). https://doi.org/10.1039/c0nj00623h

142. Gao, E., Wang, W., Shang, M., Xu, J.: Synthesis and enhanced photocatalytic performance of graphene-Bi 2WO6 composite. Phys. Chem. Chem. Phys. **13**(7), 2887–2893 (2011). https://doi.org/10.1039/c0cp01749c

143. Xu, T., Zhang, L., Cheng, H., Zhu, Y.: Significantly enhanced photocatalytic performance of ZnO via graphene hybridization and the mechanism study. Appl. Catal. B Environ. **101**(3–4), 382–387 (2011). https://doi.org/10.1016/j.apcatb.2010.10.007

144. Li, Q., Guo, B., Yu, J., Ran, J., Zhang, B., Yan, H., Gong, J.R.: Highly efficient visible-light-driven photocatalytic hydrogen production of CdS-cluster-decorated graphene nanosheets. J. Am. Chem. Soc. **133**(28), 10878–10884 (2011). https://doi.org/10.1021/ja2025454

145. Xiang, Q., Yu, J., Jaroniec, M.: Preparation and enhanced visible-light photocatalytic H_2-production activity of graphene/C_3N_4 composites. J. Phys. Chem. C **115**(15), 7355–7363 (2011). https://doi.org/10.1021/jp200953k

146. Jia, L., Wang, D.H., Huang, Y.X., Xu, A.W., Yu, H.Q.: Highly durable N-Doped graphene/CdS nanocomposites with enhanced photocatalytic hydrogen evolution from water under visible light irradiation. J. Phys. Chem. C **115**(23), 11466–11473 (2011). https://doi.org/10.1021/jp2023617

147. Zhou, L., Mao, H., Wu, C., Tang, L., Wu, Z., Sun, H., Zhang, H., Zhou, H., Jia, C., Jin, Q., et al.: Label-Free graphene biosensor targeting cancer molecules based on non-covalent modification. Biosens. Bioelectron. **87**, 701–707 (2017). https://doi.org/10.1016/j.bios.2016.09.025

148. Li, Y., Wang, C., Zhu, Y., Zhou, X., Xiang, Y., He, M., Zeng, S.: Fully integrated graphene electronic biosensor for label-free detection of lead (II) ion based on G-quadruplex structure-switching. Biosens. Bioelectron. **89**, 758–763 (2017). https://doi.org/10.1016/j.bios.2016.10.061

149. Lee, D.H., Cho, H.S., Han, D., Chand, R., Yoon, T.J., Kim, Y.S.: Highly selective organic transistor biosensor with inkjet printed graphene oxide support system. J. Mater. Chem. B **5**(19), 3580–3585 (2017). https://doi.org/10.1039/c6tb03357a

150. Lei, Y.M., Xiao, M.M., Li, Y.T., Xu, L., Zhang, H., Zhang, Z.Y., Zhang, G.J.: Detection of heart failure-related biomarker in whole blood with graphene field effect transistor biosensor. Biosens. Bioelectron. **91**, 1–7 (2017). https://doi.org/10.1016/j.bios.2016.12.018

151. Ng, S.P., Qiu, G., Ding, N., Lu, X., Wu, C.M.L.: Label-free detection of 3-Nitro-L-Tyrosine with nickel-doped graphene localized surface plasmon resonance biosensor. Biosens. Bioelectron. **89**, 468–476 (2017). https://doi.org/10.1016/j.bios.2016.04.017

152. Choi, B.G., Park, H., Park, T.J., Yang, M.H., Kim, J.S., Jang, S.-Y., Heo, N.S., Lee, S.Y., Kong, J., Hong, W.H.: Solution chemistry of self-assembled graphene nanohybrids for high-performance flexible biosensors. ACS Nano **4**(5), 2910–2918 (2010). https://doi.org/10.1021/nn100145x

153. Maurya, J.B., Prajapati, Y.K., Singh, V., Saini, J.P., Tripathi, R.: Performance of graphene–MoS_2 based surface plasmon resonance sensor using silicon layer. Opt. Quantum Electron. **47**(11), 3599–3611 (2015). https://doi.org/10.1007/s11082-015-0233-z

154. Kanyong, P., Krampa, F.D., Aniweh, Y., Awandare, G.A.: Polydopamine-functionalized graphene nanoplatelet smart conducting electrode for bio-sensing applications. Arab. J. Chem. 1–9 (2018). https://doi.org/10.1016/j.arabjc.2018.01.001

155. Tachibana, N., Ikeda, S., Yukawa, Y., Kawaguchi, M.: Highly porous nitrogen-doped carbon nanoparticles synthesized via simple thermal treatment and their electrocatalytic activity for oxygen reduction reaction. Carbon N. Y. **115**, 515–525 (2017). https://doi.org/10.1016/j.carbon.2017.01.034

156. Kruusenberg, I., Ratso, S., Vikkisk, M., Kanninen, P., Kallio, T., Kannan, A.M., Tammeveski, K.: Highly active nitrogen-doped nanocarbon electrocatalysts for alkaline direct methanol fuel cell. J. Power Sources **281**, 94–102 (2015). https://doi.org/10.1016/j.jpowsour.2015.01.167

157. Peng X., Omasta, T.J.; Magliocca, E.; Wang, L.; Varcoe, J.R.; Mustain, W.E.: N-doped carbon CoOx nanohybrids: the first precious metal free cathode to achieve 1.0 w/cm2 peak power and 100 h life in anion-exchange membrane fuel cells. Angew. Chemie Int. Ed. 1–7 (2018). https://doi.org/10.1002/anie.201811099

158. Bi, E., Chen, H., Yang, X., Peng, W., Grätzel, M., Han, L.: A quasi core–shell nitrogen-doped graphene/cobalt sulfide conductive catalyst for highly efficient dye-sensitized solar cells. Energy Environ. Sci. **7**(8), 2637–2641 (2014). https://doi.org/10.1039/C4EE01339E

159. Ma, H., Tian, J., Cui, L., Liu, Y., Bai, S., Chen, H., Shan, Z.: Porous activated graphene nanoplatelets incorporated in TiO$_2$ photoanodes for high-efficiency dye-sensitized solar cells. J. Mater. Chem. A **3**(16), 8890–8895 (2015). https://doi.org/10.1039/C5TA00527B

Synthesis and Properties of Graphene and Graphene Oxide-Based Polymer Composites

Srikanta Moharana, Sushree Kalyani Kar, Mukesh K. Mishra and R. N. Mahaling

Abstract Graphene and graphene oxide-based polymer composites have remarkable interests over the last one decade due to their excellent mechanical, thermal and electrical properties. The nanometric synthesized fillers with polymeric matrix enhance the structural, morphological and functional properties of the composite materials, and this can be prepared by both ex situ/in situ processes. However, the presence of graphene and graphene oxide even at a very small amount of loadings can give major reinforcement to the final properties of the composites. In addition, graphene is one of the finest material of choice for electronic and energy storage applications in the form of polymer–graphene composites. This chapter reviews and explores the progresses of fabrication of graphene and graphene oxide-based polymer composites with different polymer matrixes such as poly(vinylidene fluoride) (PVDF), epoxy and polyurethane (PU) with special emphasis on their modification, surface alternation and their properties from the scientific literature.

Keywords Graphene · Graphene Oxide · Polymer composites · Surface modification · Electrical conductivity · Mechanical properties

S. Moharana (✉)
School of Applied Sciences, Centurion University of Technology and Management, Balangir Campus 767001, Odisha, India
e-mail: srikantanit@gmail.com

S. Moharana · S. K. Kar · R. N. Mahaling (✉)
Laboratory of Polymeric and Materials Chemistry, School of Chemistry, Sambalpur University, Jyoti Vihar, Burla 768019, Odisha, India
e-mail: rnmahaling@suniv.ac.in

M. K. Mishra
Dalmia Cement Research Centre [DCRC], Research Unit of Dalmia Cement Bharat Limited, Chennai 600116, India

R. N. Mahaling
Nano Research Centre, Sambalpur University, Jyoti Vihar, Burla 768019, Odisha, India

Centre of Excellence, Natural Products and therapeutics, Sambalpur University, Jyoti Vihar, Burla 768019, Odisha, India

© Springer Nature Switzerland AG 2019
S. Sahoo et al. (eds.), *Surface Engineering of Graphene*, Carbon Nanostructures,
https://doi.org/10.1007/978-3-030-30207-8_7

1   Introduction

During last few decades, the developments of graphene and graphene oxide-based polymer nanocomposites (PNCs) have attracted much attention in the modern science and technology because of their ease processing, unique properties and a wide range of applications [1–4]. The polymer nanocomposites were first exposed by Toyota and his co-workers [5], and they give a new direction in the field of material science especially, by using the inorganic nanomaterial as fillers in the preparation of polymer-filled inorganic nanocomposites and several potential applications including automotive, aerospace, construction and electronic industries [6, 7]. Generally, the conventional polymer composites are prepared with the filler size less than 100 nm. However, utilization of the nanofiller-based polymer composites is to give better properties to the pure polymer without compensating in originality and its processibility, inherent mechanical properties and lightweight [8, 9]. The major part for the fabrication of nanofiller-based polymer nanocomposites includes size and property of the nanofillers with the proper interaction between them and the polymer matrix [10]. The polymer composites with carbon nanotubes (CNTs)-based materials have been widely investigated in recent past years. However, the intrinsic bundling of CNTs with the limited accessibility of high-quality nanotubes and high cost restricts their applications [11, 12]. In recent times, graphene has attracted as a promising material for making novel types of PNCs due to its excellent properties and readily available from the carbon-based material of graphite. The incorporation of graphene into the polymer matrix helps to improve the electrical, physical, mechanical and barrier properties of the composites with very low filler contents. Further, the elements of carbon are distributed widely in nature, and its development has laid a solid foundation for plastics, synthetic rubber and fiber-based material. Graphene oxide (GO) is a monolayer of two-dimensional sp^2 and sp^3 bonded carbon atoms, and those are arranged in a hexagonal honeycomb crystal lattice with remarkable structure and has various interesting properties [13, 14]. Graphene is the basic structural unit of some allotropes of carbon including graphite, carbon nanotubes and fullerenes. Graphene had been first established theoretically in 1940, and thought to be building a block of graphite [15]. In 2004, Geim and his co-workers at Manchester University had successfully identified that graphene was a single layer of graphite, which was previously thought to be thermodynamically unstable and could not exist under ambient conditions [16–18].

GO is a two-dimensional nanostructured carbon material which has attracted considerable attention for its various applications due to its remarkable physiochemical properties such as its high thermal conductivity ($\approx$5000 W/(mK), high Young's modulus ($\approx$1 TPa), fracture strength ($\approx$130 GPa), high electrical conductivity ($\approx$6000 S/cm) and high surface area ($\approx$2600 m^2 g). [16, 19, 20]. The fascinating properties of graphene as 2D sheet of sp^2-hybridized carbon, responsible for various advanced applications of future generations such as high speed and radio frequency logic devices, energy conversion devices, electronically and thermally conducting resistant nanocomposites, sensors, electronic circuits, ultra-thin carbon film, flexible electrodes for displays, solar cell and catalysts [21–26]. However, as graphene oxide is composed of stacks of sheets with heavily oxygen-containing functional groups in the form of hydroxyl (–OH), epoxide, diols, ketones and carboxyl functional groups on the basal planes, those can be exfoliated to produce chemically modified graphene oxide nanosheets, which can interact with the polymer matrix [21]. The graphene oxide-based polymer composites exhibit superior mechanical, thermal, gas barrier and electrical properties as compared to pristine polymers [21, 27–29]. It is also revealed that the enhancements of dielectric, electrical and mechanical properties of graphene oxide-filled polymer composites are superior to clay or other carbon-based polymer composites [29–31]. On the other hand, carbon nanotubes (CNTs) exhibit comparable mechanical properties as that of graphene oxide, but graphene oxide is better nanofiller than CNT in certain contest such as it has improved thermal and electrical conductivity [32–36]. These improved properties of the nanocomposites depend on the dispersion of graphene oxide layer in the polymer matrix as well as interfacial bonding between the graphene oxide sheet and polymers. This interfacial bonding between graphene oxide and pure polymer matrix may be an important factor to catch the final properties of the composites. As a consequence of modification, additional functional groups are placed in the edge of the graphene sheets, which makes GO a strongly hydrophilic in nature and then allowed to swell and disperse in water. Therefore, the GO has much attention and acts as nanofiller for polymer composites. The GO sheets can only be dispersed in aqueous solution of water, but it is not favorable for most of the organic polymers [37, 38]. Moreover, the graphene oxide is electrically insulating which makes it unsuitable for the fabrication of conducting nanocomposites. In order to overcome these issues, the surface modification is a necessary step for obtaining a molecular level dispersion of individual graphene sheet in a polymer matrix.

Urged by scientific interest and potentially important for the graphene–polymer composites, related research has increased to a surprising scale, opening new challenges and opportunities for the materials sciences of the hybrid systems. This chapter deals with the brief overview of the synthetic techniques for the preparation of graphene and graphene oxide-based polymer composites with a particular emphasis on their properties.

2 Graphene-Based Polymer Nanocomposites

Polymer nanocomposites (PNCs) containing selected graphene materials have opened a scope of considerable interest in the recent few decades because they have wide ranges of potential applications, especially in the field of electrical and electronic devices. Graphene-based fillers are thus expected to be promising replacement to CNTs, because they are filler-based materials, but have relatively high cost as compared to GO. However, the graphene has a high surface to volume ratio as compared to GO and also more suitable for enhanced electronic, mechanical and thermal properties of polymer matrices. However, a lower content of graphene incorporation onto the polymer matrix is the key factor for the better properties of the composites. In this section, the general techniques for the preparation of graphene-based polymer nanocomposites involve are solution mixing, melt blending and in situ polymerization.

2.1 Solution Mixing

Solution mixing is the simplest technique for the preparation of polymer composites. In this technique, the nanofillers are dispersed with appropriate solvents and get incorporated into the polymer matrix through ultrasonication process. Finally, the resulting solution was placed in an oven for the evaporation of the solvent. During the solution mixing process, compatibility of solvent with the polymer and the filler particles plays a crucial role to achieve better dispersion of nanofiller. The synthetic strategy can be used for the preparation of composite films using various polymers including polyvinyl alcohol (PVA) [39, 40], poly(vinylidene fluoride) (PVDF) [41, 42], polyurethane (PU) [43], poly(methyl methacrylate) (PMMA) [44], etc. Secondly, the solvent removal is a crucial factor for the incorporation of nanofiller into the polymer matrix for improved properties of the composites. Zhao et al. [40] have reported the GO–PVA composite films by direct mixing of GO into the PVA solution at 85 °C and stirring for 6 h to ensure the uniform dispersion of the nanofiller into the polymer matrix. Moreover, the microstructural and X-ray diffraction (XRD) study reveals that most of the graphene oxide sheets are fully exfoliated and uniformly dispersed in the PVA matrix.

In addition, the chemical modification of graphene also leads to change in the properties of graphene by changing the edge characteristics of the sheets and simultaneously improving the solubility and interaction between GO and polymer matrix. Till date, there are various kinds of polar polymers such as poly(vinylidene fluoride) (PVDF), poly(methyl methacrylate) (PMMA) polyurethane (PU), and those have been successfully incorporated to the functionalized GO. Similarly, GO-modified isocyanate, amine, hydroxyl or polymer grafted GO have been synthesized via solution mixing technique [45–47]. However, the surface functionalization is one of the valuable approaches to disperse graphene oxide homogeneously in organic solvents with decreased agglomeration and to get higher loading in the composites. Furthermore, the ultrasonication technique may help to obtain a homogeneous dispersion of graphene oxide in the composites, but longtime exposure to high-power sonication may induce certain defects in graphene which may produce a negative impact into the properties of the composites [48].

2.2 Melt Blending

Melt blending is a promising technique widely used in the industry purposes for the preparation of thermoplastic polymers. This technique shows the mixing of the nanofillers (graphene oxide) with the polymer matrix at high temperature by the application of shear force with a proper dispersion of the nanofillers. This method is free from toxic solvent, but less effective to disperse graphene in the polymer matrix, generally at high filler loading [48]. One more shortcoming of this technique is buckling, rolling or even shortening of graphene sheets during mixing of the polymer solution by strong shear force with high-temperature diffusion; as a result, its aspect ratios are not suitable for better dispersion [49]. For example, Kim et al. [50] have prepared graphene–polyethylene nanocomposites using blending techniques. It is observed that the graphene oxide is fully isolated when blended into the polymer solution and also appears in phase separation and complete exfoliation for the melt-blended samples. The melt-blended composites do not show notably enhanced electrical conductivity even at 1.2 vol.% of graphene loading; whereas, solvent-blended graphene/graphene oxide can reduce the surface resistance of polymers even at low concentration (0.2 vol.%) of graphene sheets. On the other hand, the composites formed by blending techniques show enhanced tensile modulus with an increase in graphene/graphene oxide to polyethylene ratio. Bao et al. [51] have fabricated poly(lactic acid) (PLA)-based graphene oxide nanocomposites via melt-blending techniques with improved properties of the polymers. In this technique, they have used a master-batch approach for uniform dispersion of graphene/graphene oxide into the PLA matrix. It is observed that the resultant composites have uniform dispersion with improved crystallinity, the rate of crystallization, mechanical properties, electrical conductivity and fire resistance. The properties of the polymer–graphene/graphene oxide composites depend on the dispersion and also loading level of graphene/graphene oxide sheets exhibiting percolation threshold at 0.08 wt%. Till date, a wide range of composite systems has been prepared using these techniques including poly(vinylidene fluoride) (PVDF) [41, 42] and its copolymers poly(vinylidene fluoride-co-hexafluropropylene) [P (VDF-HFP)] [52], poly(methyl methacrylate) (PMMA) [44], polystyrene (PS) [53] and polypropylene (PP) [54], etc.

2.3 In Situ Polymerization

In situ polymerization is another often used technique to synthesize variety of graphene-based polymer composites, i.e., epoxy [55], polyaniline [56], polystyrene [53], poly(methyl methacrylate) [54], polyimide [57], polypyrrole (PPY) [58], etc. In this typical process for the preparation of graphene-based polymer composites, the graphene or modified graphene-based nanofiller is first swollen within the liquid monomer under high shear forces and then a curing agent is added to initiate the polymerization or either by heat or radiation. On the other hand, the polymerization for PANI is an oxidative process, thus the oxidative agent such as ammonium per sulfate is used to facilitate the polymerization. In addition, the graphene-based PANI composites may also be prepared by in situ anodic electro-polymerization methods [56]. The covalent bonding between functionalized graphene/graphene oxide sheets and polymer matrix is fabricated via various chemical reactions using in situ polymerization techniques. Generally, one of the major problems of this technique is the

increase of viscosity with the progress of polymerization that hinders the manipulation and limits volume fraction of filler contents [49]. Moreover, some of the
graphene-based composite systems stack onto the layers of the graphene together
due to the weak forces. However, they can be easily dispersed in suitable solvents
such as water, acetone, chloroform, tetrahydrofuran (THF), *N*,*N*-dimethylformamide
(DMF) or toluene. After dispersion polymer can be adsorbed into the graphene sheets
when the solvent is evaporated and the sheets look into the sandwiching between the
polymers in the form of the composites [59].

3 Surface Modification of Graphene

From the above discussion, it is observed that when there is the direct incorporation
of graphene oxide into the polymer matrix there might be the possibility of agglomeration or aggregation due to the high surface energy of GO. However, the oxidation
followed by chemical functionalization of graphene (functional groups attached to
the graphene) will help the uniform dispersion and stabilize graphene to prevent
the agglomeration [60]. Thus, the surface modification is the crucial factor for the
development of better properties for the graphene-based polymer composite systems.
The various types of modified graphene oxide in combination polymer by different
academic and industrial researchers are described as follows:

Graphene oxide layers are successfully grafted on the surface of the polymer matrix using chemical activity treatment and layer-by-layer assembly method.
The physicochemical properties of graphene oxide-modified membrane result in a
decrease in the pore size, an increase in the negative surface charge and hydrophilicity of the modified membrane [61]. These polymer membranes exhibited irreversible
fouling caused by pore blocking, but the modified GO membrane shows antifouling
behavior [62]. Similarly, Bose et al. [63] have prepared PVDF-PBSA membrane by
using graphene oxide and phosphonium derivatives for the application of antibacterial
and antifouling activities. It is observed that the poly(butylene succinate-co-adipate)
(PBSA)-modified graphene composite shows better functional sites with improving
antibacterial and antifouling properties as compared to pristine PVDF [64]. A group
of researchers have fabricated and characterized the tri-layer piezoelectric nano-
generator (PNG) by using n-type graphene (n-Gr)-modified barium titanate (BTO)
nanoparticles. It is observed that the negative charge carrier of n-Gr plays an important role to improve the energy harvesting performance by aligning the dipoles in one
direction. As a result, the fabricated tri-layer PNG shows a maximum output voltage
of 10 V along with the current of 2.5 μA at an applied force of 2 N. The fabricated
device shows good stability even after 1000 pressing releasing cycles. This tri-layer
PNG structure may be a promising material for future piezoelectric generating technologies [65]. Parsi et al. [66] have reported the octadecyl amine functionalized
graphene oxide (GO-ODA)-PVDF dual-layer flat sheet membrane by air gap membrane distillation (AGMD) techniques. It is observed that 3.5 wt% of NaCl solution
as feed at 80 °C [67], unmodified, M1 (low GO-ODA) and M2 (high GO-ODA)-
modified membrane shows water fluxes of 18.2, 13.8 and 16.7 kg/m^2, while the salt

rejection of 88.5%, 96.3% and 98.3%, respectively [68]. This improvement of the PVDF membrane was attributed to the existence of modified GO onto the top layer of the modified membrane. Furthermore, the ODA modified GO-PVDF membrane with high hydrophobicity and low thermal conductivity on the surface of modified membranes, which leads to reduce pore wetting, temperature polarization and heat diffusion across the membrane [66–69]. The surface modification of graphene oxide by the help of polyvinyl pyrrolidone (PVP) with thermoplastic polyurethane (TPU) by oxidation process shows that the graphene is easily dispersed in the TPU by the absorption of PVP on reduced graphene oxide (RGO) as a stabilizer during reduction [70]. However, these PVP-coated RGO composites have low modulus region. The modulus of TPU–GO–PVP and TPU–RGO–PVP composites significantly increases with an increase in the filler content, and the electrical percolation threshold is observed at 0.35 wt% for TPU–RGO–PVP. By comparing this work to the previous literature, the prepared nanocomposites have a good balance between the electrical conductivity and mechanical properties [71].

The modification of graphene oxide by the amine and the microemulsion of epoxy-based nanocomposites are reported, and it is found that the functionalization process effectively promotes the replacement of oxygen with amine groups and simultaneously creates defects in the graphitic structure, which results in the increase in hardness of the synthesized nanocomposites. On the other hand, the amino-functionalized graphene oxide (f-GO) was successfully synthesized and characterized by Fourier transform infrared spectroscopy (FTIR) and X-ray diffraction (XRD) technique. The results show that incorporation of 0.1 wt% of f-GO nanosheets into the epoxy coating causes a significant enhancement in the corrosion resistance of the coating through improving its ionic resistance as well as barrier properties [72, 73].

Hyper branched epoxy resin or the tannic acid epoxy resin (TAE) is synthesized by a simple polycondensation reaction using bio-based tannic acid as a branch generating moiety. In this method, the nanocomposites of TAE with different doses of 0.25, 0.5 and 1 wt% of reduced graphene oxide (RGO) are prepared by solution technique [74]. The TAE is cured with poly(amido-amine) hardener at a temperature of 100 °C followed by post curing at 120 °C. There are different studies such as FTIR, ^{1}H NMR, ^{13}C NMR, and those have been employed to support the hyper branched structure of the synthesized epoxy resin [75]. Moreover, other techniques such as transmission electron microscopy (TEM), X-ray diffractometry (XRD), scanning electron microscopy (SEM) and UV–Vis spectroscopy confirm the formation of polymer nanocomposites. These polymer nanocomposites exhibit better mechanical and thermal performance than the pristine polymers. The tensile strength, scratch hardness, impact strength and initial degradation temperature are found to be improved by the formation of nanocomposite and have potential applications for the use of surface coating [74–76].

4 Different Graphene-Based Polymer Composites

Polymer composites consist of carbon-based nanofillers including nanographite (NG), graphene oxide (GO), carbon nanotubes (CNTs) and carbon fibers (CNF), etc., have been widely investigated in recent past decades. In this chapter, we have discussed a brief overview of graphene-filled polymeric materials consisting of poly(vinylidene fluoride) (PVDF), epoxy and polyurethane (PU). The following section may be helpful to the academic and industrial researchers to develop new graphene-based polymeric composites.

4.1 Poly(vinylidene fluoride) (PVDF)–Graphene Composites

In recent years, poly(vinylidene fluoride) (PVDF) and its copolymers are the multifunctional polymeric materials having ferroelectric, pyroelectric, piezoelectric and superior dielectric properties. These polymers are widely used due to their lightweight, easy processing, flexibility and wearable applications. Poly(vinylidene fluoride) has five different crystal phases such as α, β, γ, δ and ε [77], and among these crystal phases the α polymorph phase has a trans-gauche$^+$–trans-gauche$^-$ (tg$^+$tg$^-$), β-polymorph has an all-trans conformation, and γ polymorph has a conformation between α and β phases (tttg$^+$tttg$^-$) [78]. Out of the five crystal phases of PVDF, the β-phase is very important which provides piezoelectric properties to the composites because of all-trans (TTTT) conformations and unidirectional polarization [79]. Further, this polymer is dissolved in various organic solvents including N-methyl-2-pyrrolidone (NMP), N,N-dimethylformamide (DMF), N,N-dimethyl acetamide (DMAc) and dimethyl sulfoxide (DMSO). PVDF has been widely applied in various fields such as chemical engineering; electronics and food industry [80]. Lin et al. [81] have reported the carbon-based poly(vinylidnene fluoride) composites with high thermal conductivity for the applications in electrical and electronic devices. In these studies, they have prepared the composites by using conducting nanocarbon fillers including zero-dimensional super fullerene, one-dimensional carbon nanotubes and two-dimensional graphene sheets into the polymer matrix via solution blending and compression modeling method. The result shows higher thermal conductivity of nanocarbon fillers–PVDF composites than that of the pristine PVDF matrix [81] as shown in Fig. 1. The thermal conductivity of the resultant composites reaches a maximum value of 2.06 Wm^{-1} K^{-1}, with 20 wt% of two-dimensional graphene sheet (GS) which is nearly ten times higher than that of the PVDF [82].

Interestingly, it is observed that when the graphene content is less than 5 wt%, the cross section of graphene–PVDF composite membranes is relatively smooth and homogeneously dispersed in the polymer matrix as compared to the pristine PVDF membrane (Fig. 2a–c). When graphene content reaches to 10 wt%, the laminar graphene is closely combinative and well contact with each other (Fig. 2d–e). However, when graphene content is 30 wt% (Fig. 2f), a large amount of graphene

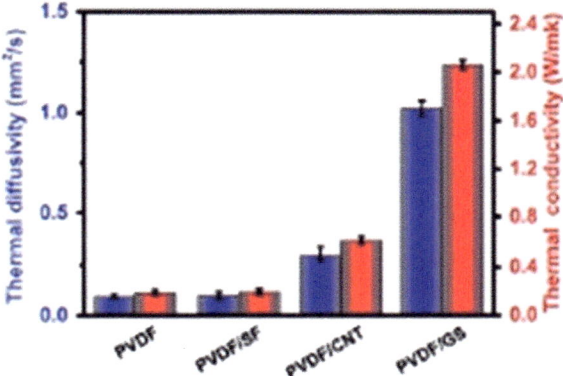

Fig. 1 Thermal conductivity and thermal diffusivity of pristine PVDF and PVDF-based composites. Reprinted with permission from Ref. [81]. Copyright 2016 The Royal Society of Chemistry

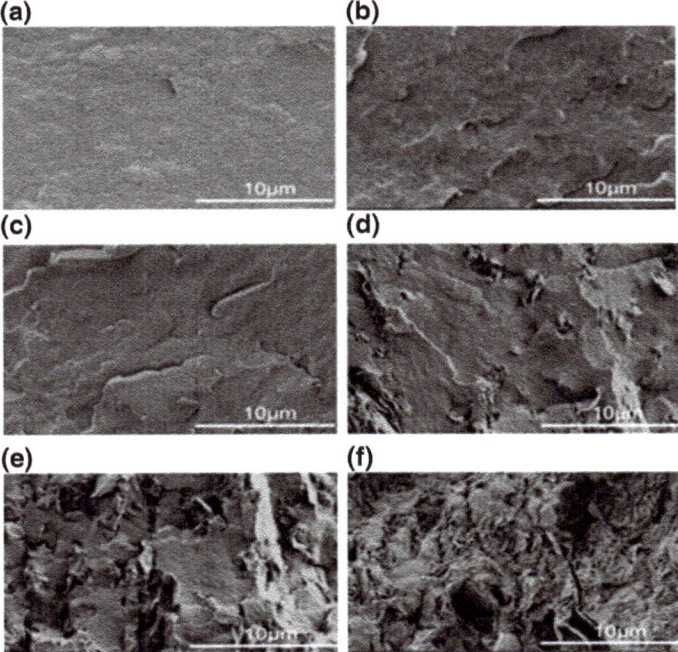

Fig. 2 SEM micrographs of graphene–PVDF composite membrane with different weight percentage of graphene and **a** pristine PVDF, **b** 1 wt%, **c** 5 wt%, **d** 10 wt%, **e** 20 wt%, and **f** 30 wt% of graphene contents, respectively. Reprinted with permission from Ref. [82]. Copyright 2017 Elsevier

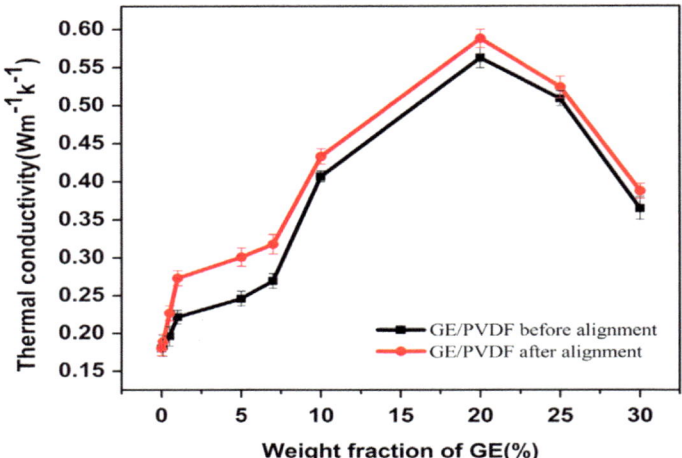

Fig. 3 Thermal conductivity of PVDF membrane and graphene-based PVDF composite membranes with different percentage of graphene contents. Reprinted with permission from Ref. [82]. Copyright 2017 Elsevier

aggregates can be observed in the cross-sectional area [82]. The thermal conductivity of the graphene-based PVDF composite membranes is found to be first increases and then decreases with increase in the graphene content. The decrease with increase in the thermal conductivity may be due to following factors such as (i) agglomeration is aggravated with high graphene content and (ii) thermal conductivity may be decreased by the interfacial thermal resistance between PVDF and aggregated graphene [82–84]. A maximum value of thermal conductivity (0.562 W m^{-1} K^{-1}) is obtained (Fig. 3) when the graphene content reaches 20 wt% [82]. The crystallinity and thermal stability of the graphene-based PVDF composite membrane are improved with the increase in graphene content due to the well dispersion of graphene into the PVDF matrix. The high thermal conductivity of the said composites is expected to be useful in the field of thermal management applications [81].

As already discussed, polyvinylidene fluoride (PVDF) is an important semicrystalline polymeric material with high piezoelectric and ferroelectric properties and has five types of crystalline structures such as α, β, γ, δ and ε [77, 78]. Among them, the β-phase is characterized by all-trans planar zigzag conformations with all the fluorine atoms located on the same side of the polymer chain with polar form and makes the polymer an eminent potential candidate having good piezoelectric and ferroelectric properties [85, 86]. The covalent functionalization of graphene oxide using novel modifier 1H, 1H, 2H, 2H-perfluorooctyltriethoxysilane (PFOES) is introduced for the preparation of PVDF/PFOES/r-GO composites via solution mixing techniques. It is observed that the three phase composite system has a higher dielectric constant than that of r-GO/PVDF composite and pristine PVDF matrix. The dielectric constant of the pristine PVDF is found to be ≈ 10 at 1 kHz, whereas the dielectric

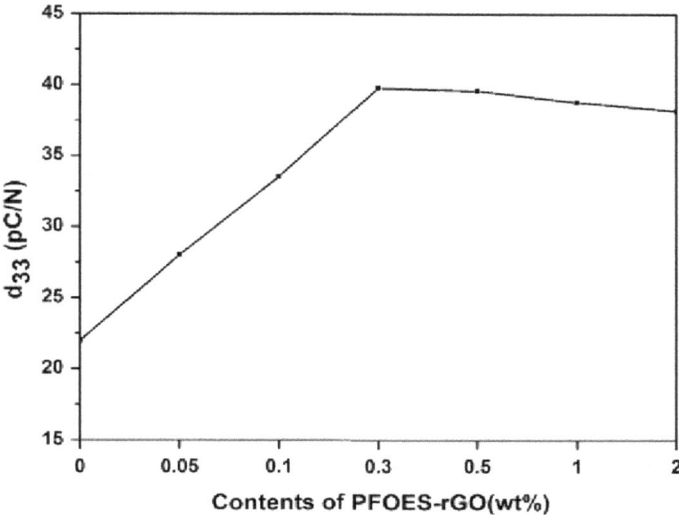

Fig. 4 Piezoelectric constant (d_{33}) of the pristine PVDF and PVDF–PFOES–r-GO (0.3 wt%) composites. Reprinted with permission from Ref. [85]. Copyright 2016 Elsevier

constant value for PVDF–PFOES–r-GO composites is ≈58 for 2.0 wt% of PFOES/r-GO contents at 1 kHz which is nearly about six times than that of pristine PVDF, while the value of the dielectric loss is 0.05. Further, the piezoelectric properties (d_{33}) of PVDF/PFOES/r-GO composite film (Fig. 4) for 0.3 wt% PFOES–r-GO contents (39.8 pC/N) are greatly improved by 370 and 80.9% as compared to the pristine PVDF matrix (22 pC/N). It is confirmed that the modification method for graphene oxide used in this study is very effective for future advanced applications [85]. The graphene oxide nanosheets (GOn)–PVDF nanocomposite films are prepared by solution casting method with various percentages of GOn contents. GOn are obtained via sonication of bulk graphite oxide using N,N-dimethylformamide (DMF) as a solvent. These graphene oxide nanosheets (GOn) can also be used as a nucleating agent for PVDF to produce high performance nanocomposite materials. The selected approach for the preparation and high compatibility between GOn and PVDF results in the formation of purely piezoelectric β-polymorph at 0.1 wt% of GOn content because below this content a mixture of β and α-polymorph is observed. In this study, the Young's modulus and tensile strength of PVDF composites are found to be higher than that of the pristine PVDF matrix and exhibit a gradual increase with increase in GO contents. Besides, the value of Young's modulus for 1 and 2 wt% of GOn (Fig. 5a, b) are found to be 2984 and 3467 MPa, respectively [86]. However, the thermal stability of PVDF polymer is also increased considerably with increasing of GOn content for the preparation of flexible nanocomposite films with low GOn contents, and these improved properties of the materials can be used in the field of piezoelectric applications [86].

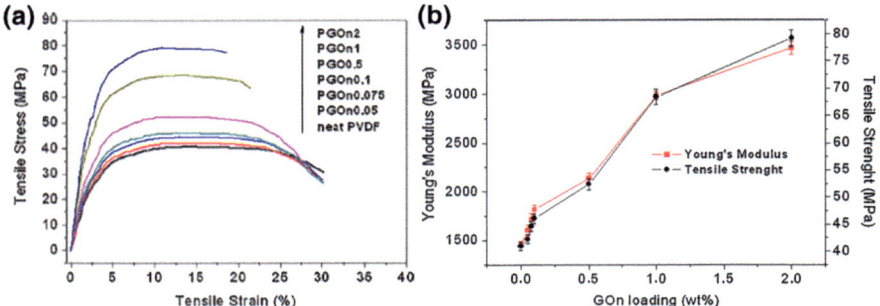

Fig. 5 **a** Stress–strain curves of PVDF–GOn composites and **b** Young's modulus and tensile strength versus various percentages of GOn content. Reprinted with permission from Ref. [86]. Copyright 2012 Elsevier

Seo et al. [87] have reported the piezoelectric properties of PVDF-based carbon materials such as multiwall carbon nanotube composites with high β-phase formation in PVDF. The unit cells of PVDF have two all-trans chains packed with their dipoles pointing toward the same direction and the molecular dipole in the β-phase are entirely aligned in one direction. Thus, this crystal form generates the largest spontaneous polarization and exhibits strong ferroelectric and piezoelectric properties with respect to carbon-based materials. Moreover, the composites made by PVDF and MWCNTs can be used as a smart material not only because of its piezoelectric properties, but also due to the higher level of β-phase formation [88].

Plastic foams or cellular plastics have many attractive properties such as low density, flexibility, thermal insulation, impact damping and good structure stability. Polymer foams are found virtually everywhere in the modern world and are used in a wide variety of applications including packing of food, cushioning of furniture, protective equipment, medical devices, transportation and in thermal insulation [89–93]. Hence, the new types of foam-based composite systems comprising functionalized graphene (f-G) and poly(vinylidene fluoride) (PVDF) are prepared nowadays, and their electrical conductivity and electromagnetic interference (EMI) shielding efficiency have been studied with the different mass fraction of the f-G. It is found that the electrical conductivity of the composites increases with increase in the concentration of f-G with respect to the PVDF matrix. It is observed that the value of the electrical conductivity is about 10^{-4} S m^{-1} for 0.5 wt% of f-G reinforced PVDF composite; whereas, it is 10^{-16} S m^{-1} for pristine PVDF which can be attributed to the formation of the conductive network by the conducting f-G nanofiller in the polymer [94]. However, the electromagnetic interference (EMI) shielding of ≈20 dB for 5 wt% of functionalized graphene reinforced foam composite in X-band region demonstrates its use for lightweight EMI shielding with lower loadings of the nanofillers into the polymer matrix [95]. Poly(vinylidene fluoride) is the most well-known ferroelectric polymer which has attracted increasing attention due to its amazing electro-active properties and a wide range of potential applications [77, 78]. It is used for the preparation of conducting composites by the end use of the self-regulated heaters, over current protections, antistatic and conducting electrodes for lithium batteries. Here, in this work, the functionalized graphene sheets (FGS) with exfoliated graphite (EG) and poly(vinylidene fluoride) matrix are prepared by solution process followed by compression modeling techniques, and their electrical, mechanical and thermal properties have been studied. The experimental results show that the 2 wt% of FGS-PVDF composite shows a lower percolation threshold as compared to EG–PVDF composite having 5 wt% of filler loading [27].

As we know most of the fluorinated hydrocarbons including poly(vinylidene fluoride) (PVDF) and its copolymers such as poly(vinylidene fluoride-co-hexafluoropropylene) (PVDF-HFP) are hydrophilic in nature, but instead of such merits, the dense skin layers of PVDF membranes make it difficult to prepare super hydrophilic surface [96]. The addition of graphene into the PVDF matrix makes it possible to tune the physical properties of the polymer composite. The composites are formed by the diffusion of a non-solvent vapor, either methanol or water into the PVDF–graphene–DMF suspension, followed by freeze drying, to obtain porous materials. Besides, the PVDF–GO composites are fabricated by solution mixing followed by either hot press molding or solution casting techniques. The strong interaction between carboxyl functional groups of the GO and fluorine groups of the PVDF matrix shows the homogeneous dispersion of the GO nanosheets within the PVDF matrix. It is revealed that the PVDF–GO composites with 0.1 and 0.5 wt% of GO exhibit lower conductivity than that of the PVDF–GO composites [97]. Further, the comparison of α-PVDF and β-PVDF–GO composites using FTIR analysis shows that β-PVDF–GO is more stable in solution casting approach than hot press molding technique due to phase transition because β-phase PVDF is a piezoelectric material, whereas α-PVDF is non-piezoelectric in nature [98].

Sulfonated poly(ether–ether–ketones) functionalized graphene oxide (SPG)/poly(vinylidene fluoride) (PVDF) composites have been prepared by a solvent evaporation technique. It is observed that the $-SO_3H$ group of SPG interact with $>CF_2$ dipole of PVDF and disperse graphene uniformly with the PVDF matrix. It also reveals that there is a formation of fiber like spherulitic crystallites with the incorporation of SPG to the PVDF matrix. Moreover, the mechanical properties of the synthesized composite show simultaneous improvement of the stress and the strain at breaking point indicating enhanced toughness of the SPG/PVDF composites as compared to pristine PVDF [99, 100]. The Young's modulus of these composite increases by 160%, and the oxygen permeability coefficient decreases by 91% in 3 wt% of SPG–PVDF film as compared to pristine PVDF [101].

Chiu et al. [102] have prepared binary and ternary graphene nanoplatelet (GNP)-poly(vinylidene fluoride) (PVDF) and PVDF/poly(methyl methacrylate) (PMMA) blended by solution mixing technique. In this work, they compare the binary and ternary composites on the basis of dispersion factor. It is found that the GNP is more randomly dispersed in binary composites than that of the ternary composites. It also exhibits that GNP has higher nucleation efficiency for PVDF crystallization in ternary composites than in binary composite. However, the storage modulus of ternary composites is 53.9% at 25 °C, whereas the binary composites possess 23.1%. In addition, the poly(methyl methacrylate)-functionalized graphene (MG)-poly(vinylidene fluoride) (PVDF) composites show stress at break 157% and Young's modulus 321% for 5% of MG. This work also shows there is random orientation of the graphene sheets (binary system) which changes to parallel orientation (ternary system) for a high volume ratio of graphene content into the polymer matrix. The MG–PVDF composite films exhibit percolation threshold at 3.8% of MG, and the variable range hopping model suggests that the conductivity is contributed from the inter-grain tunneling and hopping between the grains [99].

Pan et al. [103] have fabricated reduced graphene oxide (r-GO)-based material and converted it to high-quality graphene oxide (HQG) by hot press technique and then incorporated the HQG into the PVDF matrix via spin coating technique. From the microstructure analysis and property studies, it is observed that (i) HQG neither contents any defects nor the oxygen-containing functional group on the surface and perform a homogeneous dispersion with well compatible to the PVDF; (ii) the storage modulus of HQG–PVDF composite is found to be twice (2) and eight (8) times higher than that of the r-GO/PVDF and pristine PVDF, (iii) the optimum additive amount of HQG–PVDF composite is found to be between 3 and 5 wt% for development of suitable composites [103]. Further, the dielectric properties of the r-GO-based polymer composite get enhanced by the addition of thionyl chloride to the graphene oxide surface, which causes the synergistic modifications of the structure, chemistry, charge carrier density, electrical conductivity of GO, interfacial interaction and phase of the surrounding matrix in the PVDF composite. The doping of an electron acceptor element like chlorine with various doping strategies enhances the electrical conductivities of the respective composite systems. The experimental results indicate that 0.2 wt% of Cl-doped r-GO/PVDF composite (Fig. 6b, c) has higher dielectric constant (364) and relatively lower dielectric loss (0.077) at 1 kHz as compared to r-GO/PVDF composites [104].

4.2 Epoxy–Graphene Composites

Polymers may also be used as corrosion protection coatings that have already been utilized in various fields in order to expand the life span of metals. However, there is a lack of important property such as interface adhesion to the metal substrates foils for the use of polymeric composite materials for coatings and corrosion protection purposes [105, 106]. Epoxy is a class of polymer that has been extensively used as a corrosion protective coating on various metal substrates. The literature studies show that the remarkable corrosion mitigation property of epoxy can further be enhanced by the incorporation of filler into the polymeric matrix [107]. Epoxy is a class of thermoset that is well known as a powerful adhesive due to its strong mechanical properties, low shrinkage, low density, outstanding adhesion, excellent weather, chemical resistance, high thermal stability and long terms of applications including electrical insulating materials, coating, paintings, adhesives, laminates and encapsulants [75]. However, sometimes the high viscosity, low toughness and inherent brittleness restrain their advanced applications such as solvent-free coatings

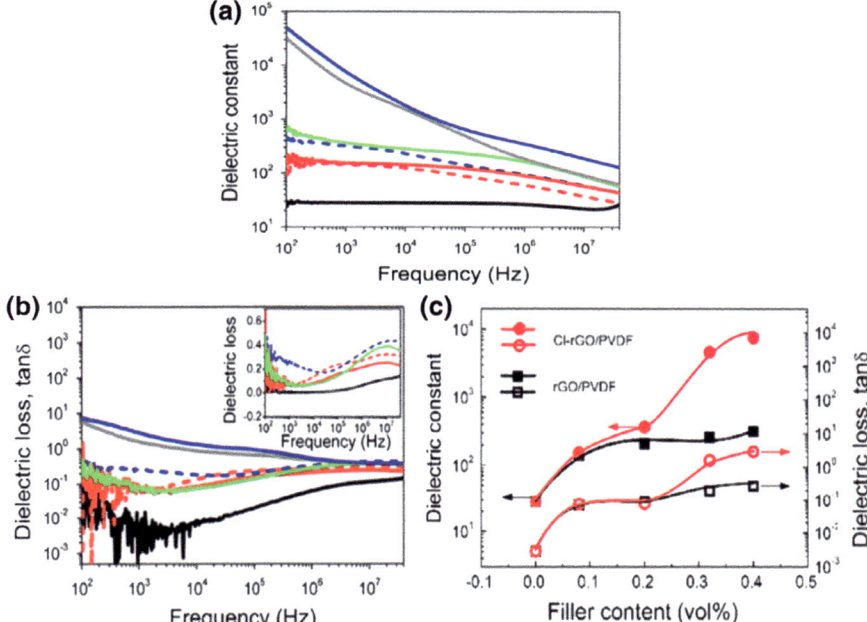

Fig. 6 Dielectric constant (**a**), dielectric loss (**b, c**) of r-GO–PVDF and Cl–r-GO–PVDF composites at 1 kHz with different filler contents. Reprinted with permission from Ref. [104]. Copyright 2015 Elsevier

and structured materials with high strength [108]. Epoxy is widely used as fiber-reinforced polymer, and its highly cross-linked structure on curing gives a remarkable improvement in certain properties such as shear strength, good toughness, reduces the coefficient of friction, excellent adhesion strength, high modulus and good performance at elevated temperatures [109–113]. Because of highly cross-linked structure of epoxy, the material has brittle and poor resistance to crack initiation. To overcome those difficulties, the nanophase is added as a matrix, and this may be in the form of rigid silica particles. Moreover, the addition of nanoparticles not only increases the toughness, but also increases the viscosity with the increase of the filler content due to its small size [113, 114].

Graphene oxide–epoxy composites are prepared, and the level of thermal expansion is studied using a thermo-mechanical analyzer [115]. The results show that the epoxy resin has very poor thermal conductivity but the incorporation of graphene sheets results in significantly enhancement conductivity. The incorporation of 1 wt% of GO to the epoxy resin exhibits improved thermal conductivity to that of 1 wt% of SWNT filler loadings. However, the thermal conductivity of the 5 wt% of GO-based epoxy resin composites reaches ≈1 W/mK, which is four times higher than that of the pristine epoxy resin. This experimental result is consistent with the previously reported literature [115–117]. It is revealed that the value of thermal conductivity increases to 6.44 W/mK by 20 wt% of filler contents. These results exhibit that the graphene-based epoxy composite is a promising thermal interface material for heat dissipation. Moreover, the thermal expansion study of the composites exhibits a similar effect on the other carbon-based materials below the glass transition temperature (T_g).

The effect of dispersion state of graphene on mechanical properties of graphene–epoxy composites has been investigated during past decades. The graphene sheets are exfoliated from graphene oxide by thermal reduction. Different dispersions of r-GO sheets are prepared with and without ball mixing. The result shows that the composites (Fig. 7a–d) which have higher dispersed r-GO show higher glass transition temperature (T_g) and strength than those with lower dispersed r-GO. In particular, the T_g is increased by nearly 11 °C with the addition of 0.2 wt% of well dispersed r-GO content into the epoxy [118]. As expected, the highly dispersed r-GO also produces one or two orders higher magnitude of electrical conductivity than the poorly dispersed r-GO [119, 120]. Since the discovery of the graphene prepared by simple scotch tape method, the composites filled with graphene have been widely investigated to achieve superior electrical, thermal and mechanical properties. The outstanding performance is attributed to the large specific surface area of graphene and its excellent properties such as electrical conductivity and Young's modulus [118]. Recently, there is

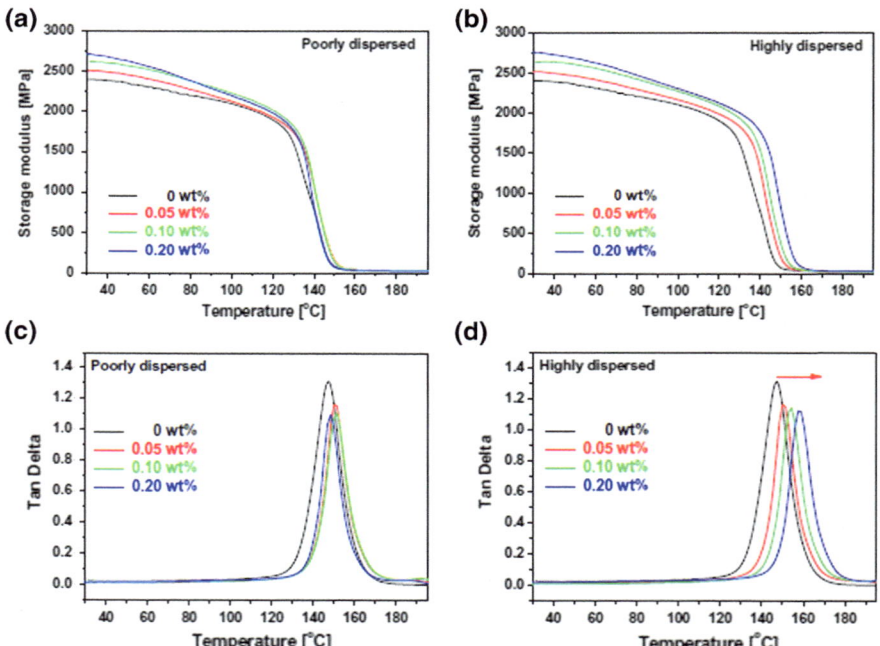

Fig. 7 Dynamic mechanical properties of epoxy-based composites comprising **a, c** poorly dispersed RGO and **b, d** highly dispersed RGO. Reprinted with permission from Ref. [118]. Copyright 2013 Elsevier

a number of research carried out on the development of graphene-based composite materials with superior properties from microscale nanosheet to macroscale bulk composites. So far, numerous studies have been reported on the improvement of the thermal, electrical and mechanical properties of the graphene-based polymer. Fire hazard in aircraft is a critical safety concern due to high flammability of polymer resins. An accidental electrical failure may produce fire, leading to potentially disastrous consequence for aircrafts. This concern has led many researchers to investigate the combustion behavior of epoxy composites using graphene and its derivatives. Liu et al. [121] have reported a reduction of 56.9% in the peak heat release rate (PHRR) of epoxy composites with a loading of 5% by the weight of graphene nanosheets (GNS) in a cone calorimeter testing. The suspension of PHRR by graphene is qualitatively explained by the barrier effect in the literature. Several researchers have revealed that graphene flakes act as blocking walls to hinder the escape of gas volatiles during pyrolysis, therefore slowing down the combustion, which results in lowering the PHRR. This explanation is qualitatively justified as graphene is reported to be impermeable even to helium and has been reported to possess efficient gas barrier properties in different polymers. Some other researchers have reported the improved char structure after adding GNP and also contribute to better combustion performance of the composites especially at lower PHRR contents [122]. The combination of graphene–epoxy gives the experimental and numerical modeling study to quantify the heat and gas barrier effect of GNP on combustion behavior of epoxy composites. Moreover, the 5 GNP–ER composites are investigated including M15 GNP loading of 0.1 wt%, 1 wt%, 3 wt% and M25 and M5 GNP loading of 1 wt%. These composites are carried out under X-ray CT scan, TGA, cone calorimeter and SEM studies. The pyrolysis of the composites is measured by Gpyro software [123]. These results show that there is no aggregation occurs, and the GNP is well dispersed on the surface of epoxy [124].

The X-ray CT scan and scanning electron microscopy (SEM) studies indicate that there are two factors those contribute to GNP effects: (i) GNP in epoxy forcing gas volatiles to move in the tortuous path inside GNP–ER composites and (ii) GNP help to improve the char structure with smaller pore diameter [124, 125]. At 3 wt% of GNP loading, the char structure appears to be compact and continuous with a pore size about one order of magnitude smaller than that of the pristine epoxy. The TGA studies indicate the similar pyrolysis process for GNP–ER composites and pristine epoxy. A method has been proposed to estimate the number of thermal and physical parameters required to simulate combustion behavior of pristine epoxy and GNP–ER composites based on available values in literature or analytical reactions with other parameters, in combination with a parameter optimization process for pristine epoxy. This study provides a new method through numerical pyrolysis modeling, to quantify these two contributions and their effects in reducing PHRR of GNP-ER [124].

In addition, the mechanical properties of graphene–epoxy composites are widely investigated for better result of the composite materials. In order to achieve the optimal enhancement in the property of graphene–polymer composites, several key issues should be resolved, i.e., improved dispersion of graphene, alignment of graphene in surface modification for good interaction. So far, there are varieties of processing methods those have been proposed to disperse graphene in the polymer matrix, such as in situ polymerization, melt mixing and solution blending techniques. The different dispersion states of graphene can lead to a different impact on the mechanical properties of the composites. The r-GO sheets show a bridge between micro-crack and delaminate during the fracture process due to the poor filler–matrix and filler—filler interfaces, which may be the key element of the toughening effect. In this work, the r-GO/epoxy composites are prepared via a ball mixing technique. It is observed that the r-GO shows better dispersion in the epoxy matrix and forms more tortuous and fine river-like structures on the fracture surface to consume fracture energy in comparison with poor dispersed r-GO [118].

4.3 Polyurethane–Graphene Composites

The fabrication of graphene-based nanocomposites attracted a lot of attention to make materials with high barrier properties specifically toward oxygen, carbon dioxide, water vapor and nitrogen. Graphene is poorly water-soluble material due to the presence of few hydrophilic groups and the attractive van der Waals forces between graphene sheets cause graphene to reaggregate. For this reason, graphene oxide (GO) is chosen to replace the graphene. The other reason is that GO is homogeneously dispersed in an aqueous solution which is attributed to the existence of a large number of oxygen-containing functional groups such as hydroxyl, carboxyl, carbonyl and epoxy group [125]. The extraordinary properties of GO and its ability to disperse in various polymeric matrices have provided an opportunity to develop a new class of polymeric nano composites.

Polyurethane (PU) is thermoplastic resilient elastomers, which consists of both soft and hard segments. It possesses a wide range of desirable properties such as elastomeric resistance to abrasion, and excellent hydrolytic stability, due to which it is used as a polymer matrix for the fabrication of nanocomposites film [126]. Generally, PU is used for coatings, adhesives and biomedical applications due to its flexibility and wide applicable temperature range. Again, it is a suitable polymer that can form chemical bonding with the graphite oxide nanoplatelets (GONPs) via reaction between isocyanate groups at the end of the PU chains and oxygenated groups on the GONps. Nowadays, PU materials have been widely used as versatile industrial materials for surface coatings of various substrates, and PU–GONPs composites are much stronger and more protective [127].

The study of the mechanical and thermal properties of the graphene-based polyurethane composites is becoming increasingly popular in both academia and industry. Cho et al. [128] have developed polyurethane-based nanocomposites using functionalized carbon nanotubes with higher thermal and mechanical properties. These composites can be used for enhanced applications in the field of smart actuators. Similarly, Xiao et al. [129] have reported that the incorporation of graphene into epoxy-based polymers enhances the scratching resistance as well as the thermal heating capability of the nanocomposites [130].

PU has been commonly used in a wide range of applications due to its high flexibility and good UV resistance. Many attempts have been made to improve the corrosion resistance and mechanical properties of PU coating by the addition of nanomaterial. The graphene oxides nanosheets are covalently functionalized by (3-glycidyloxypropyl) trimethoxy silane and introduce into the PU matrix to enhance the corrosion resistance and mechanical properties. The modified graphene oxide (f-GO) and unmodified graphene oxide nanosheets (GO) are characterized by Fourier transform infrared (FT-IR) spectroscopy [131, 132], X-ray diffraction (XRD), Field emission-scanning electron microscopy (FE-SEM), UV–visible and thermal gravi-metric analysis (TGA). The physio-mechanical properties of the PU coatings rein-forced with GO and f-GO nanosheets are characterized by dynamic mechanical–ther-mal analysis (DMTA) and tensile stress. The influence of GO and f-GO nanosheets on the fractured surface morphology of the PU coating after the tensile stress is stud-ied by scanning electronic microscopy (SEM) analysis. In addition, the corrosion protection properties of the mild steel panels coated with PU coatings are analyzed by salt spray test and electrochemical impedance spectroscopy (EIS). The results show that the physio-mechanical and anti-corrosion properties of the composites get enhanced after incorporation of f-GO nanosheets into PU. The tensile stress, energy at break, loss factor and storage modulus values are significantly increased by the addition of f-GO nanosheets. The stability of f-GO and dispersion in the PU matrix is also improved after the modification with (3-glycidyloxypropyl) trimethoxy silane and the interfacial bonds between the polyurethane coating and f-GO nanosheets [133].

During the preparation of high-performance polyurethane (PU) nanocomposites using functionalized graphene nanoplatelets (f-GNP) via in situ polymerization tech-niques, it is observed that the f-GNP/PU nanofibers exhibit improved of mechanical, thermal and shape recovery properties. The modulus and thermal stability of the f-GNP/PU nanofibers at 2 wt% of graphene nanoplatelets are ten times greater and 30 °C higher than that of the pristine PU matrix [130]. Polyurethane (PU) can show the shape memory effect by employing crystalline or amorphous segments in the soft segments [134, 135]. Shape memory polyurethane has many advantages such as good flexibility, high deformation, low density and easy processing relative to shape memory alloys, but has certain disadvantages such as shape recovery force and shape recovery rate. To overcome these issues, polyurethane composites can be made by nanomaterials such as carbon nanotubes, graphene, nanoclay and metal particles. The enhancements in the actuating properties of shape memory polymer composites using carbon nanotubes (CNT) and graphene nanocarbon have already been reported in many literature [136–138]. Koerner et al. [137] have reported the enhancement of electrical, thermal and optical stimuli in the CNT incorporated polymer compos-ites. Liang et al. [139] have reported sulfonated graphene-based polymer compos-ites and found that they show good mechanical properties and excellent infrared-triggered and light-triggered actuation due to good solubility and largely restored aromatic networks. However, the poly(ε-caprolactone) (PCL)-based shape memory polyurethane (PU) nanofibers with three kinds of graphene-based materials including graphene oxide (GO), PCL-functionalized graphene (f-GO) and reduced graphene (r-GO) show advanced mechanical and shape memory properties. It is because the incorporation of graphene-based materials to the PU nanofibers increases the modulus and breaking stress as compared to that of pristine PU nanofibers. In the shape memory test, f-GO or r-GO-incorporated PU nanofibers show much faster actuation speed than that of pristine PU nanofibers. The shape recovery time of 1 wt% f-GO or r-GO nanofibers is found to be 8 s, whereas PU nanofibers and GO-incorporated nanofibers were 27 and 13 s, respectively. From this study, it is observed that the incorporation of f-GO or r-GO into shape memory PU nanofibers may effectively enhance both high-speed shape recovery and high mechanical strength. Further, these nanofibers (Fig. 8) show improved shape recovery to the extent of 96%, which is quite significant and offers a wide range of applications of these materials [140]. Nowadays, most of the researchers have attentions toward the improvement of the thermal and mechanical properties of organic polymers by incorporating nano-sized inorganic particles [141]. In this study, the authors have introduced allyl isocyanate into GO and then UV cured with acrylate termini of WPU to introduce GO into the polymer chains and found better dispersion of GO in waterborne polyurethane (WPU). The mechanical, dynamic mechanical and thermal properties of the WPU isocyanated GO (i-GO) nanocomposites are then analyzed and compared with those of physical blends of WPU and GO [142].

The mechanical properties, i.e., the stress–strain behavior of the cast films indicate that the polyurethane (PU) is cross-linked by the 2-hydroxyethyl acrylate (HEA) termini and show typical rubber-like behavior. The physical blending of waterborne polyurethane (WPU) with GO increases the modulus and strength over the polyurethane due to the effects of the fillers. When the GO is introduced into the main chain of WPU, there is further increase in modulus and strength, and this may be due to the dual effects of i-GO particles performing multifunctional chemical cross-links and reinforcing fillers. These multifunctional cross-links provide high cross-link density and elasticity in nature, whereas the fillers augment the rigidity of the nanocomposites [142]. The most remarkable increase is noted with 1% of i-GO,

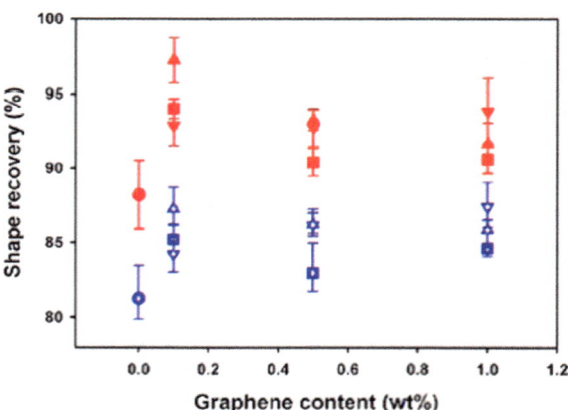

Fig. 8 Shape recovery of the pristine PU and PU-graphene nanofibers. Reprinted with permission from Ref. [140]. Copyright 2014 American Chemical Society

due to its decreased modulus and strength, and it is revealed that the optimum loading is often reported to be around 1% with various nanoparticles [143–145]. The rigid polyurethane foams have excellent features as insulating materials and thus they can be extensively used in building, transportation insulation, refrigeration systems and pipelines. The improved thermal and mechanical properties of rigid polyurethane (RPU) foams may be enhanced by the incorporation of diverse fillers into the PU matrix. The addition of nanoparticles even in a low amount gives rise to promising improvements in the field of physical properties, thermal, mechanical and barrier properties. During last few times, graphene-based materials have gathered a lot of attentions from the day of its discovery (graphene), since graphene is considered as "the thinnest material in the universe" with different kinds of properties [146].

There are various types of parameters those affect the mechanical properties of graphene-based nanocomposites including the structure of graphene, preparation method, dispersion of the filler in the matrix, filler–matrix interaction, the orientation of the fillers and many others, those have already been discussed on above and some of them are going to be discussed. Here, the current status of the intrinsic mechanical properties of the graphene family with that of the preparation and properties of the bulk graphene-based nanocomposites are examined thoroughly. Here, the filler loading is ranging from 0.05 to 0.94 vol.%. The results show that the functionalized graphene (f-GO) enhances the mechanical properties of a composite which is better than that of the reduced graphene oxide (r-GO), due to which their chemical interaction and intrinsic mechanical properties get improved [147]. Wang et al. [148] have reported that the preparation of reduced graphene oxide-polyurethane nanocomposites foam (for solar steam generation driven by local hot spots) is a useful technique for the development of solar energy equipment. In this technique, the reduced graphene nanosheets are covalently cross-linked to the PU matrix with better stability and broad optical absorption. Moreover, the hydrophilic segments and the interconnected pores of the r-GO/PU composite can be worked as water channels for the replenishment of surface water evaporation. Thus, the r-GO/PU composite (Fig. 9) shows the highest evaporation rate, i.e., 3. 4 and 11.7 times higher than the pure water and 1.2 and 1.3 times higher than the GO–PU composites at the densities of 1 and 10 kW m^{-2}, respectively [148]. The dynamic mechanical, dielectric and rheological properties of the PU–GO nanocomposites have been determined and compared to those of the other modified PU composite. It is revealed that the dispersion of GO in the PU matrix is enhanced by introducing 4, 4′-methylenebis (phenyl isocyanate) (MDI). This isocyanate (NCO) group at the end of the linear PU interacts with the oxygen groups on the GO surface. The solution mixing method is employed in order to achieve better dispersion of the GO in the PU matrix where the concentration of the filler varies between 0 and 3 wt%. The dielectric constant of the PU composites (Fig. 10a, b) is significantly increased with increase in frequency by the addition of GO-modified nanosheets (m-GO), because of their high surface area. It is observed that there is a significant increase in dielectric constant value from

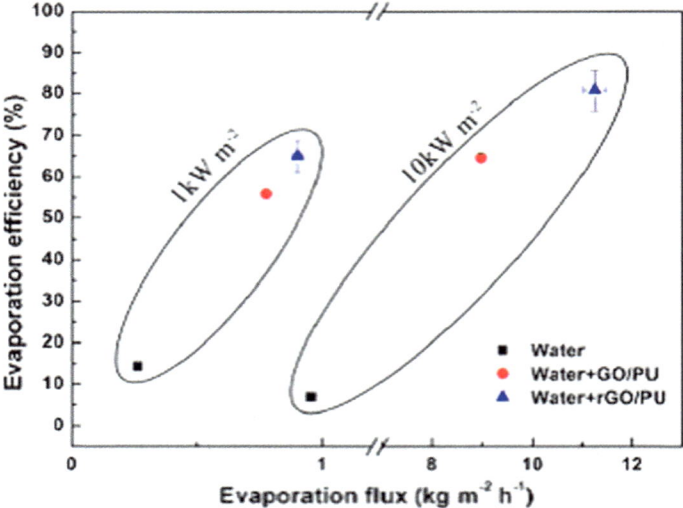

Fig. 9 Evaporation efficiency and evaporation flux of the r-GO–PU composite is compared with water and GO-PU composites under 1 and 10 kW m^{-2} solar illumination. Reprinted with permission from Ref. [148]. Copyright 2017 American Chemical Society

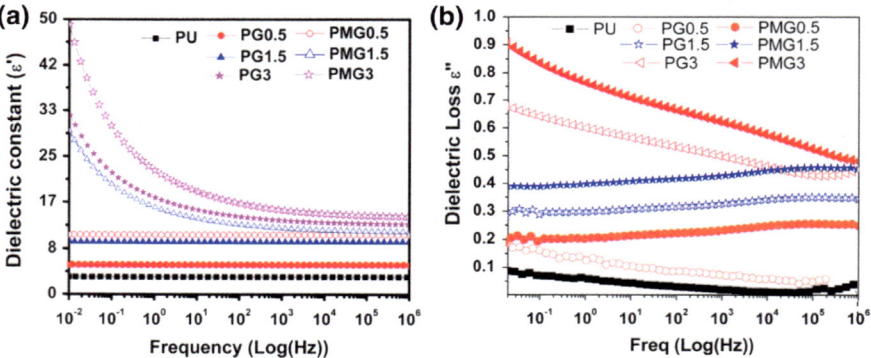

Fig. 10 Dielectric constant (**a**) and dielectric loss (**b**) for pristine PU and PU-based composites. Reprinted with permission from Ref. [149]. Copyright 2014 Elsevier

PG0.5, PG1.5, PG3, PmG0.5 and PmG1.5 to PmG3 films, respectively, at different frequencies. This increase in the dielectric constant of the nanocomposites is due to the motion of free charge carriers because of the continuous conductive paths present on the surface of modified GO nanosheets [149].

5 Summary

In summary, the graphene-based polymeric materials show most challenging and interesting properties for the development of advanced applications with low cost and possibility of mass production. The surface functionalization of graphene is well defined, and the polymers do have specific control over molecular weight and have the ability of better dispersion on graphene surface aroused immense scope for the progress of new and improved strategies in the technological world. This would be useful to make tailor-made materials based on their structure–property relationships. However, most of the graphene-based polymer composites are superior than that of the polymer matrix as well as other carbon filler-based composites. These enhanced properties of the composites are obtained at a low weight percentage of graphene contents which depends mainly on the dispersion rate of the filler phase in the polymer matrix. Thus, the graphene oxide reinforced into the polymer may exhibit undesirable properties as compared to pristine graphene-based polymer composites. Further, the graphene-filled polymer composites have enhanced thermal stability which makes them potential candidates for the applications in the field of electrical and electronic industries.

Acknowledgements The authors gratefully acknowledge the financial support obtained from the DST-FIST and UGC-DRS grant for the development of research work in the School of Chemistry, Sambalpur University and project grant of DST Govt. of Odisha, India. One of the authors SM thanks UGC, New Delhi, for financial support through BSR Research fellowship.

References

1. Szabo, T., Szeri, A., Dekany, I.: Carbon **43**, 87 (2005)
2. Wu, C., Li, F., Zhang, Y., Guo, T.: Thin Solid Films **544**, 399 (2013)
3. David, D., Evanoff, J., Chumanov, G.: Chem. Phys. Chem. **6**(7), 1221 (2005)
4. Li, Y., Huang, X., Hu, Z., Jiang, P., Li, S., Tanaka, T., Appl, A.C.S.: Mater. Interfaces **3**, 4396 (2011)
5. Li, B., Cao, X.H., Ong, H.G., Cheah, J.W., Zhou, X.Z.: Adv. Mater. **22**, 3058 (2010)
6. Robinson, J.T., Perkins, F.K., Snow, E.S., Wei, Z., Sheehan, P.E.: Nano Lett. **8**, 3137 (2008)
7. DeArco, L.G., Zhang, Y., Schlenker, C.W., Ryu, K.: ACS Nano **4**, 2865 (2010)
8. Wu, J., Pisula, W., Mullen, K.: Chem. Rev. **107**, 718 (2007)
9. Berger, C., Song, Z., Li, X., Wu, X., Brown, N., Naud, C., Mayou, D., Li, T., Hass, J., Marchenkov, A.N., Conrad, E.H., First, P.N., de Heer, W.A.: Science **312**, 1191 (2006)
10. Terrones, M., Martin, O., Gonzalez, M.: Adv. Mater. **23**, 5302 (2011)
11. Verdejo, R., Bernal, M.M., Romasanta, L.J., Manchado, M.A.L.: J. Mater. Chem. **21**, 3301 (2011)
12. Liang, J., Wang, Y., Ma, Y., Liu, Z., Cai, J., Zhang, C., Gao, H., Chen, Y.: Carbon **47**, 922 (2009)
13. Kesavan, S., Kumar, D.R., Baynosa, M.L., Shim, J.J.: Mater. Sci. Eng., C **85**, 97 (2018)
14. Lee, J., Kim, J., Kim, S., Min, D.H.: Adv. Drug Deliv. Rev. **105**, 275 (2016)
15. Wallace, P.R.: Phys. Rev. **71**, 622 (1947)

16. Novoselov, K.S., Geim, A.K., Morozov, S.V., Jiang, D., Zhang, Y., Dubonos, S.V., Grigorieva, I.V., Firsov, A.A.: Science **306**, 666 (2004)
17. Zhou, J., Wang, Q., Sun, Q., Chen, X.S., Kawazoe, Y., Jena, P.: Nano Lett. **9**, 3867 (2009)
18. Mermin, N.D.: Phys. Rev. **176**, 250 (1968)
19. El-Kady, M.F., Shao, Y., Kaner, R.B.: Nat. Rev. Mater. **1**, 16033 (2016)
20. Yang, N.J., Swain, G.M., Jiang, X.: Electroanalysis **28**, 27 (2016)
21. Kuilla, T., Bhadra, S., Yao, D., Kim, N.H., Bose, S., Lee, J.H.: Prog. Polym. Sci. **35**, 1350 (2010)
22. Allen, M.J., Tung, V.C., Kaner, R.B.: Chem. Rev. **110**, 132 (2010)
23. Sundaram, R.S., Navarro, C.G., Balasubhramaniam, K., Burghard, M., Kern, K.: Adv. Mater. **20**, 3050 (2008)
24. Geim, A.K., Novoselov, K.S.: Nat. Mater. **6**, 183 (2007)
25. Kim, K.S., Zhao, Y., Jang, H., Lee, S.Y., Kim, J.M., Kim, K.S., Ahn, J.H., Kim, P., Choi, J.Y., Hong, B.H.: Nature **457**, 706 (2009)
26. Wu, J., Agrawal, M., Becerril, H.A., Bao, Z., Liu, Z., Chen, Y.: ACS Nano **4**, 43 (2010)
27. Ansari, S., Giannelis, E.P.: J. Polym. Sci. Part B: Polym. Phys. **47**, 888 (2009)
28. Lee, Y.R., Raghu, A.V., Jeong, H.M., Kim, B.K.: Macromol. Chem. Phys. **210**, 1247 (2009)
29. Eda, G., Chhowalla, M.: Nano Lett. **9**, 814 (2009)
30. Liang, J., Xu, Y., Huang, Y., Zhang, L., Wang, Y., Ma, Y.: J. Phys. Chem. **113**, 9921 (2009)
31. Kim, H., Macosko, C.W.: Polym. **50**, 3797 (2009)
32. Liang, J., Huang, Y., Zhang, L., Wang, Y., Ma, Y., Guo, T.: Adv. Funct. Mater. **19**, 2297 (2009)
33. Balandin, A.A., Ghosh, S., Bao, W., Calizo, I., Teweldebrhan, D., Miao, F.: Nano Lett. **8**, 902 (2008)
34. Scarpa, F., Adhikari, S., Phani, A.S.: Nanotechnology **20**, 065709 (2009)
35. Yu, M., Lourie, O., Dyer, M.J., Kelly, T.F., Ruoff, R.S.: Science **287**, 637 (2000)
36. Li, Y., Wang, K., Wei, J., Gu, Z., Wang, Z., Luo, J., Wu, D.: Carbon **43**, 31 (2005)
37. Marcano, D.C., Kosynkin, D.V., Berlin, J.M., Sinitskii, A., Sun, Z., Slesarev, A., Alemany, L.B., Lu, W., Tour, J.M.: ACS Nano **4**, 4806 (2010)
38. Du, J., Cheng, H.M.: Macromol. Chem. Phys. **213**, 1060 (2012)
39. Mo, S., Peng, L., Yuan, C., Zhao, C., Tang, W., Ma, C., Shen, J., Yang, W., Yu, Y., Min, Y., Epstein, A.J.: RSC Adv. **5**, 97738 (2015)
40. Zhao, X., Zhang, Q., Chen, D.: Macromolecules **43**, 2357 (2010)
41. Wang, Q., Jiang, W., Guan, S., Zhang, Y.: J. Inorg. Organomet. Polym. **23**, 743 (2013)
42. Wang, D., Zhou, T., Zha, J.W., Zhao, J., Shi, C.Y., Dang, Z.M.: J. Mater. Chem. A **1**, 6162 (2013)
43. Vaithylingam, R., Ansari, M.N.M., Shanks, R.A.: Polym. Plastics Tech. Eng. **56**, 1528 (2017)
44. You, F., Li, X., Zhang, L., Wang, D., Shi, C.Y., Dang, Z.M.: RSC Adv. **7**, 6170 (2017)
45. Huang, X., Qi, X., Boey, F., Zhang, H.: Chem. Soc. Rev. **41**, 666 (2012)
46. Goncalves, G., Marques, P.A.A.P., Timmons, A.B., Bdkin, I., Singh, M.K., Emami, N., Gracio, J.: J. Mater. Chem. **20**, 9927 (2010)
47. Park, S., An, J., Piner, R.D., Jung, I., Yang, D., Velamakanni, A.V., Nguyen, S.T., Ruoff, R.S.: Chem. Mater. **21**, 6592 (2008)
48. Singh, V., Joung, D., Zhai, L., Das, S., Khondaker, S.I., Seal, S.: Prog. Mater Sci. **56**, 1178 (2011)
49. An, X., Simmons, T., Shah, R., Wolfe, C., Lewis, K.M., Wasington, M., Nayak, S.K., Talapatra, S., Kar, S.: Nano Lett. **10**, 4295 (2010)
50. Kim, H., Kobayashi, S., Abdur Rahim, M.A., Zhang, M.J., Khusainova, A., Hillmyer, M.A., Abdala, A.A., Wacosko, C.W.: Polym. **52**(8), 1837 (2011)
51. Bao, C., Song, L., Xing, W., Yuan, B., Wilkie, C.A., Huang, J., Guo, Y., Hu, Y.: J. Mater. Chem. **22**, 6088 (2012)
52. Moharana, S., Mahaling, R.N.: Chem. Phys. Lett. **680**, 31 (2017)
53. Zhao, F., Zhang, G., Zhao, S., Cui, J., Gao, A., Yan, Y.: Compos. Sci. Tech. **159**, 232 (2018)
54. Huang, Y., Qin, Y., Zhou, Y., Niu, H., Yu, Z.Z., Dong, J.Y.: Chem. Mater. **22**, 4096 (2010)

55. Chen, L., Chai, S., Liu, K., Ning, N., Gao, J., Liu, Q., Chen, F., Fu, Q., Appl, A.C.S.: Mater. Interfaces **4**, 4398 (2012)
56. Wang, H., Hao, Q., Yang, X., Lu, L., Wang, X.: Nanoscale **2**, 2164 (2010)
57. Ma, L., Wang, G., Dai, J.: High Perform. Polym. **29**, 187 (2017)
58. Liu, Y., Zhang, Y., Ma, G., Wang, Z., Liu, K., Liu, H.: Electrochim. Acta **88**, 519 (2013)
59. Lee, W.D., Im, S.S.: J. Polym. Sci. Part B: Polym. Phys. **45**, 28 (2007)
60. Wei, T., Luo, G., Fan, Z., Zheng, C., Yan, J., Yao, C., Li, W., Zhang, C.: Carbon **47**, 2296 (2009)
61. Gohari, R.J., Halakoo, E., Nazri, N.A.M., Lau, W.J., Matsuura, T., Ismail, A.F.: Desalination **335**, 87 (2014)
62. Song, H., Shao, J., He, Y., Liu, B., Zhong, X.: J. Membr. Sci. **405**, 48 (2012)
63. Samantaraya, P.K., Madras, G., Bose, S.: J. Memb. Sci. **548**, 203 (2018)
64. Li, X., Cai, T., Chung, T.S.: Environ. Sci. Technol. **48**, 9898 (2014)
65. Yaqoob, U., Iftekhar Uddin, A.S.M.. Chung, G.S.: Appl. Surf. Sci. **405**, 420 (2017)
66. Zahirifar, J., Sabet, J.K., Moosavian, S.M.A., Hadi, A., Parsi, P.K.: Desalination **428**, 227 (2018)
67. Shaulsky, E., Nejati, S., Boo, C., Perreault, F., Osuji, C.O., Elimelech, M.: J. Membr. Sci. **530**, 158 (2017)
68. Dogu, M., Ercan, N.: Chem. Eng. Res. Des. **109**, 455 (2016)
69. Bhadra, M., Roy, S., Mitra, S.: Desalination **378**, 37 (2016)
70. Yang, L., Phua, S.L., Toh, C.L., Zhang, L., Ling, H., Chang, M., Zhou, D., Donga, Y., Lu, X.: RSC Adv. **3**, 6377 (2013)
71. Li, X., Deng, H., Li, Z., Xiu, H., Qi, X., Zhang, Q., Wang, K., Chen, F., Fu, Q.: Compos. Part A **68**, 264 (2015)
72. Ferreira, F.V., Brito, F.S., Franceschi, W., Simonetti, E.A.N., Cividanes, L.S., Chipara, M., Lozano, K.: Surf. Interfaces **10**, 100 (2018)
73. Ramezanzadeh, B., Niroumandrad, S., Ahmadi, A., Mahdavian, M., Moghadam, M.H.M.: Corros. Sci. **103**, 283 (2016)
74. Boro, U., Karak, N.: Prog. Org. Coat. **104**, 180 (2017)
75. De, B., Karak, N.: J. Mater. Chem. A **1**, 348 (2013)
76. Thakur, S., Karak, N.: Ultratough, ductile, castor oil-based. ACS Sustainable Chem. Eng. **2**, 1195 (2014)
77. Shi, K., Sun, B., Huang, X., Jiang, P.: Nano Energy **52**, 153 (2018)
78. Salimi, A., Yousefi, A.A.: Polym. Test. **22**, 699 (2003)
79. Abolhasani, M.M., Shirvanimoghaddam, K., Naebe, M.: Compos. Sci. Technol. **138**, 49 (2017)
80. Chiong, S.J., Goh, P.S., Ismail, A.F.: J. Nat. Gas Sci. Eng. **42**, 190 (2017)
81. Cao, Y., Liang, M., Liu, Z., Wu, Y., Xiong, X., Li, C., Wang, X., Jiang, N., Yu, J., Lin, C.T.: RSC Adv. **6**, 68357 (2016)
82. Guo, H., Li, X., Li, B., Wang, J., Wang, S.: Mater. Design **114**, 355 (2017)
83. Yan, H., Tang, Y., Long, W., Li, Y.: J. Mater. Sci. **49**, 5256 (2014)
84. Yu, J., Huang, X., Wu, C., Jiang, P.: IEEE Trans. Dielect. Electr. Insul. **18**, 478 (2011)
85. Huang, L., Lu, C., Wang, F., Dong, X.: J. Alloy. Compd. **688**, 885 (2016)
86. Achaby, M.E., Arrakhiz, F.Z., Vaudreuil, S., Essassi, E.M., Qaiss, A.: Appl. Surf. Sci. **258**, 7668 (2012)
87. Ahn, Y., Lim, J.Y., Hong, S.M., Lee, J., Ha, J., Choi, H.J., Seo, Y.: J. Phys. Chem. C **117**, 11791 (2013)
88. Kim, G.H., Hong, S.M., Seo, Y.: Phys. Chem. Chem. Phys. **11**, 10506 (2009)
89. Mills, N.J., Fitzgerald, C., Gilchrist, A., Verdejo, R.: Compos. Sci. Tech. **63**, 2389 (2003)
90. Lee, J., Kim, J., Kim, S.W., Shin, C.H., Hyeon, T.: Chem. Commun. **5**, 562 (2004)
91. Nikje, M.M.M.A., Tehrani, Z.M.: Des. Monomers Polym. **13**, 249 (2010)
92. John, B., Nair, C.P.R., Ninan, K.N.: Mater. Sci. Eng. A **527**, 5435 (2010)
93. Shunmugasamy, V.C., Gupta, N., Nguyen, N.Q., Coelho, P.G.: Mater. Sci. Eng. A **527**, 6166 (2010)

94. Eswaraiah, V., Sankaranarayanan, V., Ramaprabhu, S.: Macromol. Mater. Eng. **296**, 894 (2011)
95. Yang, Y., Gupta, M.C., Dudley, K.L., Lawrence, R.W.: Adv. Mater. **17**, 1999 (2005)
96. Zha, D., Mei, S., Wang, Z., Li, H., Shi, Z., Jin, Z.: Carbon **49**, 5166 (2011)
97. Chee, W.K., Lim, H.N., Huang, N.M., Harrison, I.: RSC Adv. **5**, 68014 (2015)
98. Wu, L., Alamusi, Xue, J., Itoi, T., Hu, N., Li, Y., Yan, C., Qiu, J., Ning, H., Yuan, W., Gu, B.: J. Intell. Mater. Syst. Struct. **25**, 1813 (2014)
99. Layek, R.K., Samanta, S., Chatterjee, D.P., Nandi, A.K.: Polym. **51**, 5846 (2010)
100. Yue, X., Zheng, W.T., Yu, W.X., Hua, L.G., Zhang, Y.J.: Chem. Res. Chin. **26**, 491 (2010)
101. Layek, R.K., Das, A.K., Park, M.J., Kim, N.H., Lee, J.H.: Carbon **81**, 329 (2015)
102. Chiu, F.C., Chen, Y.J.: Compos. Part A **68**, 62 (2015)
103. Yu, C., Li, D., Wu, W., Luo, C., Zhang, Y., Pan, C.: J. Mater. Sci. **49**, 8311 (2014)
104. Wu, Y., Lin, X., Shen, X., Sun, X., Liu, X., Wang, Z., Kim, J.K.: Carbon **89**, 102 (2015)
105. Bellucci, F., Nicodemo, L., Monetta, T., Kloppers, M.J., Latanision, R.M.: Corros. Sci. **33**, 1203 (1992)
106. Roy, D., Simon, G.P., Forsyth, M., Mardel, J.: Adv. Polym. Techn. **21**, 44 (2002)
107. Alhumade, H., Nogueira, R.P., Yu, A., Elkamel, A., Simon, L., Abdala, A.: Prog. Org. Coat. **122**, 180 (2018)
108. Chen, S., Xu, Z., Zhang, D.: Chem. Eng. J. **343**, 283 (2018)
109. Martini, C., Ceschini, L.: Tribol. Int. **44**, 297 (2011)
110. Samyn, P., Baets, P.D., Paepegem, W.V., Degrieck, J., Schepdael, L.V., Gerber, A., Leendertz, J.S.: Lubr. Sci. **12**, 119 (2006)
111. Samad, M.A., Sinha, S.K.: Wear **270**, 395 (2011)
112. Aderikha, V.N., Shapovalov, V.A.: Wear **271**, 970 (2011)
113. Deng, S., Ye, L., Friedrich, K.: J. Mater. Sci. **42**, 2766 (2007)
114. Christy, A., Purohit, R., Rana, R.S., Singh, S.K., Ranae, S.: Mater. Today: Proc. **4**, 2748 (2017)
115. Wang, S., Tambraparni, M., Qiu, J., Tipton, J., Dean, D.: Macromolecules **42**, 5251 (2009)
116. Yu, A., Ramesh, P., Sun, X., Bekyarova, E., Itkis, M.E., Haddon, R.C.: Adv. Mater. **20**, 4740 (2008)
117. Yu, A., Ramesh, P., Itkis, M.E., Elena, B., Haddon, R.C.: J. Phys. Chem. C **111**, 7565 (2007)
118. Tang, L.C., Wan, Y.J., Yan, D., Pei, Y.B., Zhao, L., Li, Y.B., Wu, L.B., Jiang, J.X., Lai, G.Q.: Carbon **60**, 16 (2013)
119. Zaman, I., Kuan, H.C., Meng, Q., Michelmore, A., Kawashima, N., Pitt, T., Zhang, L., Gouda, S., Luong, L., Ma, J.: Adv. Funct. Mater. **22**, 2735 (2012)
120. Zaman, I., Phan, T.T., Kuan, H.C., Meng, Q., La, L.T.B., Luong, L., Youssf, O., Ma, J.: Polymer **52**, 1603 (2011)
121. Liu, S., Fang, Z.P., Yan, H.Q., Wang, H.: RSC Adv. **6**, 5288 (2016)
122. Ran, S., Chen, C., Guo, Z., Fang, Z., Barrier, C.: J. Appl. Polym. Sci. **131**, 40520 (2014)
123. Lautenberger, C., Pello, C.F.: Fire Saf. J. **44**, 819 (2009)
124. Zhang, Q., Wang, Y.C., Bailey, C.G., Yuen, R.K.K., Parkin, J., Yang, W., Valles, C.: Compos. Part B **146**, 76 (2018)
125. Song, W., Wang, B., Fan, L., Ge, F., Wang, C.: Appl. Surf. Sci. **463**, 403 (2019)
126. Bandyopadhyay, P., Park, W.B., Layek, R.K., Uddin, M.E., Kim, N.H., Kim, H.G., Lee, J.H.: J. Membr. Sci. **500**, 106 (2016)
127. Cai, D., Yusoh, K., Song, M.: Nanotechnology **20**, 085712 (2009)
128. Cho, J.W., Kim, J.W., Jung, Y.C., Goo, N.S.: Macromol. Rapid Commun. **26**, 412 (2005)
129. Xiao, X., Xie, T., Cheng, Y.T.: J. Mater. Chem. **20**, 3508 (2010)
130. Yadav, S.K., Cho, J.W.: Appl. Surf. Sci. **266**, 360 (2013)
131. Stankovich, S., Piner, R.D., Nguyen, S.T., Ruoff, R.S.: Carbon **44**, 3342 (2006)
132. Ramezanzadeh, B., Ghasemi, E., Mahdavian, M., Changizi, E., Moghadam, M.H.M.: Chem. Eng. J. **281**, 869 (2015)
133. Haghdadeh, P., Ghaffari, M., Ramezanzadeh, B., Bahlakeh, G., Saeb, M.R.: J. Taiwan Inst. Chem. Eng. **86**, 199 (2018)
134. Liang, C., Rogers, C.A., Malafeew, E.: J. Intell. Mater. Syst. Struct. **8**, 380 (1997)

135. Lee, B.S., Chun, B.C., Chung, Y.C., Sul, K., Cho, J.W.: Macromolecules **34**, 6431 (2001)
136. Ni, Q.Q., Zhang, C.S., Fu, Y., Dai, G., Kimura, T.: Compos. Struct. **81**, 176 (2007)
137. Koerner, H., Price, G., Pearce, N.A., Alexander, M., Vaia, R.A.: Nature Mater. **3**, 115 (2004)
138. Jung, Y.C., Kim, J.H., Hayashi, T., Kim, Y.A., Endo, M., Terrones, M., Dresselhaus, M.S.: Macromol. Rapid Commun. **33**, 628 (2012)
139. Liang, J., Xu, Y., Huang, Y., Zhang, L., Wang, Y., Ma, Y., Li, F., Guo, T., Chen, Y.: J. Phys. Chem. C **113**, 9921 (2009)
140. Yoo, H.J., Mahapatra, S.S., Cho, J.W.: J. Phys. Chem. C **118**, 10408 (2014)
141. Jeong, H.M., Lee, S.H.: J. Macromol. Sci. Part B Phys. **42**, 1153 (2003)
142. Yoon, S.H., Park, J.H., Kim, E.Y., Kim, B.K.: Colloid Polym. Sci. **289**, 1809 (2011)
143. Kyung Lee, S., Yoon, S.H., Chung, I., Hartwig, A., Kim, B.K.: J. Polym. Sci.: Part A: Polym. Chem. **49**, 634 (2011)
144. Jang, M.K., Hartwig, A., Kim, B.K.: J. Mater. Chem. **19**, 1166 (2009)
145. Jung, D.H., Jeong, H.M., Kim, B.K.: J. Mater. Chem. **20**, 3458 (2010)
146. Calvo, M.S., Blasco, V., Ruiz, C., Paris, R., Villafane, F., Ángel, M., Perez, R.: Eur. Polym. J. **97**, 230 (2017)
147. Papageorgiou, D.G., Kinloch, I.A., Young, R.J.: Prog. Mater Sci. **90**, 75 (2017)
148. Wang, G., Fu, Y., Guo, A., Mei, T., Wang, J., Li, J., Wang, X.: Chem. Mater. **29**, 5629 (2017)
149. Sadasivuni, K.K., Ponnamma, D., Kumar, B., Strankowski, M., Cardinaels, R., Moldenaers, P., Thomas, S., Grohens, Y.: Compos. Sci. Tech. **104**, 18 (2014)

Application of Reduced Graphene Oxide (rGO) for Stability of Perovskite Solar Cells

Bhim P. Kafle

Abstract Rapid increase in performance of methyl ammonium lead halide perovskite solar cells (PSCs) has been observed in the last decade, reaching overall power conversion efficiency up to 23%. This made them the serious alternative to the silicon-based solar cells. However, there are still several challenges to address before commercialization of this kind of solar cell technology. For example, PSCs showed very poor tolerance against moisture, oxygen, temperature, and UV illumination. The graphene and its derivatives [in particular, graphene oxide (GO) and reduced graphene oxide (rGO)] demonstrate several key features that may address above-underlined issues prevailing in PSCs and also in organic photovoltaic solar cells (OPVs), leading to enhance the energy conversion efficiency of these third-generation photovoltaic devices. In this context, this review highlighted on the key features of graphene, GO, and rGO and also provides overview of very latest successful examples of their applications as TCO, electron transport layer or hole transport layer mainly in PSCs. Finally, the potential issues and the perspective for future research in graphene-based materials for PSC applications are presented.

Keywords Graphene · Reduced graphenen oxide · Transparent conducting oxide · Electron transparent conducting oxide · Hole transport layer · Perovskite solar cell

1 Brief Review on Perovskite Solar Cells

1.1 Progress in Devising Efficient and Stable Perovskite Solar Cells

Solar cell technology based on thin film of organic–inorganic metal halide perovskite materials is considered to be one of the most promising for harnessing solar energy to fulfill the energy needs in a sustainable manner [1–5]. One of the main reasons to

B. P. Kafle (✉)
Department of Chemical Science and Engineering, School of Engineering, Kathmandu University, Dhulikhel, Nepal
e-mail: bhim@ku.edu.np

© Springer Nature Switzerland AG 2019
S. Sahoo et al. (eds.), *Surface Engineering of Graphene*, Carbon Nanostructures,
https://doi.org/10.1007/978-3-030-30207-8_8

consider perovskite-based photocells promising, specifically, is due to more intensely absorption of solar radiation over a broad region of the solar spectrum than other light absorbers commonly used in solar cells [6, 7]. Besides, it exhibits very long electron–hole diffusion length so that most of the photogenerated charge carriers are collected [8, 9] and also, more importantly, the solar cells of this kind can be fabricated with inexpensive precursor materials with the simple solution-processing at low temperatures [1]. As a result, achievement of high efficiency of perovskite solar cells (PSCs) is possible, even with the thinner than other types of solar cells (due to which material cost is reduced). With the use of this crystalline material as a sensitizer, the photo-conversion efficiency (PCE) increased from of 3.8% [1] in 2009 to 23.3% in late 2018 in single-junction architectures [10] exceeding the maximum efficiency achieved in single-junction silicon solar cells. Solar cells of this kind are therefore the fastest-advancing solar technology to date within the time period of 10 years.

1.2 Components of Perovskite Solar Cells

In the PSCs, primarily, a thin layer of the methyl ammonium lead halide perovskite, $CH_3NH_3PbX_3(X = Cl, Br, I)$, molecule is sandwiched between electron transport layer (ETL), and hole transport layer (HTL). When perovskite thin film is exposed to solar radiation, as elucidated in Fig. 1, it absorbs the solar radiation and the electron–hole pairs (known as excitons) are generated, which then separated and migrated to opposite directions: the photoelectron migrates toward ETL and hole migrates toward HTL. To collect electrons from these electron–hole pairs created by photon interaction at a perovskite/ETL interface, usually mesoporous TiO_2 nanoparticles (a versatile inorganic compound semiconductor with a wide band gap) have been used. While on the other hand, mostly, the holes are transported via organic p-type semiconducting mediators (e.g., Spiro-MeOTAD) [2–5].

Fig. 1 Schematic view of a thin-film perovskite solar cell having a flat layer of perovskite sandwiched between n-type and p-type charge selective contacts. The generation of electron and hole pairs and extraction process is also demonstrated

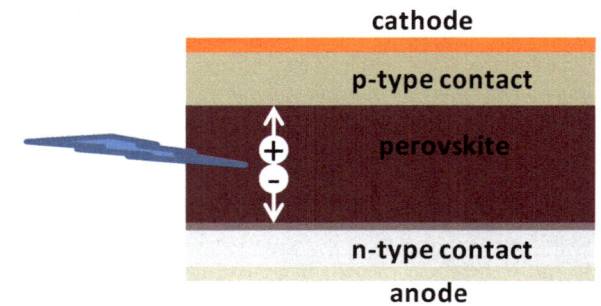

1.3 Problem in Organic p-type Semiconductor

Long term stability of this class of solar cell that limits commercial viability is, however, a key question that still remains open to date. Degradation is understood to be due to the fact that PSCs are highly susceptible to intense solar radiation (degradation due to induced stress under long term solar light illumination), high temperature, and humidity. Moreover, the use of Spiro-OMeTAD as a HTL for PSCs has also some limitations [11]. Because the doped Spiro-OMeTAD suffers degradation, which negatively affects the cell performance (PCE degrades gradually with time and the cell dies in a couple of days). For example, when exposed to ambient conditions with relative humidity around 40% and temperature around 25 °C, the dopant (Li-TFSI) used in Spiro-OMeTAD is deliquescent and tends to dissociate from the Spiro-OMeTAD, thus consequently reducing its (dopant's) effect. The other most used dopant, i.e., tert-butyl pyridine (TBP) also possesses severe limitations. Therefore, to simultaneously enhance both the stability and efficiency, scientists from solar cell community have been extensively revisiting the components of the conventional PSC designs (glass/TCO/TiO$_2$/perovskite/Spiro-MeOTAD/metal contact). Specifically, they are not only modifying composition of perovskite itself but also altering charge transport materials (both ETL and HTL) and transparent conducting oxide (TCO) layer as well.

1.4 Alternative Approach: Metal Oxides as Charge Transport Layers

In this context, studies from last 5 years inferred us that substitution of organic charge transport materials with inorganic one (for both the ETL and HTL) could be the most appropriate approach to address the underlined issues of PSCs. Because, charge transport layers from inorganic materials found to possess high chemical stability, high charge mobility and are low cost, when compared with organic counterparts. Indeed, since its discovery in 2009, numerous important contributions have been published which clearly demonstrated the possibility of application of inorganic semiconducting materials for collecting photogenerated charge carriers. For example, CuI, CuSCN, and NiO$_x$ have been tested for hole transport layer and found that these materials helped to enhance product's stability significantly, maintaining reasonable efficiency. You et al. [12], utilizing p-type NiO$_x$ nanoparticles as HTL, demonstrated improvement in photon to current conversion efficiency (PCE $= 16.1\%$) and stability against water and oxygen degradation, when compared with devices with organic charge transport layers. After 60 days storage in air at room temperature, their devices retain about 90% of their original efficiency, which is a significant progress as the photocells with organic p-type semiconductors usually degrade within a few days.

In summary, above study revealed that the inorganic charge transport layer, besides its own higher stability, it also plays a key role in protecting the perovskite photoactive layer from exposure to such environments, thus allowing to achieve highly stable perovskite-based photovoltaic cells.

1.5 Alternative Approach: Graphene-Based Materials for Perovskite Solar Cells

Alternatively, carbon-based materials (which are metal-free) has also been considered a promising alternatives to both the organic- and metal-based charge transport materials: Indeed, evidences have shown that graphene-based products can be used as transparent conducting oxide, ETL, HTL and back contact in PSCs [13–19], organic solar cells (OSCs) [20] and dye-sensitized solar cells (DSCs) [21 and references therein], and polymer-based solar cells [22, 23]. Recently published review article highlights possibilities of all kinds of components in these kinds of solar cells [24].

For example, reduced graphene oxide, rGO, as HTL (hole receiving layer and electron blocking layer) found to be quite effective, although the observed efficiency of PSCs with this material as HTL ($\eta =$ ca. 16%) is less by about 30% when compared to best efficiency of PSCs (ca. 23%) with PEDOT:PSS as HTL. Besides, they also can withstand against environmental parameters and also protect perovskite layer, being an overcoat layer [16–19].

The reasons for adaption of these graphene-based materials for charge transport application is mainly due to their possession of unique electronic, optical (high optical transparency: transmittance above 90%, tunable band gap, from zero to insulator level) and electrical properties (sheet resistance of undoped graphene films is in the range of 150–500 Ω/sq. and high conductivity (10^4 S/cm)) [25, 26], processability with green-chemistry methods and low costs [24, 27]. Also, to note, within the graphene-based materials (GBMs) family, rGO stands out for its good electronic properties, processability, low-costs, possibility to use green-chemistry methods for its production, and suitability for large scale applications [19, 23–26, 28].

In this context, this chapter mainly aims to cover the properties (optical-, electronic- and electrical-properties), most commonly used synthesis methods, characterization techniques of GO and rGO, and their applications in perovskite and organic photovoltaic solar cells.

2 Properties of Graphene-Based Materials

2.1 Properties of Graphene, GO, and rGO

Graphene is the base material for building all the graphite derivatives including GO and rGO. These graphene-based materials have attracted the interest of material scientists and chemists because of its heterogeneous chemical and electronic structures, along with the fact that it can be processed in solution in substitution of expensive techniques such as chemical vapor deposition and epitaxial growth and high-temperature thermal decomposition [29]. For example, crystalline graphene is highly conducting, whereas its derivative GO can be considered as insulating and disordered. Moreover, as shown in Fig. 2b, c, GO consists of the attachment of functional groups such as hydroxyl, carboxyl, and epoxide [24, 29, 30] due to which it is highly soluble in water, allowing exfoliation into individual sheets. Moreover, the availability of these oxygen-containing functional groups on the basal plane and the sheet edge allows GO to interact with a wide range of organic and inorganic materials in non-covalent, covalent and/or ionic manner so that functional hybrids and composites with unusual properties can be readily synthesized [19, 22, 23, 30, 31].

To be noted, due to the presence of covalently bound oxygen (see Fig. 2b, c) and the displacement of the sp³-hybridized carbon atoms slightly above and below the original graphene plane, the graphene oxide sheets are expected to be "thicker" than the pristine graphene sheet, which is atomically flat with a well-known van der Waals thickness of about 0.34 nm (It is reported, however, that depending on the

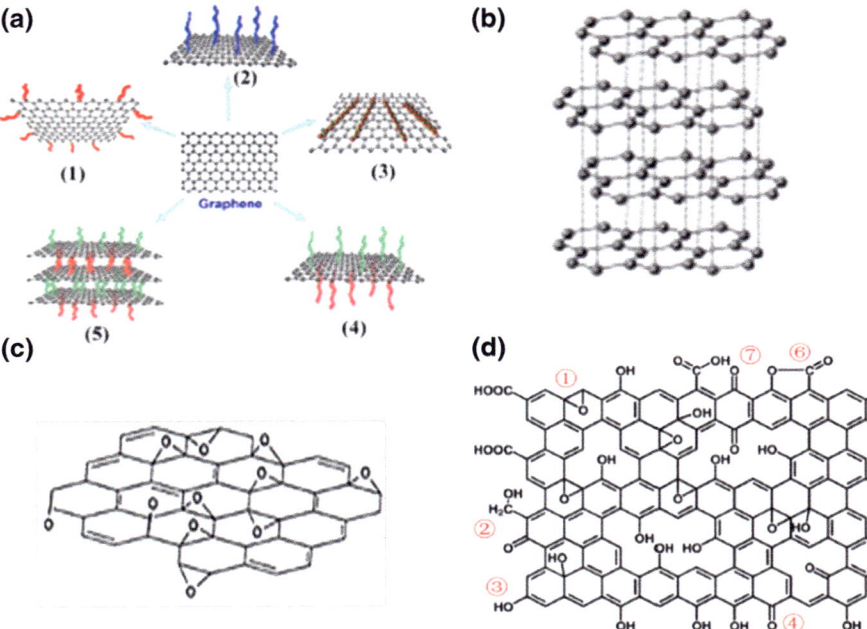

Fig. 2 a Functionalization possibilities for graphene: (1) edge-functionalization, (2) basal-plane functionalization, (3) non-covalent adsorption on the basal plane, (4) asymmetrical functionalization of the basal plane, and (5) self-assembling of functionalized graphene sheets. b–d Structural models of graphite, graphene oxide sheet (only with epoxy functional group), and with varieties of other oxygen-containing functional groups, respectively

synthesis method, the thickness and its size of the graphene sheet varies from several nanometers to centimeters) [24–27, 29, 30].

In connection to structural and electrical properties, GO possesses both conducting π-*states* from sp^2 carbon sites and a large energy gap between the σ-*states* of its sp^3-bonded carbons. Therefore, by varying the ratio of the sp^2 and sp^3 fractions (reduction either by chemical or thermal means), one can tune its band gap, and therefore, controllably to transform GO from an insulator to a semiconductor and to a graphene-like semi-metal [30, 32, 33]: The new product after reduction of GO is known as reduced graphene oxide (rGO). For example, employing vacuum deposition method, Eda et al. [32] demonstrated that very thin films (1–2 layers) of rGO were semiconductor in nature, whereas the thicker films were semi-metal. Also, these authors showed that the sheet resistance of the films can be tuned over six orders of magnitude and the transparency from 60 to 95%. Besides, it is also shown that thermally reduced (at 1100 °C in Ar/H$_2$ atmosphere) GO yields transparent and conducting thin films with transparency of ca. 60% and sheet resistance of about 600 Ω/sq. [29 and reference therein].

However, although GO can be chemically or thermally reduced to achieve graphene-like properties, it still possesses residual defects such as oxygen atoms lying along off plane of the graphene's basal plane, Stone–Wales defects (pentagon–heptagon pairs) and holes due to loss of carbon (in the form of CO or CO$_2$) from the basal plane which limit the electronic quality. But, recent advances in efficient reduction maintain high transparency and electrical conductivity (e.g., besides above-mentioned achievements, charge mobility values in field-effect devices found to be 5000 cm^2 V^{-1} s^{-1} [34]).

Owing to these unique characteristics, interest on GO and rGO have been increased, for its possibilities in a wide landscape of applications ranging from plastic electronics, optical materials, and solar cells to biosensors. Therefore, for customizing (manipulating) as per application requirements, it is crucial for understanding the unique optoelectronic properties of GO.

Means of probe: As indicated above, there is wide variability in the type and coverage of the oxygen-containing functional groups on GO, primarily arising from differences in preparation processes (the functional groups on the basal plane consist mainly of hydroxyls and epoxies). A consequence of the non-uniform coverage by oxygen-containing functional groups of the graphene basal plane is that ordered small (2–3 nm) sp^2 clusters isolated within the sp^3 C–O matrix can be readily observed also by Raman spectroscopy, besides scanning tunneling microscopy, high-resolution transmission electron microscopy (HRTEM), and transport studies. The reduction of GO leads to creation of new sp^2 clusters, through removal of oxygen, which provide percolation (infiltration) pathways between the 2 and 3 nm sp^2 domains already present and the newly formed isolated and eventually percolating sp^2 clusters mediate the transport [30, 34].

3 Methods for Synthesis, Exfoliation, and Reduction of GO

3.1 Synthesis Methods for Graphite Oxide

For synthesizing GO both the "*top-down*" and "*bottom-up*" approaches have been employed. The *top-down* approach was originally developed by Hummers' et al. in 1958 [35] and later modified by Fujiwara et al. [36], *known as modified Hummers' method.* This method involves production of graphite oxide mostly by exfoliation of bulk graphite. In particular, with Hummers' method, GO is synthesized through the oxidation of graphite using oxidants including concentrated sulfuric acid, nitric acid, and potassium permanganate. GO, produced by this oxidative treatment of graphite, still retains a multiple-layered structure of graphene (that is, the material is still in the graphite oxide form), but is much lighter in color than graphite due to the loss of electronic conjugation brought about by the oxidation [37].

Although many other methods for synthesis of graphite oxide have been developed, Hummers' method (and *modified Hummers' method*) remains key point of interest because it is an easy method of producing large quantities for industrial applications.

3.2 Methods for Exfoliation of Graphite Oxide to GO and Thin-Film Preparation

Due to hydrophilic nature of the oxygenated graphite oxide, it is readily exfoliated in water and forms colloidal suspensions of thin sheets in water. However, forming homogeneous and stable dispersion of very thin sheets of graphene oxide (or single sheet of graphene oxide); ultrasonic treatment of colloidal suspension with sufficient dilution (at concentrations 1 mg/mL in water yields sheets with uniform thickness of nearly 1 nm) is necessary. Also, to note, these sheets are reported to be different from graphitic nanoplatelets or pristine graphene sheets due to their low electrical conductivity [37–39]. Then thus obtained suspension is spin casted or spray casted on desired substrates followed by thermal treatment for moisture removal and also for reduction.

GO film can also be prepared alternatively by vacuum filtration technique which is recently developed method by Eda et al. [27]. Briefly, for deposition of uniform film from a GO suspension, this method employs filtration technique of GO suspension supported by a vacuum pump (filtration of the suspension containing the GO sheets is carried out through a 25 nm pore size mixed cellulose ester membrane). As the liquid is filtrated, the GO sheets block some of the pores where the permeation rate of the liquid is dramatically reduced. Therefore, in regions, where the pores are not blocked, the permeation rate is enhanced until covered by GO sheets. Once the GO films are deposited on the ester membrane, it is necessary to transfer them onto substrates. The transfer process is performed by placing the ester membrane with the film side down onto the substrate (the material onto which film to be casted such as glass and plastics. The films are well adhered to glass and plastic substrates). The ester membrane is then etched away by sequential acetone washes. For other techniques, readers are referred to Refs. [25–27, 29, 38, 40].

3.3 Reduction of GO

Moisture and the functional groups attached on the surface or at the edge of GO can be removed by either thermal or chemical reduction routes [41–45] (see in the Sect. 4.2). The former method is highly effective, but usually needs a temperature above 800–1000 °C under inert gas atmosphere [41, 43–45]. This reduction produces films with better electrical conductivity than that of chemical reduction [41]. Whereas, on the other hand, the chemical reduction process can be realized at a temperature lower than 100 °C, which is extremely important for practical applications, since graphene films are commonly supported on substrates such as plastics that cannot stand high temperature. Strong alkali agents, hydrazine and sodium borohydride (NaBH$_4$) have been well accepted as effective chemical reducing agents for the de-oxygenation [41]. During the hydrazine reduction of graphene oxide sheets dispersed in water, the brown-colored dispersion turns black and the reduced sheets aggregate and eventually precipitates. The formation of precipitate of the reduced sheets is due to their progression from hydrophilic to more hydrophobic as a result of oxygen removal and thus increased incompatibility with the aqueous medium. As solvation of the reduced sheets decreases, the inter-sheet hydrophobic interactions cause them to aggregate. But, it is reported that the rGO films prepared by GO suspension after reduction with hydrazine and also with sodium borohydrate possess rather low electrical conductivity. Alternatively, GO can also be reduced with hydroiodic acid (HI). The obtained rGO film prepared by reduction by this method found to show much higher electrical conductivity and C/O atomic ratio than those of films reduced by other chemical methods, while maintaining very good flexibility with a much higher tensile strength than that of the original GO film [44].

4 Characterization Techniques of GO and rGO

4.1 Surface Area Analysis of the rGO

Surface area measurement of the reduced GO sheets is carried out via nitrogen gas absorption. The measurement carried by Stankovich and coworkers yielded a BET value of 466 m^2/g [37]. This high specific surface area has been considered to be an indication of GO exfoliation prior to the reduction. However, it is still lower than the theoretical specific surface area for completely exfoliated and isolated graphene sheets (2620 m^2/g [37]). This low value is assigned to agglomeration of the graphene oxide sheets upon reduction.

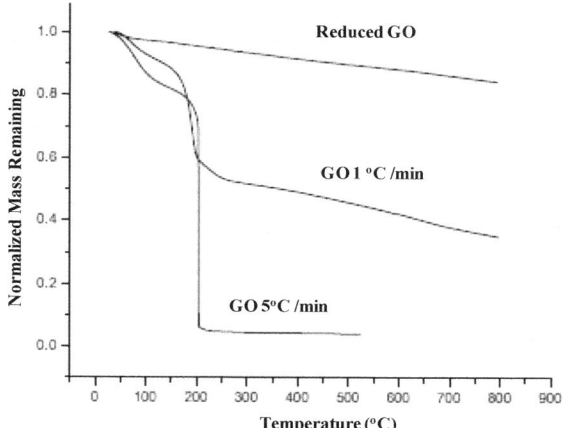

Fig. 3 Normalized TGA plots for thermal reduction process of GO and the chemically reduced rGO (upmost curve) Fig. 3, adopted from Ref. [37]

4.2 Thermal Gravimetric Analysis (TGA)

TGA is performed under a nitrogen flow (100 mL/min) using a TA Instruments and the mass is recorded as a function of temperature. For example, TGA measurement performed by Stankovich and coworkers employed TA Instruments (TGA-SDT 2960) on sample sizes from 5 to 6 mg, and the mass was recorded as a function of temperature. The samples were heated from room temperature to 800 °C at 5 °C/min. To avoid thermal expansion of the GO due to rapid heating, GO samples were also heated from room temperature to 800 °C at 1 °C/min. As shown in Fig. 3, the reduction of oxygen in GO starts at even below 100 °C, but the major loss occurs from 200 °C. This is due to pyrolysis of the loosely bound oxygen-containing functional groups, yielding CO, CO_2, and steam [37, 41]. The thermal decomposition of GO assumed to be accompanied by a vigorous release of gas, resulting in a rapid thermal expansion of the material, causing both a large volume expansion and a larger mass loss during a more rapid heating regime. On the other hand, the removal of the oxygen functional groups by rGO, reduced by chemical reduction, results in better thermal stability (see in Fig. 3): Apart from a slight mass loss below 100 °C, which can be attributed to the loss of adsorbed water, no significant mass loss occurs when this material is heated up to 800 °C [37, 38].

4.3 Optical Spectroscopy

A key requirement for material science research is the ability to identify and characterize all the members of a product's family and their applications, both at the laboratory- and at mass-production scale, which applies to the graphene family as well. As we discussed in the previous sections, graphene family has already many members (i.e., its derivatives) such as GO, rGO, metal/inorganic element doped GO, graphene quantum dots, and graphene-based composites and knowing fundamental characteristics of individual species of this class or carbon is very crucial for both the fundamental and application points of view. However, to be appealing, a characterization tool must be non-destructive, fast, with high resolution, and give the maximum information on structural and electronic behaviors. Raman-, FTIR-, UV-Vis-, photoluminescence spectroscopies are some of the most widely used tools used in material characterization. Below we discuss some of these techniques to characterize graphene family.

4.4 Raman Spectroscopy

Indeed, among the spectroscopic methods, Raman spectroscopy has been becoming the standard tool in the fast-growing field of material science and engineering. It is due to the fact that this technique is a fast, non-destructive, and is a high-resolution tool for characterization of the lattice structure, electronic properties, optical properties, and phonon properties of nanomaterial-based materials (including graphene family). Similar to other materials, the roles of the Raman spectroscopic study of materials this class of carbon may be summarized as follows: (1) It enables us to identify the vibrational modes using laser excitation (the external perturbations to graphene flakes could affect their lattice vibrations and also band structures [46–49]). (2) Through Raman imaging, it provides spatially distributed information concerning the number and quality of layers (confinement in graphene nanostructures), local stress within such a multilayer graphene structure, doping level and nature of defects in graphene (point defects, line defects, and edges on the D, G, and 2D modes), and growth mechanism of graphene flakes. (3) Besides, it also enables to monitor the property modification of three-dimensional (3D) diamond, graphite, 2D graphene, carbon dots, epitaxial graphene on SiC, GO, and rGO prepared by various methods. For example, in the process of making graphene, be it from mechanical cleavage, epitaxial growth, chemical vapor deposition, chemical exfoliation, all kinds of carbon species can in principle prevail. Moreover, while shaping graphene and graphene derivatives into devices, unwanted by-products and structural damage can also be created. Therefore, it is necessary to have a structural reference for which Raman spectroscopy has been employed as a common well-accepted technique.

As an example, here, we describe the Raman scattering processes of the entire first- and second-order modes in the intrinsic graphite, GO and rGO. In Fig. 4, we present the Raman spectra measured by three different research groups [11, 37, 50]. As demonstrated in the panels of (a), (b), and (c), we can clearly notice differing features of graphite, GO, rGO, and the GO films of monolayer and multilayer graphene (Fig. 4c). Specifically, as depicted in Raman spectra in Fig. 4a recorded by Stankovich et al. [37], the significant structural changes were observed during the chemical processing from pristine graphite to GO, and then to the rGO: Raman spectrum of the pristine graphite (in Fig. 4a: top panel) displays a prominent G peak as the

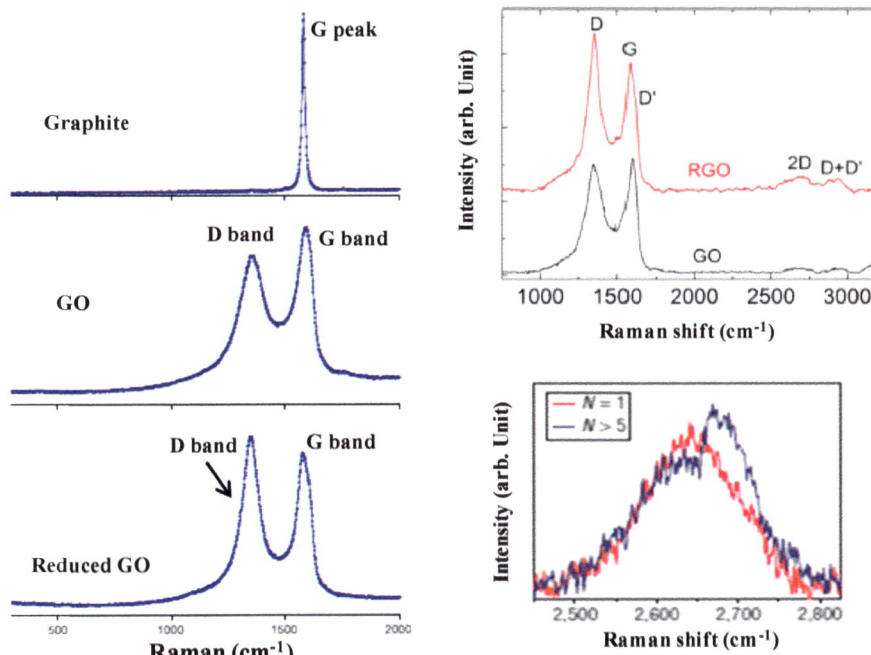

Fig. 4 Panel (**a**: left) shows the Raman spectra of SP-1 graphite (top), GO (middle), and the reduced GO (bottom) [37]. **b** (right-upper) Raman spectrum of GO (black line) and RGO (red line) (Adopted from Ref. [11]). The main peaks are labeled. **c** (right-bottom) The Raman spectra of single layered ($N = 1$) and multilayered ($N > 1$) GO (adopted from Ref. [50])

only feature at 1581 cm^{-1} which corresponds to the first-order scattering of the E_{2g} mode of graphite [37]. Whereas in the Raman spectrum of GO (a: center panel), the G band was broadened and slightly shifted to 1594 cm^{-1}. In addition, the D band at 1363 cm^{-1} becomes prominent, indicating the reduction in size of the in-plane sp^2 domains, which is understood to be due to the oxidation of the graphite. Moreover, the Raman spectrum of the rGO (a: bottom panel) also contains both G and D bands (at 1584 and 1352 cm^{-1}, respectively) which were prevailed in GO spectrum; however, with an increased intensity ratio of D peak to G peak (D/G ratio) compared to that in GO. This change suggests a decrease in the average size of the sp^2 domains upon reduction of the exfoliated GO, and this is due to creation new graphitic domains that are smaller in size to the ones present in GO before reduction, but more numerous in number [37]. Moreover, the spectra presented in Fig. 4b were measured by Palma et al. [11] from GO (black line) and rGO (red line) layers in the range of 500–3500 cm^{-1} of GO. Similar to spectra recorded by Stankovich et al. [37] in Fig. 4a, the spectrum of rGO (red line in Fig. 4b) shows D peak at $(D) = 1351$ cm^{-1}, and G peak $(G) = 1581$ cm^{-1}. It is understood that the narrower FWHM of D peak (FWHM(D) = 115 cm^{-1}) and down-shifted frequency $((G) = 1581$ cm$^{-1})$

with respect to GO (FWHM(D) = 160 cm^{-1} and $\overline{(G)}$ = 1591 cm^{-1}, respectively, is an indication of partial sp^2 restoration [51, 52]. Moreover, there is an increase in the $I(D)/I(G)$ ratio passing from GO($I(D)/I(G)$ = 0.95) to RGO ($I(D)/I(G)$ = 1.18) indicating a medium level of defects and presence of both crystalline and amorphous carbons [52]. The G can be roughly considered constant as a function of disorder, be in related to the relative motion of sp^2 carbons, while an increase of $I(D)$ is directly linked to the presence of sp^2 rings [51]. Thus, an increase of the $I(D)/I(G)$ ratio is considered as the restoration of sp^2 rings. Moreover, it is also reported that $2D$ peak (not shown here) is highly intense in single layer defectless graphene but with stacking of the layers, interaction increases, and this $2D$ peak splits into multiple peaks making it shorter and wider [53]. Figure 4c-bottom also represents Raman spectrum recorded by Eda et al. [27] rGO layers (number of layer, N = 1 and N = 5) in the range of 2500–2800 cm^{-1}. One can clearly notice the shift of the wave number in the peak position and variation in the peak intensity.

4.5 FTIR Analysis

Being complement of Raman, Fourier-transform infrared spectroscopy (FTIR) has also been employed for further chemical characterization (presence of functional groups that are IR active). Moreover, this technique has also been utilized to study the degree of reduction of graphene oxide films by analyzing the number of oxygen-containing functional groups present on the surface of rGO by observing the results for various modes of vibration [54].

For example, the peaks of GO originated at wave numbers ~1720 cm^{-1}, 1410 cm^{-1}, 1226 cm^{-1}, and 1050 cm^{-1} in FTIR spectrum (Fig. 5: left) are related to the stretching vibration of C=O, C–O–H, C–O (epoxy) and C–O (alkoxy) functional groups, respectively [45, 55]. Also, reported were other two peaks at about ~1558 cm^{-1} and 3600 cm^{-1} which have been assigned to C=C and O–H stretching, respectively [45]. After reduction process, the loss (partially) of oxygen-containing

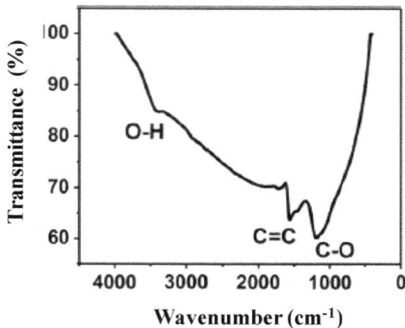
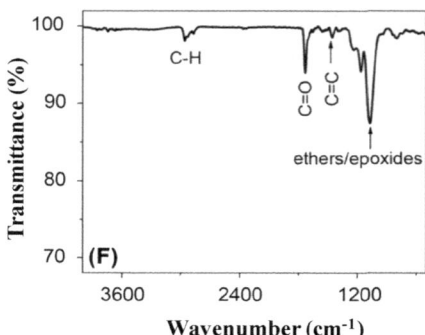

Fig. 5 FTIR spectrum of rGO produced from Ref. [45], of rGO bilayer film produced from Ref. [45]

functional groups is expected as observed (see Fig. 5 right) by Kumar et al. [55]. FTIR spectrum of graphite possesses peak associated strong OH vibration, which mainly expected to come from humidity and/or adsorbed water and this OH signal changes its shape and becomes wider upon graphite oxidation. But, becomes narrower after reduction of GO [55].

4.6 UV-Vis Spectroscopy (For Optical and Electronic Properties)

The optical properties of GO and rGO thin films can be tuned by varying the film thickness, chemical composition, average flake size, and film morphology: The films can be made insulating, semiconducting, or semi-metallic, while maintaining their optical transparency, by appropriately tuning the deposition and reduction parameters [45]. Further, preparing composite of rGO sheets with other materials allow manipulation of rGO-host interactions, offering additional degrees of freedom in device functions [55]. It is well known that an ideal sheet of graphene is a zero-gap semiconductor [54]. That is, the valence and the conduction bands of graphene meet at the Fermi energy, where the density of states (DOS) of the two bands vanishes linearly [56]. In multilayered graphene, overlap of the two bands gives rise to finite DOS at the Fermi level, rendering it semi-metallic [57]. In contrast to the large body of work devoted to understanding the electronic structure of graphene and graphite [51], some guanine contribution has been made to understanding the electronic structure of graphene on GO and rGO and their multilayers. The fundamental properties of GO and rGO such as the energy band gap, transmittance as a function of film thickness will be illustrated. In the following sections, experimental results providing insight into the electronic structure and transport properties of GO and rGO are highlighted, while discussing their manifestation in device properties.

A colloidal solution (suspension) of GO dissolved in water yields dark brown to light yellow color, depending on the concentration, whereas that of rGO appears black, signifying appreciable differences in the electronic properties [40]. Also, it is understood that individual sheets of GO and rGO (in liquid medium) exhibit nearly identical properties [55]. Similar changes in the physical appearance are also observed in thin films. Figure 6a, b represent the UV-Vis absorption spectrum of GO dispersed in water, dye methyl formamide (DMF), NMP, ethylene glycol, THF, and ethanol measured by Paredes et al. Similarly, Fig. 6b is the corresponding Tauc plot (produced from one of the spectrum in Fig. 6a) [54]. The spectra are plotted in the wavelength range from 200 to 1000 nm, except for DMF and NMP, for which data appear at ≥ 265 nm as a result of the impossibility of properly compensating for the strong absorption of both solvents (DMF and NMP) at smaller wavelengths. The formation of stable graphene dispersions in these solvents enables them to measure and understanding the optical absorption nature, correlating the observed peaks with electronic transitions as follows.

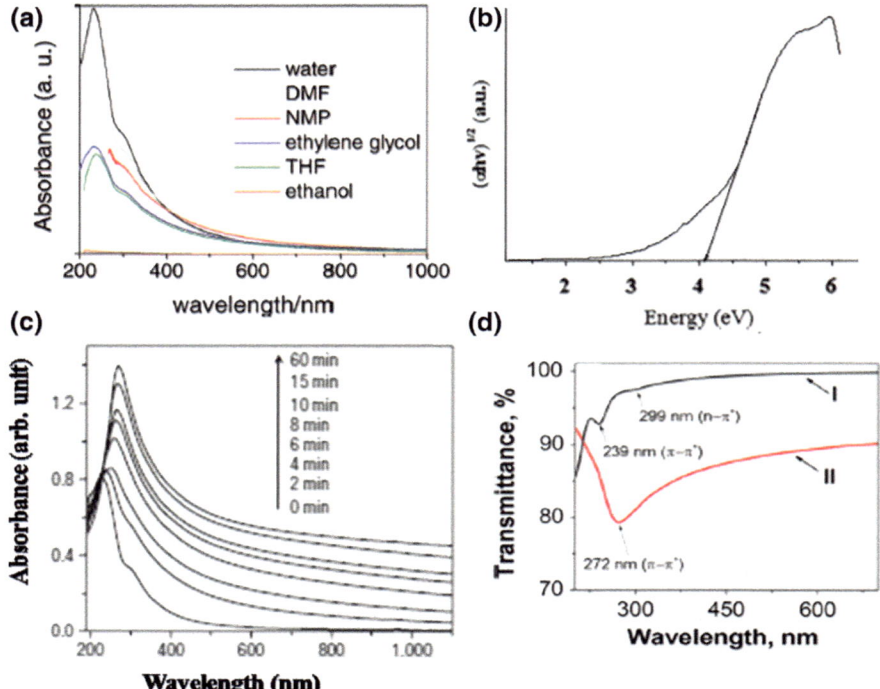

Fig. 6 a UV-Vis absorption spectra of as-prepared graphite oxide dispersed in different solvents by means of 1 h ultrasonication. **b** Tauc plot derived from the measured absorbance curve (adopted from Ref. [40]). **c** GO reduction as a function of time. The absorption peak of the GO dispersion at 231 nm gradually shifted toward longer wavelength (redshift) to 270 nm and the absorption in the whole spectral region (>231 nm) increases with reaction time, suggesting that the electronic conjugation within the graphene sheets is restored upon hydrazine reduction [59]. **d** UV-Vis transmittance spectra of GO bilayer film encapsulated with nanometer-thick copolymer (GO-P) (I), and rGO-P bilayer film (II) (produced from Ref. [40])

4.6.1 Peaks Associated with Electronic Transition

Most of the spectra presented in Fig. 6a exhibits a peak at about 230 nm which is ascribed to $\pi \rightarrow \pi^*$ transitions of the aromatic C=C bond (it is also characterized by the $\pi-\pi^*$ plasmon peak [29]). Moreover, the spectra in Fig. 6a also consists of a shoulder at 300 nm of n $\rightarrow$ π^* transitions of the C=O bond [54, 56]. From this UV-Vis-IR spectroscopic study, it can be inferred that the optical absorption of GO is dominated by $\pi-\pi^*$ transitions, which typically give rise to an absorption peak between 225 and 275 nm (4.5–5.5 eV). The contribution of conduction electrons is minimal in the visible/near-UV photon energy range [49].

4.6.2   Effect of Solvent in Disperability and Transition Peaks

Besides nature of electronic transition, the UV-Vis spectra also have been used to test the dispersability in a particular solvent. For example, among the spectra in Fig. 6a, recorded in different solvents, the spectrum of the ethanol which exhibits almost no absorption provides the negative case, i.e., no dispersion of GO in this solvent. The spectra recorded in water, ethylene glycol, DMF, NMP, and THF confirm that the as-prepared graphite oxide material was successfully dispersed in these solvents (among the five successful solvents, water displays the best dispersibility because it provided the highest absorption intensity and therefore the largest amount of suspended graphite oxide, followed closely by DMF and NMP. Ethylene glycol and THF exhibited very similar dispersibility toward as-prepared graphite oxide, although they are noticeably smaller than those of the other three solvents [58]. In any case, the concentration of dispersed graphite oxide in all of these solvents is estimated to be in the range of a few tenths of 1 mg/mL, and it can be increased through further sonication (Such concentration values are often considered for most practical uses of this material). Note also the peak positions of in spectra for these solvents hint redshift due to solvent effect. Moreover, the optical band gap determined from measured absorbance (with the aid of *Tauc plot*) found to be 4.1 eV (see Fig. 6b) [57].

Figure 6c shows the change of GO dispersions as a function of reaction time (Absorbance curves were recorded from just the initiation of reduction process of GO till 1 h, after hydrazine addition). The absorption peak of the GO dispersion at 231 nm gradually shifted toward longer wavelength (redshift) to 270 nm and the absorption in the whole spectral region (> 231 nm) increases with reaction time, suggesting that the electronic conjugation within the graphene sheets is restored upon hydrazine reduction [59]. Similar features and trends were observed by Savchak et al. for GO films deposited on substrates [40] (Fig. 6d), of which measurement conditions and observations are described below. With the aim of gaining insight about the absorbance behavior in solid form, UV-Vis spectrum for the bilayer GO and rGO films (encapsulated and cemented with nanometer-thick copolymer during film deposition) were measured by Savchak et al. [40] and is included in Fig. 6d. It is shown that the rGO bilayer has an average transparency of ~89% at 550 nm (in comparison with bare quartz). Moreover, when rGO films are sufficiently thin (<30 nm), they are semitransparent, while much thicker films appear opaque with graphite-like luster. The optical transmittance of GO and rGO films can be continuously tuned by varying the film thickness or the extent of reduction [29]. Regarding peak position, as observed in the solution phase, unreduced GO exhibits a maximum of adsorption at 231 nm (which corresponds to the $\pi \rightarrow \pi^*$ transitions of aromatic C$-$C bonds) and a shoulder at $\sim$300 nm (which corresponds to the n $\rightarrow$ π^* transitions of C=O bonds). After reduction, the redshift found to occur to about 272 nm.

4.7 Photoluminescence Measurement

Both the GO and rGO suspensions also found to show unique photoluminescence (PL) features and therefore has been utilized this technique to characterize both of these materials. Figure 7 shows photoluminescence under illumination by visible and UV light sources of GO suspensions in water (represented with l-GO) as well as in films (represented with s-GO). To note, two distinct types of PL have been reported so far [29]. The first type is a broad PL covering visible to near-IR range often exhibiting maximum intensity between 500 and 800 nm (1.55–2.48 eV; Fig. 7: left panel). The second type is blue emission, centered around 390–440 nm (2.82–3.18 eV) and is observed upon excitation with UV light. The origin of the two different kinds of PL is still being debated but it has been suggested that the type of PL could be related to the state of dispersion. No absorption features (That is, no PL was observed near the absorption peak position shown in Fig. 7a in the PL-energy range, and thus the PL couldn't be directly correlated with the band gap of the material. Nevertheless, the energies of visible to near-IR PL coincide with the band gap values of graphite oxide estimated from diffuse reflectance measurements, which range from 1.7 to 2.4 eV depending on the degree of oxidation [29 and Ref. therein].

Interestingly, the PL spectrum is found to be independent of the GO sheet size and no obvious peak shift is observed even when the GO sheets are cut down to few nanometers in size. This observation indicated that the PL in GO is ascribed to the atomic-scale structure of the material and that the size of the sheet does not define the electron confinement, in contrast to the case of SWNTs where PL wavelength is strongly dependent on the tube diameter [60]. Thus, GO is expected to possess a range of local energy gaps across the sheet, giving rise to the broad PL. The behavior of the PL upon gradual reduction of GO is also distinctly different in the two systems, suggesting that the origin of the two types of PL is also different [61]. The common trend is that the PL intensity is quenched upon extensive reduction.

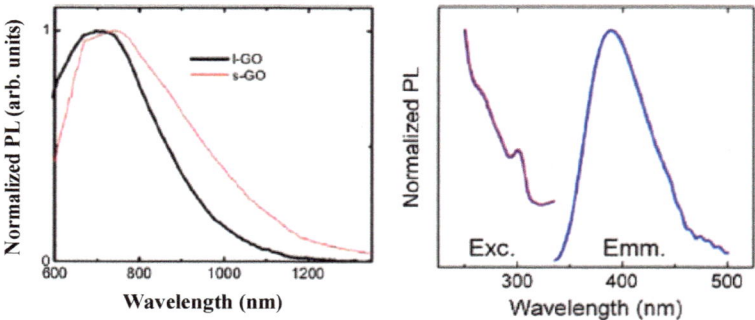

Fig. 7 Left panel, photoluminescence (PL) spectra of a GO suspension in water (l-GO) and a GO film on a solid substrate (s-GO), excited at 500 nm right panel, PL emission spectrum for excitation at 325 nm and excitation spectrum for emission at 388 nm for a GO thin film reduced by hydrazine for 3 min (reproduced from Ref. [57])

4.8 XRD Crystallographic Measurement

The XRD patterns of graphite, GO and rGO found to be quite different to each other and, therefore, has been used as a tool to differentiate them (see in Fig. 8: adapted from Ref. [55]). For example, Kumar et al. [55] found that GO films prevailed a large interlayer distance (ca. 0.8 nm) due to the formation of hydroxyl, epoxy, and carboxyl groups during chemical oxidation. However, after reduction the interlayer distance decreased to nearly 0.38 nm fairly close to the graphite values (0.34 nm), due to the removal of oxygen-containing functional groups which is mainly associated with the ring-opening of the epoxides (C–O). In other word, these observations are linked to the removal of oxygen-containing functional groups situated on the surface of each layer of GO film during the reduction process (the conjugated graphene network (sp^2 carbon) is re-established and that the graphene layers tend to aggregate). Moreover, X-ray photoelectron spectroscopy (XPS) and nuclear magnetic resonance (NMR) spectroscopy have also been employed to confirm and distinguish the characteristic features of GO and rGO. In order to gain insight about these characterization techniques for these species, readers are referred to Stankovich et al. [37] and Tiwari et al. [62].

5 Application of Graphene and Its Derivatives in Perovskite and Organic Solar Cells

As highlighted in the recent reviews [19, 24, 57, 63, 64], rapid increment of number of publications in recent years which mostly concern on applications of graphene-based

Fig. 8 XRD spectra of graphite, graphite oxide, and reduced GO produced from Ref. [55]

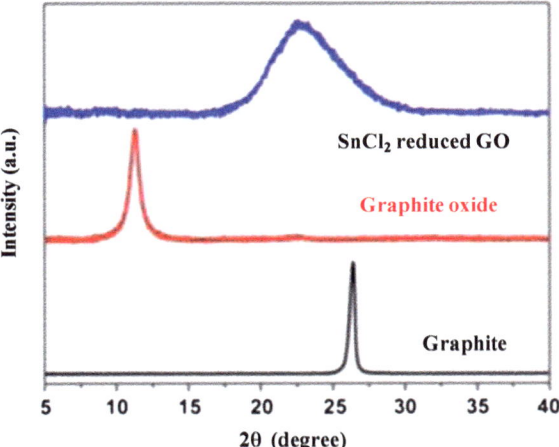

materials in nanomaterial-based photovoltaic (third-generation solar cells) technologies clearly indicate their emergence as promising materials for future energy-related technologies. Besides, energy-related devices, these materials have been integrated in other applications as well (readers are referred to the literatures [29, 31, 62]). Here, we focus on applications in perovskite solar cell technology only.

Consideration of these materials, as discussed in the Sect. 1, is due to the fact the thin film made from graphene and its derivatives can be prepared at low temperature, can sustain against the intense solar radiation and harsh environment conditions: For example, the conventional structural design of the perovskite solar cells require high-temperature processing for preparing metal oxides thin films such as TiO_2, NiO_x as charge transporting materials, is a big challenge for the market entry. Alternatively, the organic charge transporting material such as PCBM and PEDOT: PSS have also been employed in PSCs as charge transport material (as electron- and hole-transfer materials, respectively). However, these compounds also found to be sensitive to the humidity, air, and intense solar radiation due to which the efficiency of the PSCs fall immediately and, as a result, the cells sustain only a couple of days.

Graphene, GO and rGO as transparent conducting oxide (TCO) layer, interface (facilitating) layer, ETL and HTL has been successfully implemented. Below will discuss some of the recent contribution, with the priority of focus on rGO as the HTL.

5.1 Application of rGO in Perovskite Solar Cells

5.1.1 Transparent Conducting Oxide, TCO Layer

Beside window to photoactive layer, TCO layer serves as an ohmic contact in third-generation photovoltaic solar cells such as OPVs, DSSCs, QDSCs, and PSCs. To be a suitable candidate as TCO, the high optical transmittance (about 90% or above) and low sheet resistance (≤ 10 Ω/sq.) are required. ITO or FTO meets these criteria and, therefore, have been commonly employed as a TCO material in this class of solar cell technologies. However, these materials are reported to be expensive, less stable, and less flexible [63]. Graphene owing to its high transparency and low sheet resistance overcomes the demerits ITO and FTO. Moreover, due to hydrophobic nature of its surface, it can provide protection against air and moisture from the surrounding medium (which can improve the stability of these classes of devices). Therefore, is considered one of the best materials to be used as transparent electrode to enhance the PV performance.

Although optimization of the sheet resistance and transmittance is still in progress, graphene-based transparent conductors have already been successfully implemented in all kinds of above-mentioned solar cells [31, 63, 64]. For example, application of graphene as a TCO was first reported by You et al. using the chemical vapor deposition (CVD) to produce graphene film [65]. Thus, produced films exhibited high sheet resistance (~1050 Ω sq^{-1}), but the authors further optimized the device by depositing a thin layer (~20 nm) of PEDOT:PSSS solution doped with fluoro surfactant (Zonyl-FS300) together with D-sorbitol on top of the graphene surface. With these modifications the photo cell showed efficiency up to 12.37%, which is relatively high compared to that of the semitransparent TCO-free PSCs [65 and Ref. therein]. The superior performance of this graphene laminated device originated from the low sheet resistance and the high transmittance of about 90%, of the thin films in the visible spectral region of graphene electrode after it was coated with PEDOT:PSS. On the other hand, Yoon et al. found that coating of about 2 nm molybdenum trioxide film (MoO$_3$, as a HTL) on top of the graphene electrode surface (which acts as a TCO), the device performance could be improved (efficiencies of up to 17.1% was achieved) [66]. This dramatic increase in the device performance was attributed to the use of MoO$_3$ HTM, which provided hydrophilicity to the graphene surface and the formation of desirable energy-level alignment between the MoO$_3$/graphene electrode and PEDOT:PSS. In 2018, Savchak et al. [40] have demonstrated a rather easy and technology-friendly dip-coating method for the production of highly conductive and transparent rGO monolayer and bilayers on nonconductive substrates. The method was based on encapsulating individual GO sheets in a nanometer-thick molecular brush copolymer layer that allowed for the nearly perfect formation of the GO layers via dip coating in aqueous medium. By thermal reduction, the bilayers (cemented by a carbon-forming polymer linker) were converted into highly conductive and transparent reduced GO films with a high conductivity up to 10^4 S/cm and optical transparency on the level of 90%. This electrical conductivity value is the highest reported so far for thermally reduced nanoscale GO films and is close to the conductivity of ITO currently in use for PSCs. These observations, thus, make layers of graphene and graphene oxide intriguing candidates for replacement of ITO films [40].

5.1.2 Interface Layer

As discussed in the introductory section, in the DSSCs and PSCs, the photoactive layer (radiation absorber layer) is sandwiched between an electron donor and an electron acceptor. This unique stage of charge donor-acceptor, comprising minimum three layers of different materials, is referred as bulk heterojunction (BHJ). This BHJ provides continuous pathway for carriers toward the front contact and back contact electrodes for efficient charge collection [57]. The direct contact between the electrodes (front or back contact layer) and with the charge acceptor (the electrons or holes receiving layers), and the charge acceptor layer with the donor layer (i.e., radiation absorber layer) cause recombination through current leakage. Thus, to minimize the current leakage requires functional layers at the interface of absorber layer/acceptor layer or acceptor layer/contact layer to support charge extraction and collection. These additional layers (also called as, buffer layers or interface layers) are highly selective to pass a specific type of charge (electron or hole) and block the other one.

Graphene and its derivatives have been utilized as an interlayer also in the perovskite solar cell, including the interfaces: (a) between transparent conductive oxide and electron transport material; (b) between the electron transport material and perovskite; (c) between the perovskite and hole transport layer; (d) between the hole transport layer and electrode, and (e) between the electrode material and atmospheric environment [67]. For example, Tavakoli et al. [67] synthesized a hollow structured 3D scaffold of graphene and employed it as an interface layer between TiO_2 electron transfer and absorber layers in PSCs. The measured photovoltaic properties indicated that rGO scaffold (rGS) lead to improve the carrier transportation, which eventually resulted improvement in device performance by about 27% (incident photon to current conversion efficiency (IPCE) was found to be as high as 17.2%). Moreover, the devices also exhibited lower hysteresis effects and longer stability than that of the photocells constructed without rGS layer. In another contribution, Fan et al. fabricated PSCs by placing graphene quantum dots (GQDs) between mesoporous TiO_2 and $MAPbI_3$ layers [68]. In this study, the role of GQD was facilitating electron transfer from $MAPbI_3$ to TiO_2 and, indeed, with such arrangement, the device's IPCE was boosted from 8.81 to 10.15% along with higher values in J_{sc} and V_{oc}. In particular, the V_{oc} enhancement relied on faster electron extraction through GQDs interlayer. IPCE also found to boost significantly in presence of GQDs. Besides, according to transient absorption measurement, the speed of the electron extraction increased by a factor of about 3 when employing GQDs (changes from 260 to 307 ps to 90–106 ps for devices without and with GQD, respectively) [68].

5.1.3 Electron Transport Layer, ETL

This layer serves as facilitating layer for electron collection from conduction band of the perovskite absorber layer. Two of the major requirements for ETL layer are high electron mobility of material and matching of band gap with the absorber material: These properties guarantee photoelectron produced in absorber layer to effectively transfer to ETL layer followed by migration to TCO (front contact). Graphene and its derivatives have exhibited high mobility and conductivity and, therefore, considered to be a suitable candidate for this application as well. For example, Wang et al. developed a mesostructured PSC with ETL composed of graphene and anatase-TiO_2 NPs (diameter ~25 nm) [69]. Panels (a and c) of Fig. 9 represent the comparative study of the J–V curve and Panels (b and d) show external quantum efficiency (EQE) and photon to current conversion efficiency (PCE) of the PSCs with and without graphene, respectively. Their device showed an IPCE of 15.6% with J_{sc} of 21.9 mA/cm^2, V_{oc} of 1.05 V, and FF of 73% (panel (d) of Fig. 9). In contrary, PSCs with ETL layers of only TiO_2, only graphene, and high-temperature-annealed graphene incorporated TiO_2 showed PCEs of 10, 5.9, and 14.1%, respectively. This research also showed that all the components of this kind of solar cells can be fabricated at temperature below 150 °C, as the incorporation of graphene during PSC fabrication does not need high-temperature annealing in contrast to preparing TiO_2 film (which requires annealing at about 450 °C).

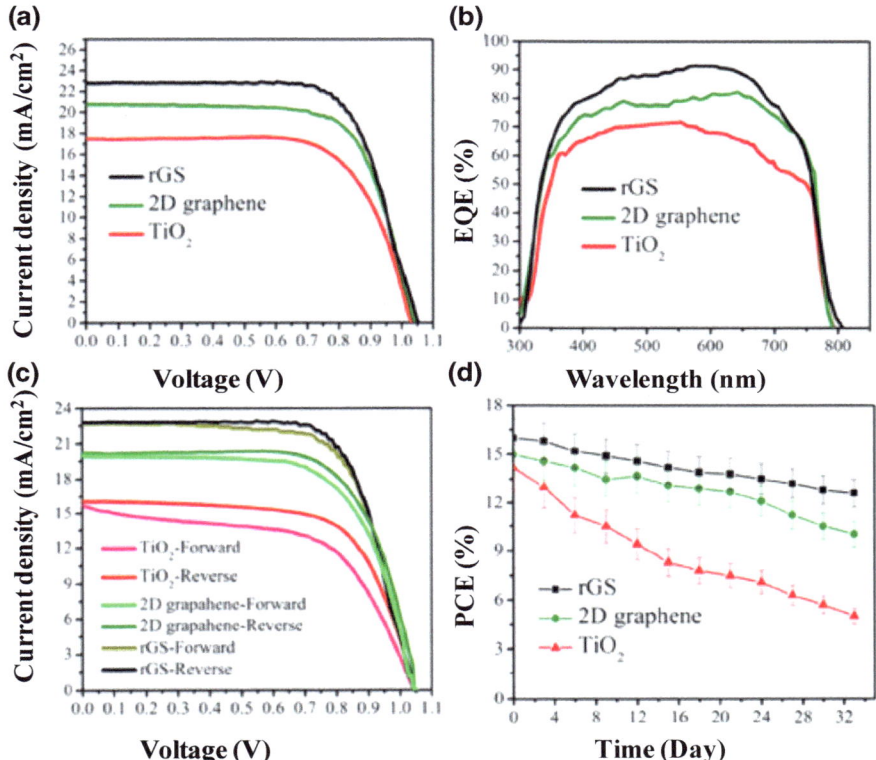

Fig. 9 **a** $J-V$ curves of the rGS-based perovskite solar cells, **b** EQE spectra, **c** $J-V$ measurement of devices in forward and reverse scan directions, and **d** Stability of the devices in an ambient environment after encapsulation using UV-cured epoxy (adopted from Ref. [69])

Furthermore, in another very recent contribution, Xie et al. have adopted tin dioxide (SnO_2): GQDs as an electron-transporting layer in planar PSCs. With this innovation, the problem of numerous electron trap states in SnO_2 film was solved by treating the film with adding GQDs. GQDs could effectively transfer the photogenerated electrons to the conduction band of the SnO_2 and improve the electron extraction efficiency as well as the conductivity of SnO_2, which was beneficial to reduce the recombination at the interface of ETL/perovskite [70]. Besides, their device also showed also with very little hysteresis, which is very common problem in this kinds of solar cells.

5.1.4 Hole Transport Layer, HTL

For polymer-based OPVs, DSSCs, QDSCs, and PSCs, the HTL with a wide band gap p-type material and high level of work function (WF) is required to block the electron and make an Ohmic contact with donor materials (polymer, dye or perovskite). For this purpose, both the inorganic and organic p-type semiconductors are employed. Some of the inorganic p-type semiconductors are MoO_3, V_2O_5, NiO, and Cu_2O [71, 72]. Whereas, the most widely used organic one is poly (3,4-ethylenedioxithiophene) (PEDOT) doped with poly (styrene sulfonic acid) (PSS). As discussed in the Sect. 1, it is reported, however, that its acidity can cause etching of the ITO (on top of hygroscopic nature of PSS allows absorption of water) facilitates to abrupt degradation of photocells shortening its lifetime. Therefore, the replacement of the PEDOT:PSS with another hole transporting material, which may lead to high efficiency and high stability of this kind of solar cells has been eagerly sought.

Among the suitable candidates, graphene-based material, rGO, has been considered one of the most suitable candidates as it possesses high electron blocking capability maintaining good hole transporting characteristics [19, 24, 57, 63, 64]. This is due to the fact that the graphene-based materials can be converted to p-type semiconductor by doping of inorganic elements such as Au-, Pt-, S-, and B [73–75]. Likewise, immobilization of p-type metal oxides (e.g., MoO_3, NiO, and CuO) on graphene leads to p-type graphene composite, respectively [76, 77]. These approaches allow scientists to adjust and modify physical properties of graphene-based 2D materials. During the doping process, dopants (i.e., molecules and ions) interact with the carbon atoms of host material and transfer charges, resulting in p-type (or n-type) doping of graphene. Some of the interesting observations from the latest contributions in PSC (and also partly covers in OVP) are highlighted below. Palma et al. used rGO as hole transport material in mesoscopic PSCs and observed a significant enhancement in devices lifetime (their photocells with rGO survived 83 days), whereas, the Spiro-OMeTAD-based solar cells showed a decrease in efficiency of about 1 order of magnitude [11]. Feng et al. incorporated ammonia modified GO into PEDOT:PSS for preparing PEDOT:PSS-GO:NH_3 films and employed it as HTL in solution-processed planar PSCs [78]. In this cell configuration, their photocells found to show 30% elevation of photon to electron conversion efficiency, which is mainly due to introduction of ammonia modified GO into PSCs which lead to keep perovskite structure stability intact. Specifically, these improvements are attributed to the better crystallization and preferred orientation order of perovskite structure and better energy-level alignment at the perovskite interface [78]. In 2017, Nouri et al. investigated perovskite solar cells employing GO for both the ETL and HTL with an inverted p–i–n architecture [79]. Specifically, in their design (shown in Fig. 10), GO was used as a hole transporting material and Li-modified GO as an electron-transporting material, while Al was used as a counter electrode. A maximum solar conversion efficiency of 10.2% was achieved by adding a Ti-based sol on the top of the Li-modified graphene oxide layer. This leads to possibility of producing solar cells at significantly low cost.

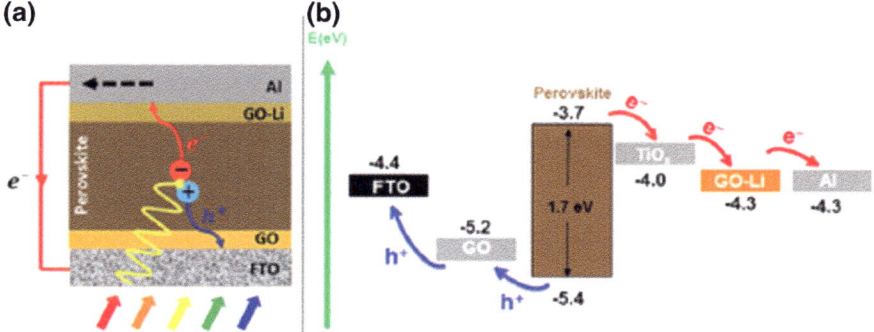

Fig. 10 **a** Schematic illustration of the inverted PSC structure employed in the Ref. [79] and **b** energy-level diagram showing possible charge transfer in the inverted planar PSC device with GO as the hole transport layer and Li doped GO as the electron transport layer

5.2 Application of rGO in Organic Photovoltaic Solar Cells, OPV

The rGO films have been used as the hole transport and electron blocking layer also in OPVs by incorporating them between the photoactive poly (3-hexylthiophene) (P3HT):phenyl-C61-butyric acid methyl ester (PCBM) layer and ITO. Uniform 2 nm thin films of rGO were deposited on top of ITO coated glass by spin coating to achieve OPV efficiencies of 3.5% PCE, which is comparable to devices fabricated with 30 nm of PEDOT:PSS as the hole transport layer that has efficiency values of around 3.6%. The rGO films could, therefore, be a simple solution-processable alternative to PEDOT:PSS as the effective hole transport and electron blocking layer in OPV and light-emitting diode devices [57]. This phenomenon depicts chemically derived rGO thin films as promising alternatives with single sheet rGO (1-1.4 nm) being particularly promising as opposed to a bilayer of rGO (2 nm), It was also indicated that with increasing the transparency by fine tuning its fabrication process and controlling of the rGO layers, one would expect greater OPV efficiency. Moreover, Yun et al. have employed GO, rGO, and pr-GO (reduced graphene oxide with a novel p-TosNHNH$_2$ reductant) as electron blocking layer in ITO/HTL/P3HT:PCBM/Ca/Al device. The devices efficiency with pr-GO was increased from PCE = 0.72% for only ITO cell to PCE = 3.7% by introducing pr-GO as HTL between the electrode and polymer photoactive layer with enhancement in FF, J_{sc}, and V_{oc}. The results were highly comparable with PEDOT:PSS (PCE = 3.6%) [74]. Similarly, Li et al. demonstrated the use of GO as HTL in ITO/GO/P3HT:PCBM/Al device [80]. To note, in regard to influence of GO film thickness on the OPV, a clear trend of decreasing power conversion efficiency with increasing GO film thickness was found. Thinnest film yielded the best results, due to the increase in serial resistance resulting in lower J_{sc} and FF and slightly lowers the transmittance of the films with decreased thickness. Beside the highly comparable device performance with PEDOT:PSS device, longer recombination lifetime was also reported with GO, by transient open-circuit voltage decay (TOCVD) measurements. Specifically, the decay lifetimes for ITO-only, PEDOT:PSS HTL, and thin GO HTL devices were found to be 8.1, 9.6, and 11.6 μS, respectively [80].

In very recent contribution by Rafique et al. [28], it was found that when GO combined it with PEDOT:PSS, it worked an effective hole transport layer. This was understood as matching of work function of GO/PEDOT: PSS (GO/PEDOT: PSS = 4.9 eV/5.1 eV) with PCDTBT (5.3 eV) facilitated the hole transportation to ITO or FTO. The improved performance attributed to decreased R_s which are highly desired for carrier transportation and collection. In this case, GO is assumed to become able as effective electron blocking layer due to its large band gap of 3.6 eV.

Role of rGO on the stability of the solar cells: Furthermore, the photovoltaic performance decayed as a function of exposure time in air, without encapsulation the PCE of the PEDOT:PSS devices dropped rapidly to about 0% after air exposure for 6 days but the pr-GO-based cells still showed high performance (up to about 54% of initial PCE) even after 13 days, suggesting that the GO suppress the issue of ITO electrode erosion via PEDOT:PSS acidity.

6 Conclusion and Outlook

In this chapter, we highlighted the prospect of utilization of graphene and its derivative (GO and rGO) as TCO and charge transport materials (electron transport and hole transport materials) in perovskite and OPV solar cells, including their synthesization and characterization techniques. Observations of low electrical resistivity, high transmittance, high carrier mobility, and easy production method (chemical route) allowed to scientist to consider this material as a promising material in such devices. As a conductive electrode, graphene is a promising substitute for commercial ITO and FTO. Moreover, GO and rGO are capable of working (mediating) as charge selective and transport layer when inserted in the perovskite photocells. In this context, achievement of maximum solar conversion efficiency of about 10.2%, employing GO and Li doped GO as ETL and HTL, respectively, in perovskite solar cells, gives hope toward producing all inorganic-based solar cells [79]. Since, utilizing GO-based charge carriers, both the electron and hole carriers, likely to replace expensive organic or metal oxide charge transporting layers.

Also in OPV, it has been shown that when GO combined with PEDOT:PSS it works effective hole transport layer and electron blocking layer, leading to PCE 4.3%. Moreover, the improvement due to atmospheric degradation of the PSC devices with rGO encapsulation was achieved due to its stable highly packed 2D structure: The photocells sustained for about 2000 h, maintaining its initial efficiency intact.

However, as demonstrated by Li et al. in Ref. [80], thickness of the films of graphene, GO or rGO products critically determine the performance of the devices.

This highlights the need of controlling mechanism for the fabrication of nanometer-scale thickness during thin preparation. Indeed, contributions similar to Ref. [40] by Savchak et al., which successfully produced highly conductive and transparent reduced graphene oxide nanoscale films via thermal conversion of polymer-encapsulated graphene oxide sheets, will definitely fulfill needs and accelerate to meet the target of producing the efficient, stable, and low-cost perovskite solar cells. Since the properties of GO/rGO (mainly determined by defect densities, type of functional groups and their positions in graphene sheets) critically depend to their fabrication process, a sensible selection of methods for thin-film preparation is essential for targeted applications (CVD, spin coating are widely used).

Yet, for controlling properties of the GO or rGO sheets as per application requirement, a simple, accurate and reproducible method needs to be developed especially for functionalization, band energy tuning and size control. This could lead to realize environmentally stable, lighter, and low-cost solar cell technology.

Acknowledgements Author would like to thank "The Word Academy of Science for developing countries (TWAS)" (Grant No. 12-165 RG/CHE/AS_I; UNESCO FR: 12-165 RG/CHE/AS_I/2013) and National Innovation Center (NIC), Kathmandu Nepal, for supporting this research project.

References

1. Kojima, A., Teshima, K., Shirai, Y., Miyasaka, T.: J. Am. Chem. Soc. **131**, 6050 (2009)
2. Kim, H.S., Lee, C.R., Im, J.H., Lee, K.B., Moehl, T., Marchioro, A., Moon, S.J., Baker, R.H., Yum, J.H., Moser, J.E., Gratzel, M., Park, N.G.: Sci. Rep. **2**, 591(1–7) (2012)
3. Lee, M.M., Teuscher, J., Miyasaka, T., Murakami, T.N., Snaith, H.J.: Science **338**, 643 (2012)
4. Gratzel, M.: Nat. Mater. **13**, 838–842 (2014)
5. Park, N.-G.: J. Phys. Chem. Lett. **4**, 2423 (2013)
6. Kim, H.-S., Lee, C.-R., Im, J.-H., Lee, K.-B., Moehl, T., Marchioro, A., Moon, S.-J., Humphry Baker, R., Yum, J.-H., Moser, J.E., et al.: Sci. Rep. **2**, 591(2012)
7. Stranks, S.D., Eperon, G.E., Grancini, G., Menelaou, C., Alcocer, M.J.P., Leijtens, T., Herz, L.M., Petrozza, A., Snaith, H.J.: Science, **342**, 341 (2013)
8. Jang, T.D.M., Park, K., Kim, D.H., Park, J., Shojaei, F., Kang, H.S., Ahn, J.-P., Lee, J.W., Song, J.K.: Nano Letters **15**(8), 5191 (2015)
9. Hutter, E.M., Gelvez-Rueda, M.C., Osherov, A., Bulovic, V., Grozema, F.C., Stranks, S.D., Savenije, T.J.: Nat. Mater. **16**, 115–120 (2017)
10. https://www.nrel.gov/pv/assets/pdfs/pv-efficiencies-07-17-2018.pdf
11. Palma, A.L., Cinà, L., Pescetelli, S., Agresti, A., Raggio, M., Paolesse, R., Bonaccorso, F., Carlo, A.D.: Nano Energy **22**, 349 (2016)
12. You, J., Meng, L., Song, T.-B., Guo, T.-F., Yang, Y.(M.),Chang, W.-H., Hong, Z., Chen, H., Zhou, H., Chen, Q., Liu, Y., Marco, N.D., Yang, Y.: Nat. Nanotechnol. **11**, 75 (2016)
13. Yeo, J.-S., Kang, R., Lee, S., Jeon, Y.-J., Myoung, N., et al.: Nano Energy **12**, 96–104 (2015)
14. Agresti, A., Pescetelli, S., Cina, L., Konios, D., Kakavelakis, G., Kymakis, E., Carlo, A.D.: Adv. Funct. Mater. **26**(16), 2686 (2016)
15. Sung, H., Ahn, N., Jang, M.S., Lee, J.-K., Yoon, H., Park, N.-G., Choi, M.: Adv. Energy Mater. **6**(3), 1501873 (2016)
16. Yang, Q.-D., Li, J., Cheng, Y., Li, H.-W., Guan, Z., Yu, B., Tsang, S.-W.: J. Mater. Chem. A **5**, 9852 (2017)

17. Luo, H., Lin, X., Hou, X., Pan, L., Huang, S., Chen, X.: Nano-Micro Lett. **9**, 39 (2017)
18. Gatti, T., Casaluci, S., Prato, M., Salerno, M., Di Stasio, F., Ansaldo, A., Menna, E., Carlo, A.D., Bonaccorso, F.: Adv. Funct. Mater. **26**, 7443 (2016)
19. Gatti, T., Lamberti, T., Topolovsek, P., Abdu-Aguye, M., Sorrentino, R., Perino, L., Salerno, M., Girardi, L., Marega, C., Rizzi, G.A., Loi, M.A., Petrozza, A., Menna, E.: Sol. RRL 1800013 (2018)
20. Lee, B.H., Lee, J.H., Kahng, Y.H., Kim, N., Kim, Y.J., Lee, J., Lee, T., Lee, K.: Adv. Funct. Mater. **24**, 1847–1856 (2014)
21. Ju, M.J., Jeon, I.Y., Kim, J.C., Lim, K., Choi, H.J., Jung, S.M., Choi, I.T., Eom, Y.K., Kwon, Y.J., Ko, J.: Adv. Mater. **26**, 3055–3062 (2014)
22. Liscio, A., Veronese, G.P., Treossi, E., Suriano, F., Rossella, F., Bellani, V., Rizzoli, R., Samori, P., Palermo, V.: J. Mater. Chem. **21**, 2924 (2011)
23. Cho, H.-W., Liao, W.-P., Lin, W.-H., Yoshimura, M., Wu, J.-J.: J. Power Sources **293**, 246 (2015)
24. Mahmoudi, T., Wang, Y., Hahn, Y.-B.: Nano Energy **47**, 51–65 (2018)
25. Bae, S., Kim, H., Lee, Y., Xu, X.F., Park, J.S., Zheng, Y., Balakrishnan, J., Lei, T., Kim, H.R., Song, Y.I., Kim, Y.J., Kim, K.S., Ozyilmaz, B., Ahn, J.H., Hong, B.H., Iijima, S.: Nat. Nanotechnol. **5**, 574 (2010)
26. Kobayashi, T., Bando, M., Kimura, N., Shimizu, K., Kadono, K., Umezu, N., Miyahara, K., Hayazaki, S., Nagai, S., Mizuguchi, Y., Murakami, Y., Hobara, D.: Appl. Phys. Lett. **102**, 023112 (2013)
27. Eda, G., Lin, Y.-Y., Miller, S., Chen, C.-W., Su, W-.F., Chhowalla, M.: Appl. Phys. Lett. **92**, 233305 (2008)
28. Rafique, S., Abdullah, S.M., Shahid, M.M., Ansari, M.O., Sulaiman, K.: Nat.: Sci. Rep. **7**, 39555 (2017)
29. Eda, G., Chhowalla, M.: Adv. Mater. **22**, 2392 (2010)
30. Loh, K.P., Bao, Q., Eda, G., Chhowalla, M.: Nat. Chem. **2**, 1015 (2010)
31. Hu, C., Liu, D., Xiao, Y., Dai, L.: Mater. Int. **28**, 121 (2018)
32. Eda, G., Mattevi, C., Yamaguchi, H., Kim, H., Chhowalla, M.: J. Phys. Chem. C **113**, 15768 (2009)
33. S.-S. Li, K.-H. Tu, C.-C. Lin, C.-Wei, and M. Chhowalla, ACS NANO.**4**, 3169 (2010)
34. Wang, S., et al.: Nano Lett. **10**, 92 (2010)
35. Hummers, W.S., Offeman, R.E.: J. Am. Chem. Soc. **80**(6), 1339 (1958)
36. Hirata, M., Gotou, T., Horiuchi, S., Fujiwara, M., Ohba, M.: Carbon **42**, 2929 (2004)
37. Stankovich, S., Dikin, D.A., Piner, R.D., Kohlhaas, K.A., Kleinhammes, A., Jia, Y., Wu, Y., Nguyen, S.T., Ruoff, R.S.: Carbon **45**, 1558 (2007)
38. Zhu, Y., Murali, S., Cai, W., Li, X., Suk, J.W., Potts, J.R., Ruoff, R.S.: Adv. Mater. **22**, 3906 (2010)
39. Stankovich, S., Dikin, D.A., Dommett, G.H.B., Kohlhaas, K.M., Zimney, E.J., Stach, E.A., et al.: Nature **442**(7100), 282 (2006)
40. Savchak, M., Borodinov, N., et al.: ACS Appl. Mater. Interfaces. **10**(4), 3975 (2018)
41. Gao, X., Jang, J., Nagase, S.: J. Phys. Chem. C **114**(2), 832 (2009)
42. Williams, G., Seger, B., Kamat, P.V.: ACS Nano **2**, 1487 (2008)
43. Stankovich, S., Piner, R.D., Chen, X., Wu, N., Nguyen, S.T., Ruoff, R.S.: J. Mater. Chem. **16**(2), 155 (2006)
44. Songfeng, P., Zhao, J., Du, J., Ren, W., Cheng, H.-M.: Carbon **48**, 4466 (2010)
45. Ahammad, A.J.S., Islam, T., Hasan, MdM, Mozumder, M.N.I., Karim, R., Odhikari, N., Pal, P.R., Sarker, S., Kim, D.M.: J. Electrochem. Soc. **165**(5), B174–B183 (2018)
46. Lerf, A., He, H., Forster, M., Klinowski, J.: J Phys Chem B **102**(23), 4477 (1998)
47. Tuinstra, F., Koenig, J.L.: J. Chem. Phys. **53**(3), 1126 (1970)
48. Reserbat-Plantey, A., Marty, L., Arcizet, O., Bendiab, N., Bouchiat, V.: Nat. Nanotechnol. **7**, 151 (2012)
49. Dresselhaus, M.S., Jorio, A., Saito, R.: Ann. Rev. Condens. Matter Phys. **1**, 89 (2010)
50. Eda, G., Fanchini, G., Chhowalla, M.: Nat. Nanotechnol. **3**, 270 (2008)

51. Ferrari, A.C.: Solid State Commun. **143**, 47–57 (2007)
52. Eda, G., Mattevi, C., Yamaguchi, H., Kim, H., Chhowalla, M.: J. Phys. Chem. C **113**, 15768–15771 (2009)
53. Venezuela, P., Lazzeri, M., Mauri, F.: Phys. Rev. B **84**, 035433 (2011)
54. J.I. Paredes, S. Villar-Rodil, A. Martı́nez-Alonso, J.M.D. Tasco´n, Langmuir **24**, 10560 (2008)
55. Kumar, N.A., Gambarelli, S., Duclairoir, F., Bidan, G., Dubois, L.: J. Mater. Chem. A **1**, 2789 (2013)
56. Islam, M.F., Rojas, E., Bergey, D.M., Yodh, A.J.: Nano Lett. **3**, 269 (2003)
57. T.A. Amollo, T.M. Genene, O. Nyamori Vincent, Solar Energy, **171**, 83 (2018)
58. Robertson, J.: Mater. Sci. Eng. R **37**, 129 (2002)
59. Li, D., Müller, M.B., Gilje, S., Kaner, R.B., Wallace, G.G.; Nat. Nanotechnol. **3**, 101 (2008)
60. O'Connel, M.J., et al.: Science **29**, 7593 (2002)
61. Eda, G., Lin, G.Y.-Y., Miller, S., Mathevi, C., Yamaguchi, H., Chen, H.-A., Chen, I.S., Chen, C.-W., Chhowalla, M.: Adv. Mater. **21**, 505 (2009)
62. Tiwari, S.K., Mishra, R.K., Ha, S.K., Huczko, A.: ChemNanoMat **4**, 1 (2018)
63. Iqbal, M.Z., Rehman, A.-U.: Solar Energy **169**, 634 (2018)
64. Lim, E.L., Yap, C.C., Jumali, M.H.H., Teridi, M.A. M., Teh, C.H.: Nano-Micro Lett. **10** (2018)
65. You, P., Liu, Z., Tai, Q., Liu, S., Yan, F.: Adv. Mater. **27**(24), 3632 (2015)
66. Yoon, J., Sung, H., Lee, G., Cho, W., Ahn, N., Jung, H.S., Choi, M.: Energy Environ. Sci. **10**(1), 337 (2017)
67. Tavakoli, M.M., Tavakoli, R., Hasanzadeh, S., Mirfasih, M.H.: J. Phys. Chem. C **120**, 19531 (2016)
68. Zhu, Z., Ma, J., Wang, Z., Mu, C., Fan, Z., Du, L., Bai, Y., Fan, L., Yan, H., Phillips, D.L.: J. Am. Chem. Soc. **136**, 3760 (2014)
69. Wang, D.H., Kim, J.K., Seo, J.H., Park, I., Hong, B.H., Park, J.H., Heeger, A.J.: Angew. Chem. Int. Ed. **52**, 2874 (2013)
70. Xie, J., Huang, K., Yu, X., Yang, Z., Xiao, K., Qiang, Y., Zhu, X., Xu, L., Wang, P., Cui, C.: ACS Nano **11**, 9176 (2017)
71. Irwin, M.D., Buchholz, D.B., Hains, A.W., Chang, R.P., Marks, T.J.: Proc. Natl. Acad. Sci. (USA) **105**, 2783 (2008)
72. Shrotriya, V., Li, G., Yao, Y., Chu, C.-W., Yang, Y.: Appl. Phys. Lett. **88**, 073508 (2006)
73. Hou, Y., Geng, X., Li, Y., Dong, B., Liu, L., Sun, M.: Sci. China Phys. Mech. Astron. **54**, 416 (2011)
74. Yun, J.M., Yeo, J.S., Kim, J., Jeong, H.G., Kim, D.Y., Noh, Y.J., Kim, S.S., Ku, B.C., Na, S.I.: Adv. Mater. **23**, 4923 (2011)
75. Mahmoudi, T., Seo, S., Yang, H.-Y., Rho, W.-Y., Wang, Y., Hahn, Y.-B.: Nano Energy **28**, 179 (2016)
76. Steim, R., Kogler, F.R., Brabec, C.J.: J. Mater. Chem. **20**, 2499 (2010)
77. Yip, H.L., Hau, S.K., Baek, N.S., Ma, H., Jen, A.K.Y.: Adv. Mater. **20**, 2376 (2008)
78. Feng, S., Yang, Y., Li, M., Wang, J., Cheng, Z., Li, J., Ji, G., Yin, G., Song, F., Wang, Z.: ACS Appl. Mater. Interfaces. **8**(23), 14503 (2016)
79. Nouri, E., Mohammadi, M.R., Lianos, P.: Chem. Commun. **53**, 1630 (2017)
80. Li, S.-S., Tu, K.-H., Lin, C.-C., Chen, C.-W., Chhowalla, M.: ACS Nano **4**, 3169 (2010)

Graphene and Graphene Oxide as Nanofiller for Polymer Blends

Benalia Kouini and Hossem Belhamdi

Abstract Due to its exceptionally outstanding electrical, mechanical, and thermal properties, graphene is being explored for a wide array of applications and has attracted enormous scientific and industrial interest. In this present work, recent research and development of the utilization of graphene and graphene oxide as nanofiller in the fabrication of nanocomposites with different polymer matrices are developed. Most production methods of graphene and the processing of Graphene/polymer blends are discussed. We also review the electrical, mechanical, thermal, and barrier properties of these blends and the influence of the intrinsic properties of these fillers (graphene and its derivatives) and their state of dispersion in polymer matrix on the properties.

Keywords Graphene oxide · Nanofiler · Nanocomposite · Blend · Polymer

1 Introduction

Polymer nanocomposites based on carbon black, carbon nanotubes, and layered silicates have been used for improved mechanical, thermal, electrical, and gas barrier properties of polymers [1–3]. The discovery of graphene with its combination of extraordinary physical properties and ability to be dispersed in various polymer matrices has created a new class of polymer nanocomposites. Graphene is an atomically thick, two-dimensional (2D) sheet composed of sp^2 carbon atoms arranged in a honeycomb structure (Fig. 1). It has been viewed as the building block of all other graphitic carbon allotropes of different dimensionality [4]. For example, graphite

B. Kouini (✉)
Laboratory of Caotings, Materials and Environment, M'Hamed Bougara University, Boumerdes 35000, Algeria
e-mail: kouinib@gmail.com; kouinib@univ-boumerdes.dz

H. Belhamdi
Research Unit: Materials, Processes and Environment (RU/MPE), M'Hamed Bougara University, Boumerdes 35000, Algeria
e-mail: h.belhamdi@univ-boumerdes.dz

© Springer Nature Switzerland AG 2019   231
S. Sahoo et al. (eds.), *Surface Engineering of Graphene*, Carbon Nanostructures,
https://doi.org/10.1007/978-3-030-30207-8_9

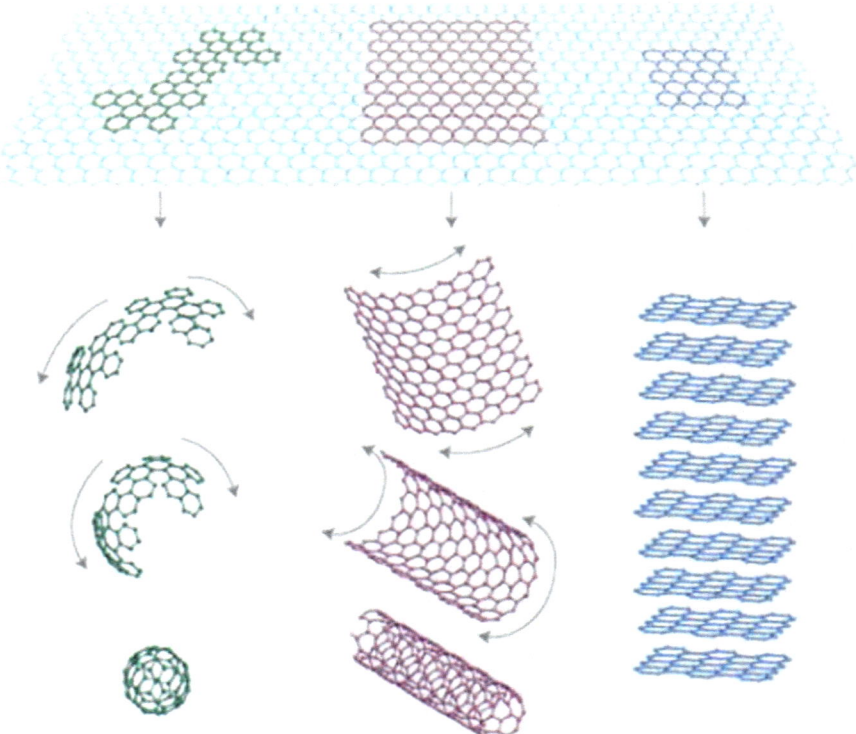

Fig. 1 Graphene is the building block of all graphitic forms. It can be wrapped up into zero-dimensional buckyballs, rolled into one-dimensional nanotubes, or stacked into three- dimensional graphite. Reproduced with permission from Ref. [4]. Copyright 2007, Springer Nature

(3D carbon allotrope) is made of graphene sheets stacked on top of each other and separated by 3.37 Å. The 0D carbon allotrope, fullerenes (buckyballs), can be envisioned to be made by wrapping a section of graphene sheet. The 1D carbon allotropes, carbon nanotubes (CNT), and nanoribbons can be made by rolling and slicing graphene sheets, respectively. In reality, however, these carbon allotropes, with the exception of nanoribbons, are not synthesized from graphene. Graphite is a naturally occurring material with the first documented deposit [5] near Borrowdale, England, in 1555, but its first use may be dated back 4000 years [6]. Single-walled CNT (SWCNT) was first synthesized in 1991 [7], following the discovery of fullerene in 1985 [8]. Although the first reported method for production of graphene nanosheets can be traced back to 1970, and [9] isolation of free-standing single-layer graphene was first achieved in 2004 when graphene was separated from graphite using micromechanical cleavage [10].

2 Production of Graphene

For years, graphene (Fig. 1) was considered an academic material that existed only in theory and presumed not to exist as a free-standing material, due to its unstable nature. A. Geim, K. Novoselov, and co-workers were among the first to successfully obtain the elusive free-standing graphene films [10], which was a remarkable achievement. The International Union of Pure and Applied Chemistry (IUPAC) defines graphene as a single carbon layer of the graphite structure, describing its nature by analogy to a polycyclic aromatic hydrocarbon of quasi-infinite size [11]. Thus, the term graphene should be used only when the reactions, structural relations, or other properties of a single layer are discussed. There has been a long and sustained effort to realize free-standing graphene films. Different ways for isolating graphene have been studied. One of the earliest documented attempts to isolate graphene was through exfoliation by physical or chemical methods. For example, graphite was first exfoliated in 1840, when C. Schafheutl tried to purify "kish" from iron smelters by treating it with a mixture of sulfuric and nitric acids [12]. Graphite oxide was first prepared by Brodie in 1859, by treating graphite with a mixture of potassium chlorate and fuming nitric acid [13, 14]. Boehm et al. described the formation of extremely thin lamellae of carbon, comprising of a few carbon layers as measured by TEM, by either "deflagration of graphitic oxide on heating or by reduction of graphitic oxide in alkaline suspension" [15]. It has been argued that sample preparation techniques for making the TEM samples resulted in the agglomeration of the otherwise single layer of graphene into the lamellae described by Boehm et al. In none of these early works was "free-standing" graphene or graphene oxide files isolated or identified as such.

Geim's group (Fig. 2) successfully isolated atomically thin graphite by using adhesive tape to peel off layers from graphitic crystal flakes and then gently rub those fresh layers against an oxidized silicon surface. They were also able to determine the thickness of this layer, which was a few angstroms' thick, using AFM. Their "Scotch tape" technique is very reminiscent of the use of adhesive tape to routinely peel

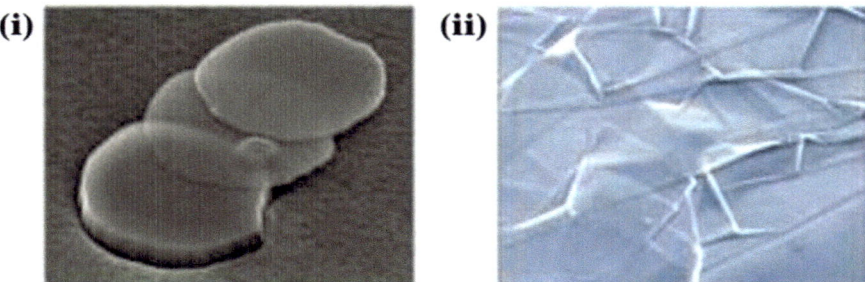

Fig. 2 One of the first photographs of isolated graphene. (**i**) They used the simple technique of ripping layers from a graphite surface (called as exfoliation) using adhesive tape. (**ii**) High resolution scanning electron micrograph image of graphene. Reproduced with permission from Ref. [20]. Copyright 2011, The Electrochemical Society

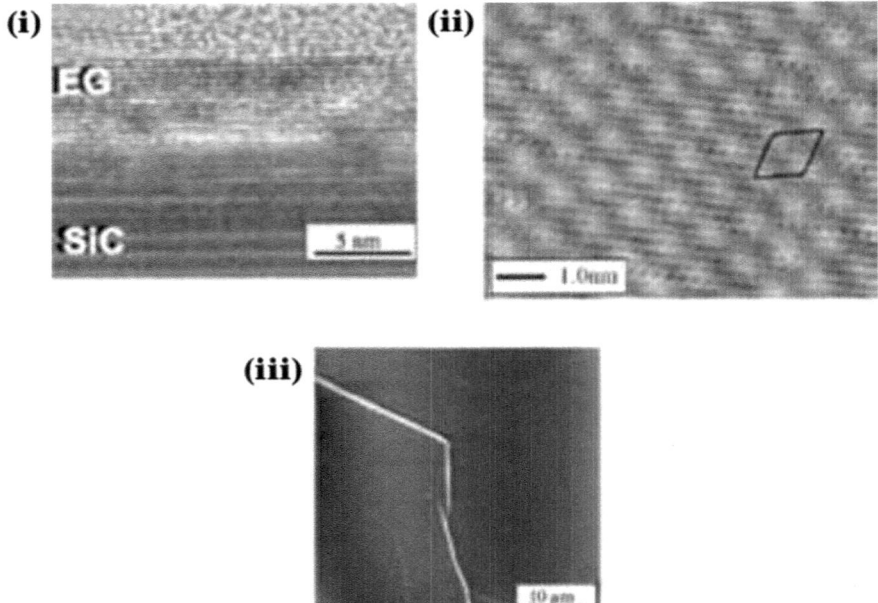

Fig. 3 Epitaxial graphene on the C-face of 4H-SiC. (**i**) TEM image of the cross section of multi-layer epitaxial graphene. (**ii**) Atomic resolution STM image showing a hexagonal lattice. and (**iii**) AFM image. Reproduced with permission from Ref. [20]. Copyright 2011, The Electrochemical Society

layered crystals (e.g., graphite, mica, etc.), held together by van der Waals forces, to expose fresh surfaces [16, 17]. In the past decade or so, the group at Georgia Tech led by Walter de Heer used the method of epitaxial growth to isolate graphene (Fig. 3). Silicon carbide was chosen as a substrate, and the group demonstrated that epitaxial graphene could be produced by thermal decomposition of SiC which can be patterned and gated [18]. Furthermore, they showed that the epitaxial graphene exhibited 2D electronic properties as well as quantum confinement and quantum coherence effects. At the same time, Philip Kim's group at Columbia University used AFM to mechanically separate graphene layers from graphite. They succeeded in isolating a multi-layer structure comprised of about ten layers [19].

3 Properties of Graphene

Graphene is a flat monolayer of sp^2 carbon atoms tightly packed into a two-dimensional (2D) honeycomb lattice, which is a basic building block for carbon-based materials (Fig. 1). In 1947, Wallace used band theory of solids with tight binding approximation to explain many of the physical properties of graphite. He makes a rather clairvoyant assumption: "Since the spacing of the lattice planes of graphite is large (3.37 Å) compared with the hexagonal spacing in the layer 1.42 Å, a first approximation in the treatment of graphite may be obtained by neglecting the interactions between the planes, and supposing that conduction takes place only in layers." This assumption makes subsequent analyses conveniently applicable to the material that we now know as graphene. The 2D system of graphene is not only interesting by itself; it also allows access to the subtle and rich physics of quantum electrodynamics in a bench-top experiment. Novoselov et al. [21] showed that electron transport in graphene is essentially governed by Dirac's (relativistic) equation. The charge carriers in graphene mimic relativistic particles with zero rest mass and have an effective speed of light, $c* \approx 106 \text{ cm}^{-1} \text{ s}^{-1}$. Their study revealed a variety of unusual phenomena that are characteristic of 2D Dirac fermions. In particular, they observed that graphene's conductivity never falls below a minimum value corresponding to the quantum unit of conductance, even when concentrations of charge carriers tend to zero.

One of the most fascinating aspects of the physics enabled by the isolation of graphene is the experimental demonstration of the so-called Klein paradox/unimpeded penetration of relativistic particles through high and wide potential barriers. The phenomenon is discussed in many contexts in particle, nuclear, and astrophysics, but direct tests of the Klein paradox using elementary particles had hitherto proved impossible. Katsnelson et al. showed that the effect can be tested in a conceptually simple condensed-matter experiment using electrostatic barriers in single- and bi-layer graphene [22]. Owing to the chiral nature of their quasi-particles, quantum tunneling in these materials becomes highly anisotropic, qualitatively different from the case of normal, non-relativistic electrons. Massless Dirac fermions in graphene allow a close realization of Klein's Gedanken experiment, whereas massive chiral fermions in bi-layer graphene offer an interesting complementary system that elucidates the basic physics involved.

4 Production of Graphene Oxide (GO)

Currently, the most promising methods for large-scale production of graphene are based on the exfoliation and reduction of GO. It is produced using different variations of the Staudenmaier [23] or Hummers methods [14].GO is generally produced by the treatment of graphite using strong mineral acids and oxidizing agents, typically via treatment with $KMnO_4$ and H_2SO_4, as in the Hummers method or its modified derivatives, or $KClO_3$ (or $NaClO_3$) and HNO_3 as in the Staudenmaier or Brodie methods [24]. These reactions achieve similar levels of oxidation (C:O ratios of approximately 2:1) [24], which ultimately disrupts the delocalized electronic structure of graphite and imparts a variety of oxygen-based chemical functionalities to the surface. While the precise structure of GO remains a matter of debate [24], it is thought that hydroxyl and epoxy groups are present in highest concentration on the basal plane, with carboxylic acid groups around the periphery of the sheets as

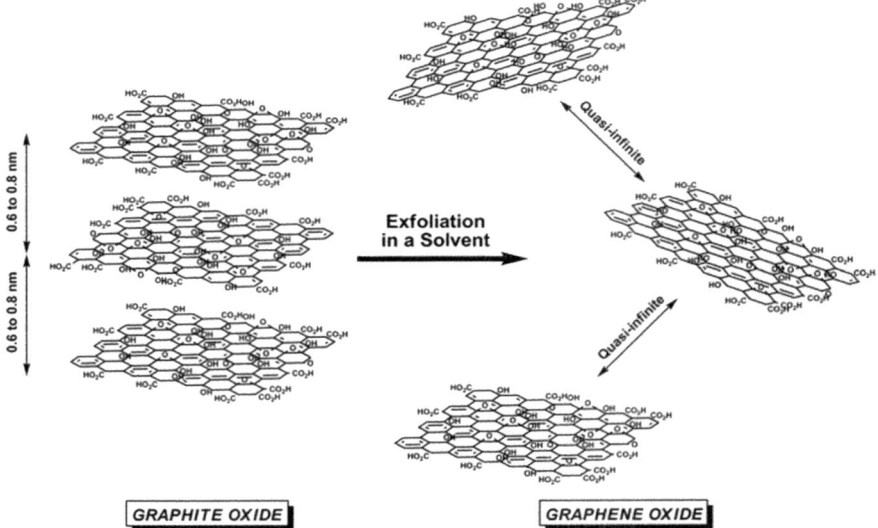

Fig. 4 Schematic illustrating the chemical structure of graphite oxide (GO) and the structural difference between layered GO and exfoliated graphene oxide (GeO) platelets. Reproduced with permission from Ref. [29]. Copyright 2010, Elsevier Ltd

shown in Fig. 4. GO has an expanded interlayer spacing relative to graphite which depends on humidity (for instance, 0.6 nm when subjected to high vacuum [25] to roughly 0.8 nm at 45% relative humidity [26]) due to intercalation of water molecules [25]. GO can be exfoliated using a variety of methods (most commonly by thermal shocking [27] or chemical reduction in appropriate media [28]), yielding a material reported to be structurally similar to that of pristine graphene on a local scale.

5 Processing of the Graphene/Polymer Nanocomposites

The properties of graphene-based polymer nanocomposites are dependent upon the processing conditions in the fabrication of graphene/polymer nanocomposites. The functionality of graphene components is critical to lower filler loading rate, make them highly dispersed and organized sheets within polymer matrix to enhance the properties of nanocomposites. In particular, the mechanical properties depend on the specific surface area, aspect ratio, organization, and loading content of graphene materials. The dispersion, interfacial strength, affinity of components, and spatial organization are all of great importance in determining the final stiffness, strength, toughness, and elongation of polymer nanocomposites under various loading conditions [30–34]. The pre-treatment procedures and the fabrication methods control the fine morphology and physical/chemical properties of graphene-based polymer nanocomposites. For various graphene/polymer nanocomposites known to date, the extent of dispersion and exfoliation of graphitic layers are controlled by the applied shear force, temperature, and solvent polarity. Effective control of restacking, wrinkling, and aggregation of graphene sheets is required for the development of functional nanocomposites with high performance. In fact, extremely flexible and high aspect ratio graphene components are prone to random wrinkling, buckling, or folding during processing, which dramatically affects the ultimate performance. In the case of the post-treatment, the degree of dispersion can be further influenced by the hydrophobic nature of reduced graphene oxide sheets and dewetting processes at the interfaces.

The choice of fabrication methods is determined by the surface functionalization of integrated graphitic sheets. Generally, traditional fabrication routines include solution-based processing [35–38] and melt-based processing [39–41]. Among most popular approaches for chemical modification and assembly are in situ polymerization, chemical grafting, latex emulsion blending, layer-by-layer (LbL) assembly, and directed assembly [42–46]. For the in situ polymerization method, intercalated monomers within expanded graphite clusters can promote their efficient exfoliation into single sheets throughout the polymer matrix caused by catalysis reactions [47].

Solution processing maximizes filler dispersion in polymer matrix by using pre-suspended single-layered graphene sheets. Different solvents (aqueous to organic) can be used to dissolve graphene materials, including graphene oxide and reduced graphene oxide materials. This approach has been widely exploited due to its high dispersion efficiency, facile and fast fabrication step, and a high level of control on component behavior. By the way, melt-based mixing is a solvent-free process in which applied mechanical shear force distributes the fillers in the polymer matrix using a screw extruder or a blending mixer [48, 49]. This method allows stacked graphite or reduced graphene oxide sheets to be exfoliated into a viscous polymer melt by suppressing unfavorable interactions and inducing component dispersion. Melt mixing is recognized as a practical approach that can be adapted to the graphene-based polymer nanocomposites.

6 Characterization of the Graphene/Polymer Nanocomposites

The excellent performances of graphene are derived from its unique 2D crystal structure. The horizontal dimension of graphene can be sufficiently extended, while the thickness is only in atomic scale. Thus, the structural characterizations of graphene need to take into account horizontal macroscopic scale as well as atomic-level analysis. Several typical structural characterization techniques, including optical microscopy, electron microscopy, scanning probe microscope, and Raman spectroscopy are presented as follows:

6.1 Microscopy

Although it is a nanomaterial, graphene can be observed directly in an optical micro-scope since a single atomic layer absorbs ~2.3% of visible light [50]. This absorption is also virtually independent of wavelength. It is also possible to distinguish flakes of graphene with different numbers of atomic layers relatively easily in a transmis-sion optical microscope [51, 52] (Fig. 5). Transmission electron microscopy has also been employed to determine the size of different grains or the atomic structure of grain boundaries, since these features are associated with the electronic [53], mag-netic [54], and mechanical [55] properties of the material [56, 57]. Novoselov et al. in their initial studies of graphene employed atomic force microscopy (AFM) in order to observe the thickness of the graphene layers and found out that mono-layer graphene possesses a thickness of 0.4 nm. Since then, many publications dealing with graphene have used AFM in order to characterize the thickness of the flakes (Fig. 5). Scanning electron microscopy and scanning tunneling microscopy have also

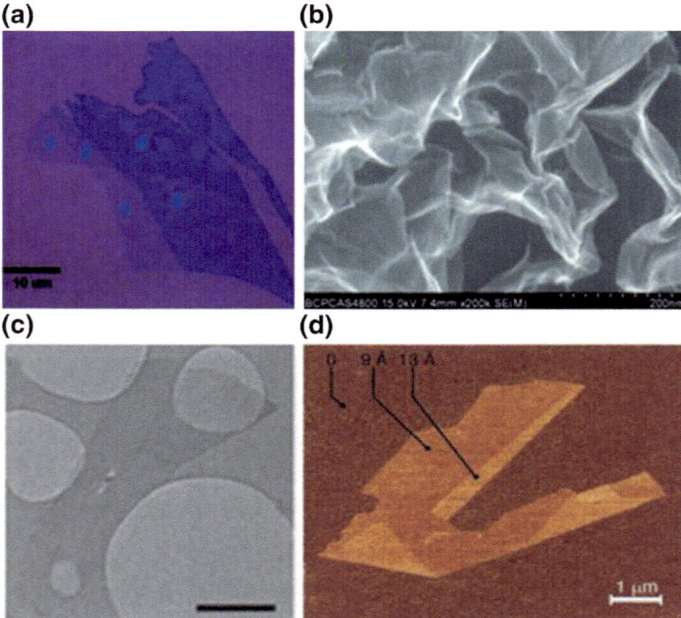

Fig. 5 Optical image of graphene with one, two, three, and four layers (Reproduced with permission from Ref. [52]. Copyright 2007, American Chemical Society), **b** SEM image of graphene (Repro-duced with permission from Ref. [62]. Copyright 2011, Royal Society of Chemistry), **c** bright-field TEM image of mono-layer graphene on a holey carbon film(Reproduced with permission from Ref. [70]. Copyright 2008, Springer Nature), **d** graphene visualized by AFM (Reproduced with permission from Ref. [4]. Copyright 2008, Nature Publishing Group)

been employed in order to observe the ripples, wrinkles, and structure of graphene sheets, which ultimately can alter the properties of the "initial" or "ideal" material [58–62].

6.2   X-ray Diffraction

X-ray diffraction (XRD) is a technique that can be utilized in order to follow the intercalation and exfoliation of graphite and the formation of graphene. The sharp Bragg reflection of graphite under normal measurement conditions found at $2\theta \approx$ 26°, becomes broader with the decreasing number of layers and ultimately disappears in measurements upon mono-layer graphene [63]. This can be undertaken upon bulk material; therefore, XRD can provide only a relative estimation regarding the average number of layers in graphene.

6.3   Raman Spectroscopy

One of the most powerful techniques, widely used in extensive studies upon graphene, has proven to be Raman spectroscopy, due to the fact that there is strong resonance Raman scattering from graphene. It is found that even measurements upon a graphene monolayer (Fig. 6) can provide a strong signal, very useful for the characterization of the material [64, 65]. There are three main characteristic bands of graphene and graphite: the D band at ~1330 cm^{-1}, the G band at ~1580 cm^{-1} , and the G′ (or 2D) at ~2650 cm^{-1}. One of the most important functions of Raman spectroscopy on the study of graphene is the accurate information it can provide regarding the number of layers of graphene [66]. Dresselhaus et al. [67] have shown that the characteristic 2D band evolves and displays differences for different numbers of graphene layers (Fig. 6). In particular, it broadens and upshifts as the number of graphene layers is increased. The D band (not present in the monolayer in Fig. 6) is related to the presence of edges and defects [68]. The intensity ratio between the D and G bands indicates the level of defects and for graphene produced by bulk preparation methods is normally higher than that of the original graphite as the result of damage during exfoliation and the formation of edges.

6.3.1   Ultraviolet/Visible (UV/Vis) Spectroscopy

UV/Vis spectroscopy can be very helpful for the identification of the characteristics of graphene produced by different methods. The UV/Vis spectrum of graphene displays a pronounced and asymmetric peak at around 4.62 eV, while at lower photon energies (0.6 eV $< E <$ 2 eV), the spectrum is flat. The number of layers does not have a pronounced effect on the UV/Vis spectrum of graphene, since bi-layer graphene shows similar excitonic effects to single-layer graphene, but with a less asymmetric optical absorption at 4.6 eV [69].

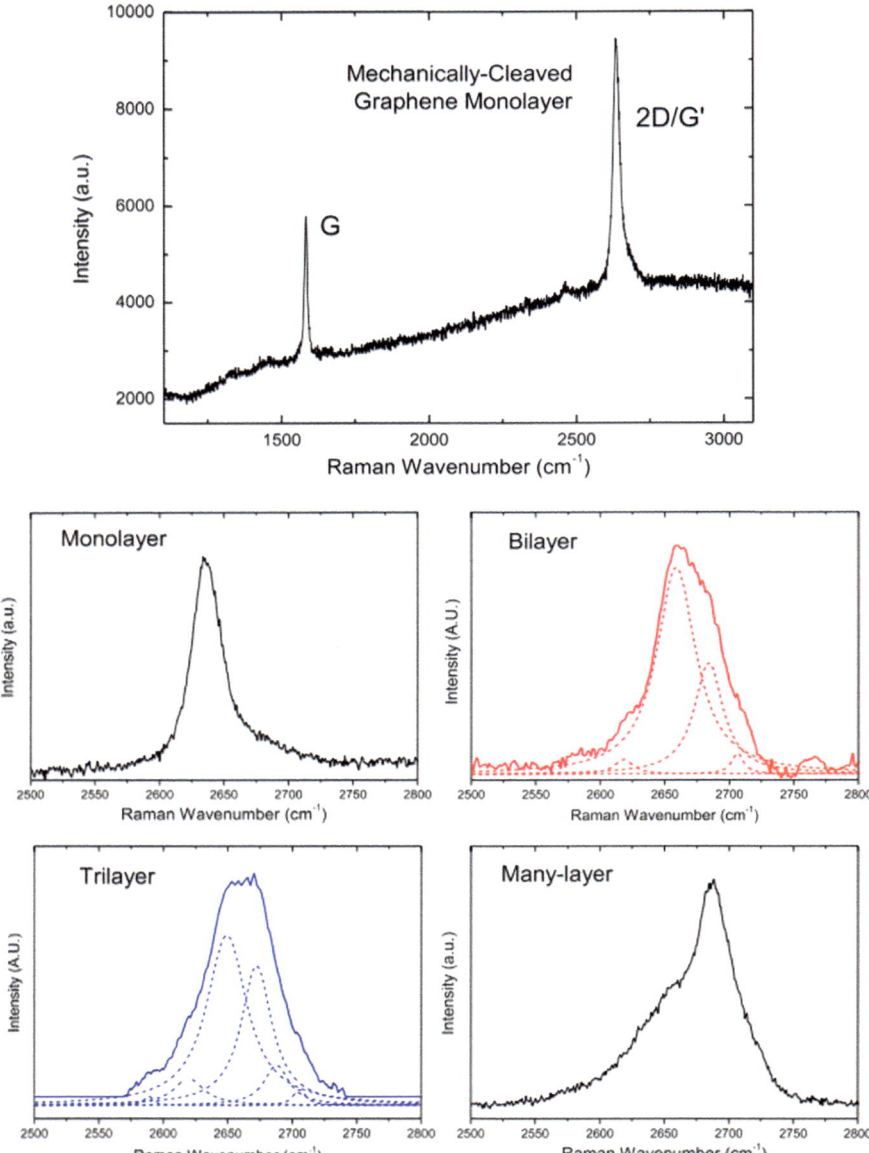

Fig. 6 Raman spectra of mono-layer graphene showing the complete spectrum with the G and 2D/G0 bands (top). Details of the 2D/G0 band for mono-layer, bi-layer, tri-layer, and many layer materials (bottom). (Reproduced with permission from Ref. [66]. Copyright 2013, Royal Society of Chemistry)

7 Properties of Graphene/Polymer Nanocomposites

In this section, the influence of the intrinsic properties of these fillers (graphene and graphene oxide) and their state of dispersion in polymer matrix on the electrical, thermal, mechanical, and gas barrier properties of graphene/polymer nanocomposites are summarized.

7.1 Electrical Conductivity

To study the effect of the graphene on the electrical properties of polymer/graphene nanocomposites, we have chosen two different works. The first paper of Hao-Bin Zhang has studied the electrical properties of PET/graphene nanocomposites. The second work of He-xin Zhang. His paper summarized the excellent electrical properties of PE/rGO nanocomposites. Figure 7 shows plots of electrical conductivity versus filler content for PET composites filled with graphene and pristine graphite. PET/graphene nanocomposites exhibit a sharp transition from insulator to semiconductor, and the inset of Fig. 6 indicates that their electrical conductivity(s) obeys the power law [71]. As shown in the inset of Fig. 6, for PET/graphene nanocomposites, the electrical conductivity quickly rises to 7.4×10^{-2} S/m from 2.0×10^{-13} with a slight increase in content from 0.47 to 1.2 vol.%. Actually, at 0.56 vol.% of the graphene content, the conductivity is 3.3×10^{-5} S/m, which is higher than

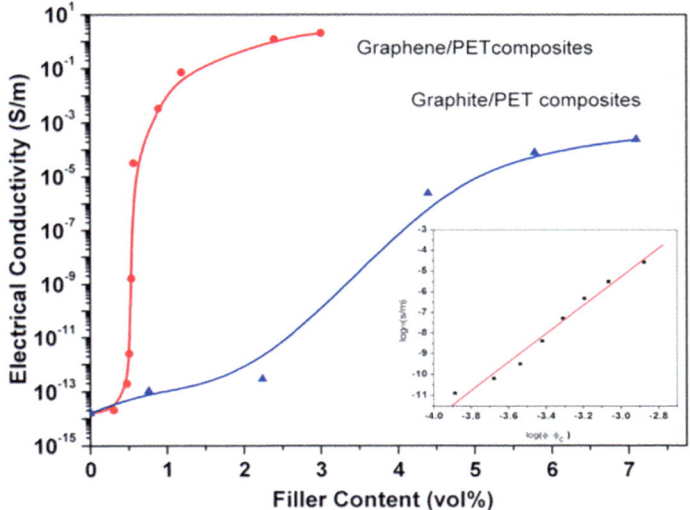

Fig. 7 Plots of electrical conductivity versus filler content for PET/graphene nanocomposites and PET/graphite composites. Reproduced with permission from Ref. [84]. Copyright 2010, Elsevier Ltd

the antistatic criterion of 10^{-6} S/m. With only 3.0 vol.% graphene, the conductivity approaches to 2.11 S/m. On the contrary, PET/graphite composites show a higher percolation threshold of 2.4 vol.% and a broad percolation transition within a range of graphite content from 2.4 to 5.8 vol.%, the conductivity of PET/pristine graphite composite with 7.1 vol.% of graphite is only 2.45×10^{-4} S/m. These results indicate the advantage of graphene nanosheets.

To investigate the percolation behavior of graphene-filled nanocomposites, aspect ratio must be taken into consideration [72]. Graphene nanosheets were evenly dispersed in matrix and covered with a thin layer of PET film, thus they can be assumed as isotropically distributed hard oblate spheroids of identical dimensions with soft shell in an insulating continuum host [73–79]. It is very important to mention that the efficiency of graphene in improving conductivity of PET is comparable to or even better than that of carbon nanotube (CNTs). Hu et al. [47] reported a low electrical percolation threshold (0.9 wt%) in PET/CNT nanocomposites fabricated via coagulation, and the electrical conductivity reached 10^{-2} S/m at higher filler loading (w5 wt%). An even lower percolation threshold was reported by Steinert and Dean [49] in PET/CNT nanocomposite films prepared using a solution mixing and casting method. It was shown that sufficient conductivity for antistatic and electrostatic dissipation purpose was achieved at 0.5 wt% CNT [49], which was reasonable. On the one hand, solution casting of a film is very efficient in forming a conductive network due to the two-dimensional distribution of CNT; on the other hand, the aspect ratio of CNT could be retained due to the less serious damage of CNT compounding like extrusion would inevitably shorten CNT and graphene, and thus reduce their aspect ratios, which is not beneficial for achieving low percolation threshold and high conductivity [80–82]. Mild condition of melt compounding is required to retain the aspect ratio of graphene sheets. The low percolation threshold obtained in the current study should be attributed to the high aspect ratio, large specific surface area, and the good dispersion of graphene in PET matrix.

The electrical conductivity of the polyethylene/reduced graphene oxide (PE/rGO) nanocomposites [83] was measured at various rGO contents, as shown in Fig. 8. The electrical conductivity generally increased as function of rGO content, demonstrating a sharp increase at approximately 3.5 wt%. This dramatic improvement occurred owing to the formation of a conducting network (at the percolation threshold), and the conductivity increased to 5.1×10^{-3} S/m, with an increase of three orders of magnitude on changing from 2.8 to 3.5 wt% rGO contents. The electrical conductivity of PE/rGO nanocomposites starts to be saturated at 5 wt% with the value of 1.2 S/m, which indicates that the electrical conductivity of the composites increased significantly once the conductive networks of rGO were formed above a certain critical concentration of rGO in the matrix. This strong increase in the conductivity at the percolation threshold could be explained by the excellent exfoliation of rGO by (3-aminopropyl) triethoxysilane and the lower number of GO layers providing better dispersion in the matrix. As the rGO contents increased to more than 5 wt%, the electrical conductivity remained steady, exhibiting single digit values.

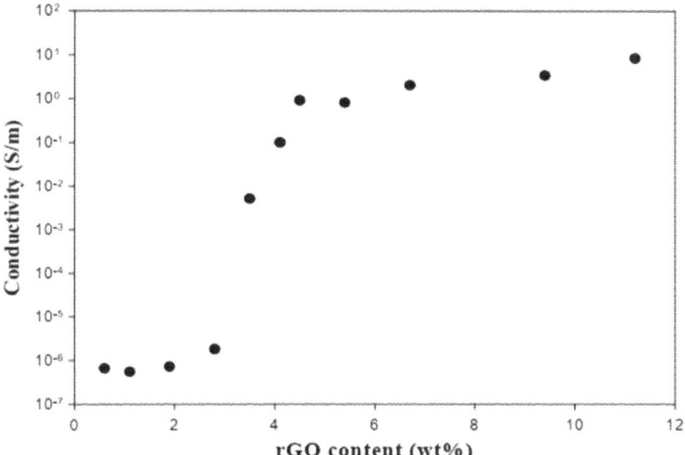

Fig. 8 Electrical conductivity of Cat-C PEs at various rGO contents. Reproduced with permission from Ref. [83]. Copyright 2010, Elsevier Ltd

7.2 Mechanical Properties

Fang [85] has developed the research article entitled "Covalent polymer fictionaliza-tion of graphene nanosheets and mechanical properties of composites" in order to show the effect of the incorporation of the graphene on the mechanical properties of polystyrene composites. He has concluded that with the addition of 0.9 wt% graphene sheets, the tensile strength and Young's modulus of the resulting PS composite film are increased by 70% and 57%, respectively. Figure 9 presents the representative stress/strain curves of the pristine PS and nanocomposite films. The pristine PS revealed a typical yield behavior with increasing stress during tension and the cor-responding Young's modulus and fracture strength are 1.45 GPa and 24.44 MPa, respectively. The addition of graphene sheets remarkably changed the tensile behav-ior of PS films: yielding became hardly discernable especially at high graphene con-tents and the elongation at break decreased gradually. However, the Young's modulus and fracture strength of nanocomposite films exhibited a remarkable increasing ten-dency. For the nanocomposite film with 0.9 wt% of graphene sheets, the Young's modulus and fracture strength increased by 2.28 GPa and 41.42 MPa, correspond-ing to increases of 57.2 and 69.5%, respectively (relative to the pristine PS film). Such mechanical improvements are significant compared to the results reported in the literature [86], which could be attributed to the efficient load transfer between graphene sheets and the PS matrix. For example, when 5 wt% of thermally reduced graphene oxide sheets were added to styrene-acrylonitrile copolymer (SAN), only 34.5% of increases in Young's modulus were reported [87]. In this regard, poor particle dispersion and interface interaction are presumably responsible for the lim-ited performance improvement. This is in contrast with the pronounced mechanical

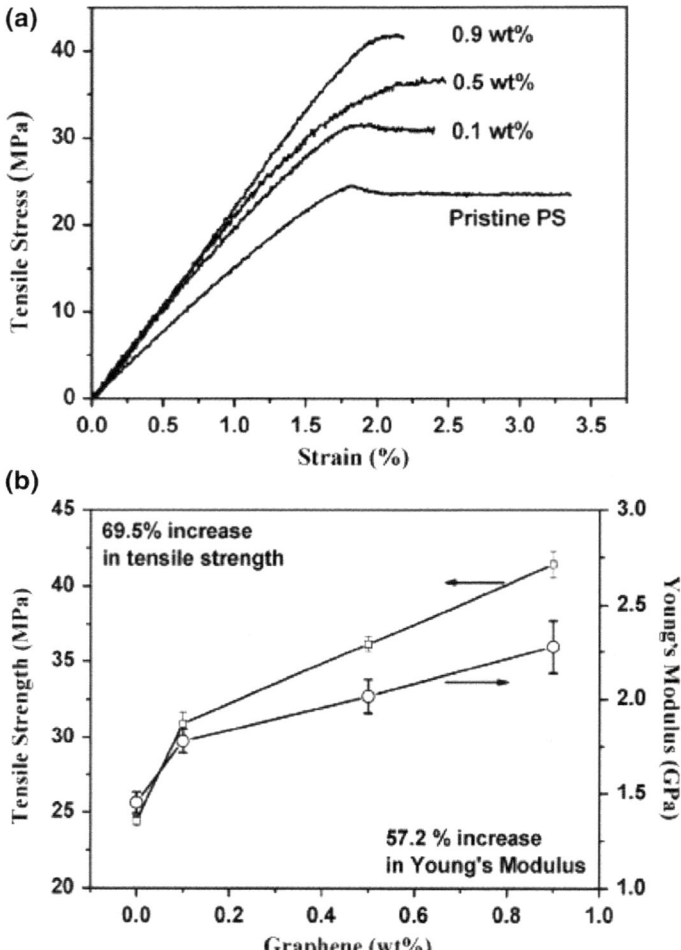

Fig. 9 **a** Representative stress/strain curves of the pristine PS and nanocomposite films with different contents of graphene sheets. **b** Young's modulus and tensile strength changes with increasing graphene content. Reproduced with permission from Ref. [85]. Copyright 2009, The Royal Society of Chemistry

enhancements observed in graphene oxide/poly(vinyl alcohol) (PVA) systems, where the PVA nanocomposite with 0.7 wt% graphene oxide sheets revealed 76 and 62% increases in Young's modulus and fracture strength. This is distinct from the case in the present graphene/PS systems (PS is a non-polar polymer), in which both dispersion and interface interaction (or load transfer efficiency) were mediated by PS chains covalently bonded to the graphene surface.

In the paper entitled "Biopolymer—Thermally reduced graphene nanocomposites: Structural characterization and properties" published by Vikas Mittal [88], the mechanical properties of nanocomposites of poly-L-lactide (PLA), bio-polyamide (PA), and poly(butylene adipate-co-terephthalate) (PBAT) with varying amounts of thermally reduced graphene were generated by melt mixing are summarized. Figure 9 shows the DMA analysis of the nanocomposites as a function of temperature. The filler incorporation enhanced the high temperature performance of the nanocomposites. Bio-polyamide (PA) and poly(butylene adipate-co-terephthalate) (PBAT) nanocomposites exhibited gradual reduction in storage modulus with temperature, whereas the PLA composites exhibited a sharp drop near the T_g and reduction of more than two orders of magnitude was observed. The T_g of PLA was also observed to increase by 5–6 °C for nanocomposite with 5% graphene content. The PA nanocomposites had the highest storage modulus especially at elevated temperatures. For instance, at 60 °C, the PA nanocomposites with 5% graphene content had storage modulus value of 4E8 Pa as compared to 5E7 Pa and 3E7 Pa for PBAT and PLA composites, respectively. Likewise, at the same temperature, the extent of increase for the PA nanocomposite was 1.8 times as compared to 1.3 and 1.1 for PBAT and PLA nanocomposites. The better interaction between the PA chains and graphene platelets was also confirmed through the tensile testing of the nanocomposites as shown in Fig. 10d and Table 1. Pure polyamide had a tensile modulus of 990 MPa,

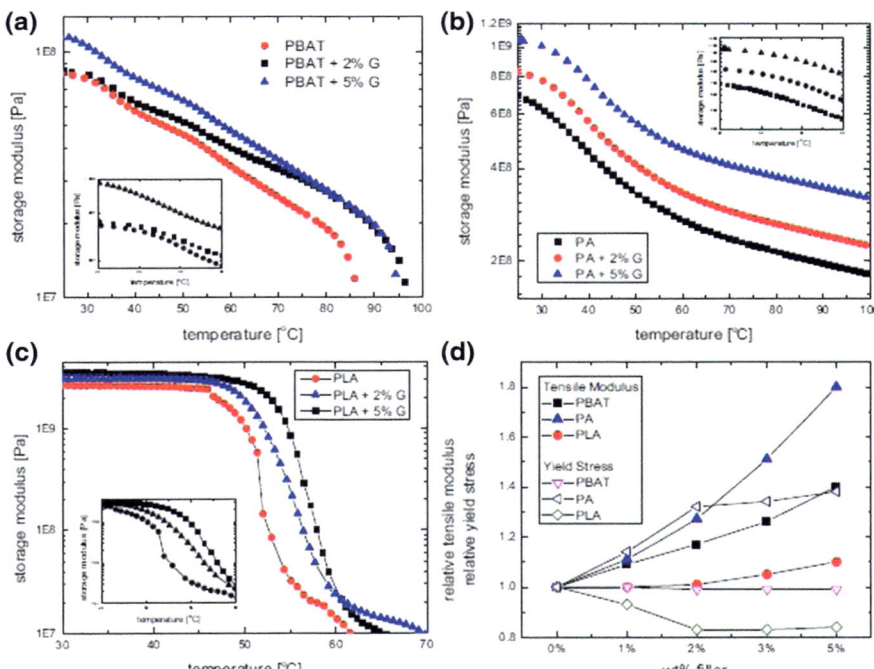

Fig. 10 Storage modulus of pure polymers and nanocomposites analyzed from DMA **a** PBAT, **b** PA, **c** PLA, and **d** relative tensile modulus and yield stress of the composites as a function of filler weight fraction. Reproduced with permission from Ref. [88]. Copyright 2014, Elsevier B.V

Table 1 Tensile properties of pure polymers and nanocomposites with graphene

Composite	Tensile modulus[a], MPa	Yield stress[b], MPa	Yield strain[a], %	Stress at break[c], MPa	Elongation[a], mm
PBAT	78	9.1	13.1	20.9	88.7
PBAT + 1%G	85	9.1	12.2	18.9	85.3
PBAT + 2%G	91	9.1	11.3	17.5	83.8
PBAT + 3%G	98	9.0	10.1	16.4	65.7
PBAT + 5%G	110	9.0	9.6	15.0	57.7
PA	990	32.5	15.3	45.1	58.0
PA + 1%G	1102	37.2	15.2	45.8	39.6
PA + 2%G	1258	42.9	15.2	46.0	25.9
PA + 3%G	1499	43.4	14.9	48.0	25.2
PA + 5%G	1780	44.7	14.3	50.5	23.2
PLA	3306	70.3	3.0	63.7	1.6
PLA + 1%G	3310	65.7	2.9	60.2	1.5
PLA + 2%G	3335	58.6	2.4	55.6	1.3
PLA + 3%G	3472	58.5	2.4	56.1	1.3
PLA + 5%G	3614	59.2	2.4	57.0	1.3

[a]Relative probable error 5%
[b]Relative probable error 2%
[c]Relative probable error 15%

which was enhanced to 1780 MPa for 5% graphene composite, thus, exhibiting an increase of 80%. For the PBAT and PLA nanocomposites, the increment in the tensile modulus for 5% graphene composites was 40% and 10%, respectively. PA composites also exhibited an increase in the yield stress on increasing the filler content from 1 to 5%. For pure PA, the yield stress was measured to ne 32.5 MPa, which enhanced to 44.7 MPa for nanocomposite with 5% graphene content. This indicated that the filler/matrix interactions and better filler dispersion led to efficient load transfer from PA matrix to the filler particles thus resulting in polymer yielding at higher stress. The PBAT nanocomposites did not show any change in the yield stress, whereas a decrease was observed in PLA nanocomposites indicating less optimal polymer/filler interaction. Stress at break also increased marginally for PA nanocomposites and yield strain was not impacted negatively on graphene incorporation. The overall elongation was decreased in all polymer systems after graphene addition indicating stiffening of the polymer chains.

7.3 Thermal Behavior

Several studies [89, 90] have demonstrated that incorporation of graphene sheets can enhance the thermal stability of the polymer to different degree under nitrogen atmosphere, while Jeong's studies [91, 92] revealed that incorporating functionalized graphene sheets accelerated the thermal degradation of the polymer under inert condition. Since the polymeric materials are commonly used under air condition, it is much more important to investigate the effect of graphene sheets on thermal oxidation stability of the polymeric materials. As shown in Fig. 11, compared to the PP matrix, both initial degradation temperature (T_i) and maximum mass loss temperature (T_{max}) are monotonously shifted up to higher temperatures with increasing

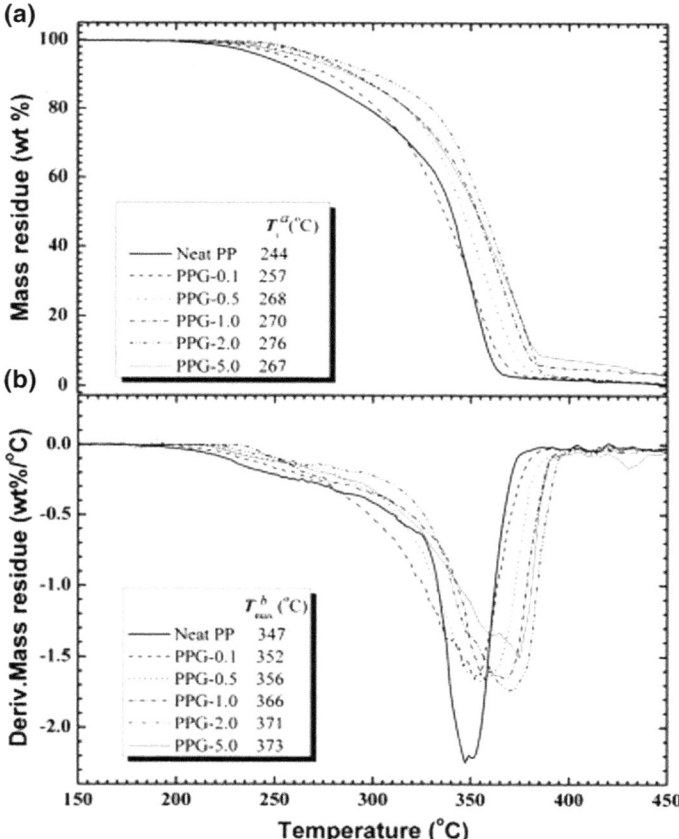

Fig. 11 **a** TGA and **b** DTG curves of PP/graphene nanocomposite with various graphene loadings, [a]T_i and [b]T_{max} represent the initial degradation temperature where 5 wt% mass loss occurs, and the maximum mass loss temperature where maximum loss rate takes place. Reproduced with permission from Ref. [94]. Copyright 2011, Elsevier Ltd

graphene loading, inferring a significant improvement of the thermal oxidation stability of PP due to the barrier effects of its sheet structure like clay planets. Neat PP starts to degrade at 244 °C (T_i) and T_{max} occurs at about 347 °C. After incorporation of 0.1 and 1.0 wt% graphene, T_i values are increased by 13 °C (257 °C) and by 26 °C (270 °C), respectively. Moreover, the T_{max} values are also enhanced by 5 °C for 0.1 wt% graphene and by 19 °C for 1.0 wt% graphene loading, respectively. Wakabayashi et al. [93] observed that addition of 0.8 wt% graphite enabled the T_i (5 wt% mass loss) of PP to be delayed by ≈15 °C in nitrogen condition, without determining the effect of graphite on the thermal oxidative degradation of PP in this study. Whatever, graphene has a great potential to enhancing the thermal stability of PP in air condition.

7.4 Barrier Properties

From the theoretical studies, the presence of impenetrable graphene nanosheets, which are homogeneously dispersed in the polymer matrix, leads to an increase of the diffusion path (tortuosity) and consequently, a decrease of the gas permeability of the graphene/polymer composites [95]. The barrier properties of graphene/polymer composites are affected strongly by the aspect ratio, dispersion, and orientation of the graphene nanosheets, the graphene nanosheets/polymer interface, and the crystallinity of the polymer matrix.

According to the article review of Yanbin Cui [96], graphene oxide can be prepared in bulk quantities by oxidation of graphite with strong oxidants containing many hydrophilic groups, such as hydroxyl, epoxy, and carboxyl acid [97]. Graphene oxide possesses the desirable characteristics of aqueous solution processability attributed to the oxygen-containing functional groups on the basal planes and edges of graphene [98]. These oxygen-containing functional groups promote complete exfoliation and homogeneous dispersion of graphene oxide sheets in polar polymer matrix and improve the interfacial bonding significantly. Graphene oxide has been compounded with various polymers [99–102]. Also Yanbin Cui cited many interesting examples, the first Kang et al. fabricated carboxylated acrylonitrile butadiene rubber (XNBR)/graphene oxide nanocomposites which have high mechanical and gas barrier properties using a simple and environment-friendly latex co-coagulation method. The addition of 1.9 vol.% of graphene oxide reduced the gas permeability coefficient of XNBR by 55% [103]. The second Zhu et al. prepared graphene oxide from graphite by using a modified Hummers method and then used graphene oxide as nanofiller to synthesize polyimide (PI)/graphene oxide by in situ polymerization (see Fig. 12). As expected, the oxygen transmission rate (OTR) decreased significantly from 377.78 of pure PI to 26.07 cm^3 m^{-2} 24 h^{-1} atm^{-1} for 30 wt% graphene oxide loaded composite, which displayed a 93% reduction compared with pure PI [104]. The third example, Huang et al. prepared high barrier poly(vinyl alcohol) (PVA)/graphene oxide nanosheets nanocomposite films by solution mixing. A more than 98% decrease in the O$_2$ permeability coefficients of PVA film from 21.17 × 10^{-15}

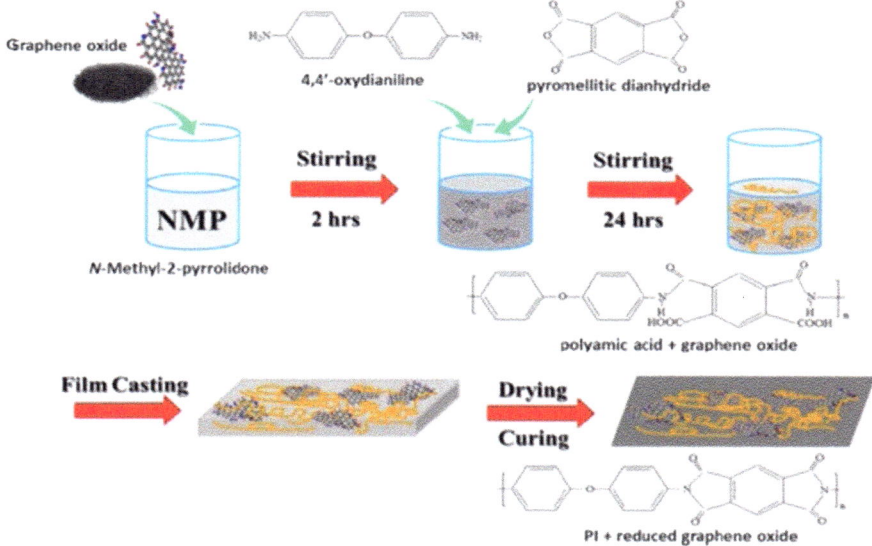

Fig. 12 The procedures of the synthesis of pure polyimide (PI) and its composite films. Reproduced with permission from Ref. [104]. Copyright 2013, Wiley Periodicals, Inc.

to 0.24×10^{-15} cm^3 cm^{-2} s^{-1} Pa^{-1} is achieved by adding only 0.72 vol.% graphene oxide nanosheets . The reduction of O2 permeability of PVA film was attributed to excellent impermeable property of graphene oxide nanosheets, their full exfoliation, uniform dispersion, and high alignment in the PVA matrix and the strong interfacial adhesion between graphene oxide nanosheets and PVA matrix [105]. Using amidation reaction and chemical reduction, dodecyl amine (DA)-functionalized graphene oxide (DA-GO) and dodecyl amine-functionalized reduced graphene oxide (DA-RGO) were produced by Ren et al. Then, high-density polyethylene (HDPE)/DA-GO and HDPE/DA-RGO nanocomposites were prepared by solution mixing method and hot-pressing process. The crystallinity, dynamic mechanical, gas barrier, and thermal stability properties of HDPE were significantly improved by the addition of DA-GO or DA-RGO. However, the performance of HDPE nanocomposites reinforced with DA-GO was almost the same as that of DA-RGO, which indicated that the reduction of DA-GO was not necessary and the interfacial adhesion and aspect ratio of graphene layers had hardly changed after reduction [106]. Morimune et al. developed an environmentally friendly technique for fabricating poly(methyl methacrylate) (PMMA)/graphene oxide nanocomposites in which PMMA was polymerized by soap-free emulsion polymerization and incorporated with graphene oxide using water as a processing medium. The addition of 1% w/w of graphene oxide to the PMMA matrix decreased the permeability by 50% and the nanocomposite with 10% w/w of graphene oxide was found to be almost completely impermeable (see Fig. 13) [98].

Fig. 13 O$_2$ gas permeability
of PMMA film and
PMMA/graphene oxide
nanocomposites.
Reproduced with permission
from Ref. [98]. Copyright
2012, American Chemical
Society

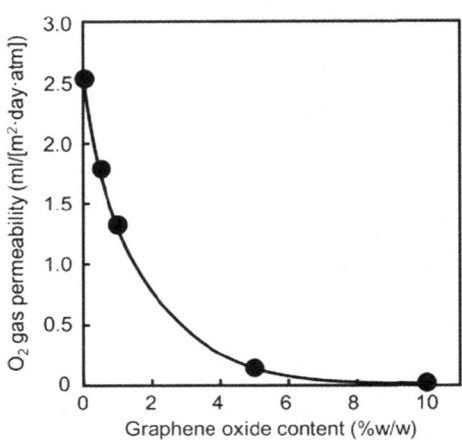

8 Examples of Graphene/Polymer Composites

8.1 Graphene/Epoxy Composites

Graphene has been incorporated into epoxy-based materials to enhance various functional properties. Graphene's excellent mechanical properties make it as a good candidate for reinforcement in nanocomposites. Bortz et al. [107] investigated the effect of graphene concentration on the mechanical properties of graphene/epoxy nanocomposites. The tensile modulus of unmodified epoxy increased by ≈12% at only 0.1 wt% loading of graphene oxide. Wajid et al. [108] used solution mixing to fabricate graphene/epoxy composites. The strength and modulus increased by 38% and 37%, respectively, at only 0.46 vol.% graphene loading. By using a simple such solution mixing technique, Galpaya et al. [109] fabricated a graphene oxide/epoxy nanocomposite with enhanced mechanical properties. At only 0.1 wt% addition of graphene oxide, fracture toughness increased by ≈50%. The graphene oxide sheets in the composite disturbed and deflected the crack propagation. The elastic modulus increased by ≈35% from the neat epoxy with only 0.5 wt% loading of graphene oxide. Shen et al. [110] investigated the tribological properties of epoxy nanocomposites reinforced with graphene oxide at low loading (0.05–0.5 wt%). The wear resistance significantly increased with the addition of graphene oxide, such that, at 0.5 wt% loading, the specific wear rate was reduced by 90.0–94.1% compared to the neat polymer.

Electrical conductivity in nanocomposites is achieved by the formation of a continuous network of the conductive fillers [111, 112]. Therefore, because graphene has a very high aspect ratio, the percolation threshold is achieved at very low graphene loading [113]. Wajid et al. [108] used sonication and shear mixing methods to disperse graphene into the epoxy matrix and achieved the percolation threshold at 0.088 vol.%.

In another study, the composites were fabricated by incorporation of functionalized graphene into the epoxy matrix by an in situ process. The percolation threshold was reached only after 0.52 wt% loading of graphene, and electrical conductivity increased from 1^{-10} to 1^{-2} S cm^{-1} by $\approx$9 vol.% loading of reduced graphene oxide. Zhao et al. [114] conducted a thorough investigation into the properties of epoxy composite filled with epoxide-functionalized graphene (G-EP). The tensile strength and Young's modulus increased by 116% and 96%, respectively, compared to the polymer at only 1 wt% loading of G-EP. The percolation threshold was reached at only 0.33 wt% loading, and electrical conductivity increased from about 1^{-17} to 1^{-2} S cm^{-1} at $\approx$2 vol.% loading. Thermal conductivity also improved by 189% at 10 wt% G-EP loading. For a more comprehensive review on graphene/epoxy nanocomposites, the reader is referred to a review by Atif et al. [115].

8.2 Graphene/Cellulose Composites

Graphene has been incorporated in cellulose composites to enhance mechanical and electrical properties. GO and cellulose were dissolved in *N*-methylmorpholine *N*-oxide (NMMO) monohydrate to prepare the composite films [116]. Despite being brittle, the produced graphene/cellulose composites showed improved thermal and electrical properties. Nguyen et al. [117] used reduced graphene oxide sheets (RGO) and amine-modified nanofibrillated cellulose (A-NFC) to produce composites with enhanced properties. These composites showed good electrical conductivity of up to 71.8 S m^{-1} and tensile strength of up to 273 MPa. Graphene cellulose paper (GCP) membranes are used to fabricate flexible supercapacitors [118]. These composites displayed high electrical conductivity stability with a decrease of only 6% after being bent 1000 times. The capacitance of these supercapacitors per geometric area is 81 mF cm^{-2}, which is equivalent to a gravimetric capacitance of 120 F g^{-1} of graphene, and these supercapacitors did not lose any capacitance after 5000 cycles. Manman et al. [119] reported cellulose/graphene composite hydrogels prepared from ionic liquids (IL). They used vitamin C to prepare RGO directly in IL. These hydrogels showed enhanced mechanical and thermal properties. At only 0.5 wt% doping of RGO in cellulose composite, the Young's modulus was improved more than four times.

8.3 PVA/Graphene Nanocomposites

Zhao et al. [120] reported the preparation of graphene nanosheets and poly(vinyl alcohol) (PVOH) via a facial aqueous solution. In their study, they showed that Graphene and flat carbon nanosheets have generated huge activity in many areas of science and engineering due to its unprecedented physical and chemical properties. With the development of wide-scale applicability including facile synthesis and high yield, this exciting material is ready for its practical application in the preparation of polymer nanocomposites. Here, they have reported that nanocomposites based on fully exfoliated graphene nanosheets and poly(vinyl alcohol) (PVA) are prepared via a facial aqueous solution. A significant enhancement of mechanical properties of the graphene/PVA composites is obtained at low graphene loading; that is, a 150% improvement of tensile strength and a nearly 10 times increase of Young's modulus are achieved at a graphene loading of 1.8 vol.%. The comparison between the experimental results and theoretical simulation for Young's modulus indicates that the graphene nanosheets in polymer matrix are mostly dispersed randomly in the nanocomposite films. Liang et al. [90] also studied poly(vinyl alcohol) (PVOH)/-graphene nanocomposites. In this study, they used water as a solvent to fuse GO into the PVOH matrix. At only 0.7 wt% GO loading, tensile strength increased by 76% and Young's modulus by 62%.

8.4 Polyurethane (PU)/Graphene Composites

Lee et al. [121] reported the study of waterborne polyurethane (WPU)/graphene sheets (FGS) nanocomposites prepared using an in situ method. The electrical conductivity of the nanocomposites was increased 10^5-fold compared with that of pure WPU. This was attributed to high dispersion and homogeneity of graphene sheets in the WPU matrix. They obtained a percolation threshold of graphene sheets at only 2 wt%. Liang et al. [122] used thermoplastic polyurethane (TPU) to fabricate nanocomposites with three differently processed graphene samples. They were able to show that the rate of thermal degradation for thermoplastic polyurethane/isocyanate modified graphene composites are much higher than that of the sulfonated graphene and reduced graphene-based TPU nanocomposites. The TPU/-graphene nanocomposites doped with only 1 wt% sulfonated graphene showed enhanced infrared-triggered actuation performance. The tests showed that the composite could contract and lift a 21.6 g weight up to 3.1 cm with 0.21 N of force when exposed to IR. The mechanical properties of this composite were also increased. The tensile strength of TPU/sulfonated graphene nanocomposites with 1 wt% graphene was increased by 75% at a strain of 100%, and the Young's modulus was enhanced by 120% [122]. However, IR-triggered actuation performance of isocyanate modified graphene/TPU nanocomposites showed inferior results.

8.5 Graphene/Polyethylene Terephthalate (PET) Nanocomposites

The melt intercalation technique was used to prepare PET/graphene composites by Zhang et al. [84]. Transmission electron microscopy analysis showed that graphene nanosheets were evenly distributed in the PET matrix. This criterion was confirmed by the high electrical conductivity values they got from the PET/graphene composites. The threshold percolation was only 0.47 vol.% loading, and with graphite 2.4 vol.%. A high electrical conductivity of 2.11 S m^{-1} was achieved with the addition of only 3.0 vol.% graphene.

8.6   Polycarbonate (PC)/Graphene Nanocomposites

Kim et al. [34] studied PC composites doped with graphite and functionalized graphene sheets (FGS) by the melt intercalation technique. In this study, they used melt rheology to study the viscoelastic properties of the composites. They annealed the composites for 10,000 s and observed that the composites manifest a solid-like state above the percolation threshold, which was around 1 wt% for the FGS, while around 3 wt% for graphite loading. In terms of electrical properties, FGS/PC showed lower percolation threshold as compared to graphite/PC composites. Both graphite and FGS fillers lead to improved PC composites stiffness and dimensional stability. The composites also exhibited good barrier properties. The permeability of nitrogen and helium of the PC composites is significantly reduced by incorporation of both FGS and graphite. Moreover, FGS showed potential for application in gas separation processes as it was more effective against molecules with large kinetic diameter.

8.7   Polystyrene (PS)/Graphene Nanocomposites

Polystyrene/graphene composites have been extensively studied for various applications. Stankovich et al. [123] reported PS/isocyanate modified graphene composites prepared using a solution blending method in DMF. They obtained a percolation threshold for the electrical conductivity at 0.1 vol.% GO in PS. The low percolation value is due to the homogeneous dispersion and extremely large aspect ratio of graphene sheets. The PS/graphene composite showed an electrical conductivity of $\approx$0.1 to $\approx$1 S m^{-1} at 2.5 vol.% loading. Liu et al. [124] prepared a PS/GNP composite in ionic liquids. The electrical conductivity increased from 10^{-14} S m^{-1} for pure PU to 5.77 S m^{-1} with the addition 0.38 vol.% GNP, which is 3–15 times higher than that of polystyrene composites filled with single-walled carbon nanotubes. They also studied the thermal stability of PS/GNP composite and pure polystyrene. The degradation temperature of the PS/GNP composite was about 50 °C higher than that of pure PS. This enhancement is due to the strong interaction of GNP and the polymer matrix at the interface, which leads to a decrease of polymer chain mobility near the interface and, hence, the increase in thermal stability.

9   Conclusions

Graphene-based polymer nanocomposites represent one of the most technologically promising developments to emerge from the interface of graphene-based materials and polymer materials. However, there are still many challenges that must be addressed for these nanocomposites to reach their full impact. In this review, many aspects of graphene have been discussed: properties, fabrication methods, processing, and properties of the graphene/polymer nanocomposites with some application in polymer composites examples. From the properties aspect of graphene, it is clear that it has a high potential for various applications including nanofiller for functional composites. The choice of fabrication methods is determined by the surface functionalization of integrated graphitic sheets. Generally, traditional fabrication routines include solution-based processing and melt-based processing. Among most popular approaches for chemical modification and assembly are in situ polymerization, chemical grafting, latex emulsion blending, layer-by-layer (LbL) assembly, and directed assembly. For the in situ polymerization method, intercalated monomers within expanded graphite clusters can promote their efficient exfoliation into single sheets throughout the polymer matrix caused by catalysis reactions. The application of graphene in polymer nanocomposite was reviewed. It is evident that GO is the most widely used graphene-based nanofiller in polymer composites, largely because of the chemical interaction between the filler and the polymer matrix that enhances homogeneous dispersion of graphene in the matrix, thus improving the overall composite performance. The electrical, mechanical, thermal, and barrier properties of these blends and the influence of the intrinsic properties of these fillers (graphene and grapheme oxide) and their state of dispersion in polymer matrix on the properties are summarized.

References

1. Huang, J.-C.: Carbon black filled conducting polymers and polymer blends. J. Poly. Proc. Inst. **21**, 299 (2002)
2. Moniruzzaman, M., Winey, K.I.: Macromol **39**, 5194 (2006)
3. Okamoto, M.: Polymer/clay nanocomposites, vol. 8. American Scientific Publishers, Stevenson Ranch, CA (2004)
4. Geim, A.K., Novoselov, K.S.: Nature Mater. **6**, 183–191 (2007)
5. Lax, A., Maxwell, R.: Natl. Trust Ann. Archaeol. Rev. 18–23 (1998–1999)
6. Weinberg, S.S.: The stone age in the Agean, vol. 1, 10th edn. Cambridge University Press, Cambridge (2007)
7. Iijima, S.: Nature **354**, 56 (1991)
8. Kroto, H.W., Heath, J.R.: Nature **318**, 1216 (1985)
9. Eizenberg, M., Blakely, J.M.: Surf. Sci. **82**, 228 (1979)
10. Novoselov, K.S., Geim, A.K., Morozov, S., Jiang, D., Zhang, Y.: Science **306**, 666 (2004)
11. Fitzer, E., Kochling, K.-H., Boehm, H.P., Marsh, H.: Pure App. Chem **67**, 473 (1995)
12. Boehm, H.-P., Stumpp, E.: Carbon **7**, 1381 (2007)
13. Brodie, B.C.: Justus Liebigs Annalen Der Chem. **114**, 6 (1860)

14. Hummers Jr., W.S., Offeman, R.E.: J. Am. Chem. Soci **80**, 1339 (1958)
15. Boehm, H.P., Clauss, A., Fischer, G.O., Hofmann, U.: Z. Naturforsch. **17**, 150 (1962)
16. Liu, C.-Y., Chang, H., Bard, A.J.: Langmuir **7**, 1136 (1991)
17. Lu, X., Yu, M., Huang, H., Ruoff, R.S.: Nanotechnology **10**, 269 (1999)
18. Berger, C., Song, Z., Li, T., Li, X., Ogbazghi, A.Y., Feng, R., Dai, Z., Marchenkov, A.N., Conrad, E.H., First, P.N., Phy, J.: Chem. B **108**, 19912 (2004)
19. Zhang, Y., Small, J.P., Amori, M.E., Kim, P.: Phy. Rev. Lett **94**, 176803 (2005)
20. Obeng, Y., Srinivasan, P.: Interface Mag. **20**, 47 (2011)
21. Novoselov, K.S., Geim, A.K., Morozov, S.V., Jiang, D., Katsnelson, M.I., Grigorieva, I.V., Dubonos, S.V., Firsov, A.A.: Nature **438**, 197 (2005)
22. Katsnelson, M.I., Novoselov, K.S., Geim, A.K.: Nat. Phys. **2**, 620 (2006)
23. Staudenmaier, L.: Beric. Deut. Chem. Gese **31**, 1481 (1898)
24. Dreyer, D.R., Park, S., Bielawski, C.W., Ruoff, R.S.: Chem. Soc. Rev. **39** (2010)
25. Buchsteiner, A., Lerf, A., Pieper, J.: J. Phys. Chem. B **110**, 22328 (2006)
26. Dikin, D.A., Stankovich, S., Zimney, E.J., Piner, R.D., Dommett, G.H., Evmenenko, G., Nguyen, S.T., Ruoff, R.S.: Nature **448**, 457 (2007)
27. Ruess, V.G., Vogt, F.: Monatsh. Chem. **78** (1948)
28. Park, S., An, J., Jung, I., Piner, R.D., An, S.J., Li, X., Velamakanni, A., Ruoff, R.S.: Nano Lett. **9**, 1593 (2009)
29. Potts, J.R., Dreyer, D.R., Bielawski, C.W., Ruoff, R.S.: Polymer **52**, 5 (2011)
30. Xu, Y., Wang, Y., Liang, J., Huang, Y., Ma, Y., Wan, X., Chen, Y.: Nano Res. **2**, 343 (2009)
31. Quan, H., Zhang, B., Zhao, Q., Yuen, R.K., Li, R.K.: Compos. Part A: App. Sci. Manuf. **40**, 1506 (2009)
32. Eda, G., Chhowalla, M.: Nano Lett. **9**, 814 (2009)
33. Liang, J., Xu, Y., Huang, Y., Zhang, L., Wang, Y., Ma, Y., Li, F., Guo, T., Chen, Y.: J. Physi. Chem. C **113**, 9921 (2009)
34. Kim, H., Macosko, C.W.: Polymer **50**, 3797 (2009)
35. Ramanathan, T., Stankovich, S., Dikin, D.A., Liu, H., Shen, H., Nguyen, S.T., Brinson, L.C., Poly, J.: Sci. Part B: Poly. Phys **45**, 2097 (2007)
36. Kim, S., Do, I., Drzal, L.T.: Macromol. Mat. Eng. **294**, 196 (2009)
37. Liang, J., Huang, Y., Zhang, L., Wang, Y., Ma, Y., Guo, T., Chen, Y.: Adv. Func. Mat **19**, 2297 (2009)
38. Zhao, X., Zhang, Q., Chen, D., Lu, P.: Macromol **43**, 2357 (2010)
39. Kalaitzidou, K., Fukushima, H., Drzal, L.T.: Comp. Sci. Tech. **67**, 2045 (2007)
40. Zheng, W., Lu, X., Wong, S.C.: J. App. Poly. Sci. **91**, 2781 (2004)
41. Zhao, Y.F., Xiao, M., Wang, S.J., Ge, X.C., Meng, Y.Z.: Comp. Sci. Tech. **67**, 2528 (2007)
42. Du, X.S., Xiao, M., Meng, Y.Z., Poly, J.: Sci. Part B: Poly. Phys. **42**, 1972 (2004)
43. Cho, D., Lee, S., Yang, G., Fukushima, H., Drzal, L.T.: Macromol. Mat. Eng **290**, 179 (2005)
44. Kim, H., Miura, Y., Macosko, C.W.: Chem. Mat. **22**, 3441 (2010)
45. Li, H., Pang, S., Wu, S., Feng, X., Müllen, K., Bubeck, C.: J. Am. Chem. Soc. **133**, 9423 (2011)
46. Cassagneau, T., Fendler, J.H.: Adv. Mater. **10**, 877 (1998)
47. Hu, H., Wang, X., Wang, J., Wan, L., Liu, F., Zheng, H., Chen, R., Xu, C.: Chem. Phys. Lett. **484**, 247 (2010)
48. Hu, K., Kulkarni, D.D., Choi, I., Tsukruk, V.V.: Prog. Poly. Sci. **39**, 1934 (2014)
49. Steinert, B.W., Dean, D.R.: Polymer **50**, 898 (2009)
50. Nair, R.R., Blake, P., Grigorenko, A.N., Novoselov, K.S., Booth, T.J., Stauber, T., Peres, N.M.R., Geim, A.K.: Science **320**, 1308 (2008)
51. Blake, P., Hill, E.W., Castro Neto, A.H., Novoselov, K.S., Jiang, D., Yang, R., Booth, T.J., Geim, A.K.: Appl. Phys. Lett **91**, 063124 (2007)
52. Ni, Z.H., Wang, H.M., Kasim, J., Fan, H.M., Yu, T., Wu, Y.H., Feng, Y.P., Shen, Z.X.: Nano Lett. **7**, 2758 (2007)
53. Yazyev, O.V., Louie, S.G.: Nat. Mater. **9**, 806 (2010)
54. Červenka, J., Katsnelson, M.I., Flipse, C.F.J.: Nat. Phys. **5**, 840 (2009)

55. Yazyev, O.V., Louie, S.G.: Phys. Rev B **81**, 195420 (2010)
56. Huang, P.Y., Ruiz-Vargas, C.S., van der Zande, A.M., Whitney, W.S., Levendorf, M.P., Kevek, J.W., Garg, S., Alden, J.S., Hustedt, C.J., Zhu, Y., Park, J., McEuen, P.L., Muller, D.A.: Nature **469**, 389 (2011)
57. Meyer, J.C., Geim, A.K., Katsnelson, M.I., Novoselov, K.S., Booth, T.J., Roth, S.: Nature **446**, 60 (2007)
58. Tung, V.C., Allen, M.J., Yang, Y., Kaner, R.B.: Nat. Nanotechnol **4**, 25 (2009)
59. Xu, K., Cao, P., Heath, J.R.: Nano Lett. **9**, 4446 (2009)
60. Decker, R., Wang, Y., Brar, V.W., Regan, W., Tsai, H.-Z., Wu, Q., Gannett, W., Zettl, A., Crommie, M.F.: Nano Lett. **11**, 2291 (2011)
61. Bao, W., Miao, F., Chen, Z., Zhang, H., Jang, W., Dames, C., Lau, C.N.: Nat. Nano **4**, 562 (2009)
62. Xu, B., Yue, S., Sui, Z., Zhang, X., Hou, S., Cao, G., Yang, Y.: Ener. Envi. Sci **4**, 2826 (2011)
63. Khanra, P., Kuila, T., Bae, S.H., Kim, N.H., Lee, J.H., Mat, J.: Chem **22**, 24403 (2012)
64. Graf, D., Molitor, F., Ensslin, K., Stampfer, C., Jungen, A., Hierold, C., Wirtz, L.: Nano Lett. **7**, 238 (2007)
65. Ferrari, A.C., Meyer, J.C., Scardaci, V., Casiraghi, C., Lazzeri, M., Mauri, F., Piscanec, S., Jiang, D., Novoselov, K.S., Roth, S., Geim, A.K.: Phys. Rev. Lett **97** (2006)
66. Young, R.J., Kinloch, I.A.: Nanoscience. Nano. Chem. R. Soc. Chem. **1** (2013)
67. Malard, L.M., Pimenta, M.A., Dresselhaus, G., Dresselhaus, M.S.: Phys. Rep. **473**, 51 (2009)
68. Eckmann, A., Felten, A., Verzhbitskiy, I., Davey, R., Casiraghi, C.: Phys. Rev B **88**, 035426 (2013)
69. Luo, Z., Yu, T., Shang, J., Wang, Y., Lim, S., Liu, L., Gurzadyan, G.G., Shen, Z., Lin, J.: Adv. Func. Mater. **21**, 911 (2011)
70. Hernandez, Y., Nicolosi, V., Lotya, M., Blighe, F.M., Sun, Z., De, S., McGovern, I.T., Holland, B., Byrne, M., Gun'Ko, Y.K., Boland, J.J., Niraj, P., Duesberg, G., Krishnamurthy, S., Goodhue, R., Hutchison, J., Scardaci, V., Ferrari, A.C., Coleman, J.N.: Nat. Nanotechnol. **3**, 563 (2008)
71. Stauffer, D.: Lond. Phil. (1985)
72. Ma, H.M., Gao, X.-L.: Polymer **49**, 4230 (2008)
73. Ambrosetti, G., Johner, N., Grimaldi, C., Danani, A., Ryser, P.: Phys. Rev. E **78**, 061126 (2008)
74. Rintoul, M.D., Torquato, S.J.: Phys. A: Math. Gen. **30** (1997)
75. Garboczi, E.J., Snyder, K.A., Douglas, J.F., Thorpe, M.F.: Phys. Rev. E **52**, 819 (1995)
76. Yi, Y.B., Tawerghi, E.: Phys. Rev. E **79**, 041134 (2009)
77. Balberg, I.: Phys. Rev. Lett. **59**, 1305 (1987)
78. Quivy, A., Deltour, R., Jansen, A.G., Wyder, P.: Phys. Rev B **39**, 1026 (1989)
79. Toker, D., Azulay, D., Shimoni, N., Balberg, I., Millo, O.: Phys. Rev B **68**, 041403 (2003)
80. Dasari, A., Yu, Z.-Z., Mai, Y.-W.: Polymer **50**, 4112 (2009)
81. Xiao, M., Sun, L., Liu, J., Li, Y., Gong, K.: Polymer **43**, 2245 (2002)
82. Zheng, W., Wong, S.C.: Compos. Sci. Tech. **63** (2003)
83. Zhang, H., Park, J.-H., Yoon, K.-B.: Comp. Sci. Tech **154**, 85 (2018)
84. Zhang, H.-B., Zheng, W.-G., Yan, Q., Yang, Y., Wang, J.-W., Lu, Z.-H., Ji, G.-Y., Yu, Z.-Z.: Polymer **51**, 1191 (2010)
85. Fang, M., Wang, K., Lu, H., Yang, Y., Nutt, S.: J. Mat. Chem. **19**, 7098 (2009).
86. Debelak, B., Lafdi, K.: Carbon **45**, 1727 (2007)
87. Steurer, P., Wissert, R., Thomann, R., Mülhaupt, R.: Macromol. Rapid Commun. **30**, 316 (2009)
88. Mittal, V., Chaudhry, A.U., Luckachan, G.E.: Mater. Chem. Phys. **147**, 319 (2014)
89. Xu, Y.X., Hong, W.J., Bai, H., Li, C., Shi, G.Q.: Carbon **47** (2009)
90. Liang, J., Huang, Y., Zhang, L., Wang, Y., Ma, Y., Guo, T., Chen, Y.: Adv. Func. Mater. **19**, 2297 (2009)
91. Raghu, A.V., Lee, Y.R., Jeong, H.M., Shin, C.M.: Macromol. Chem. Phys. **209** (2008)
92. Lee, Y.R., Raghu, A.V., Jeong, H.M., Kim, B.K.: Macromol. Chem. Phys. **210** (2009)

93. Wakabayashi, K., Pierre, C., Dikin, D.A., Ruoff, R.S., Ramanathan, T., Brinson, L.C., Torkelson, J.M.: Macromolecules **41**, 1905 (2008)
94. Song, P., Cao, Z., Cai, Y., Zhao, L., Fang, Z., Fu, S.: Polymer **52**, 4001 (2011)
95. Nielsen, L.E.: J. Macromol. Sci. Chem. **1**, 929 (1967)
96. Cui, Y., Kundalwal, S.I., Kumar, S.: Carbon **98**, 313 (2016)
97. Shim, S.H., Kim, K.T., Lee, J.U., Jo, W.H., Appl, A.C.S.: Mater. Interfaces **4**, 4184 (2012)
98. Morimune, S., Nishino, T., Goto, T., Appl, A.C.S.: Mater. Interfaces **4**, 3596 (2012)
99. Bao, C., Guo, Y., Song, L., Kan, Y., Qian, X., Hu, Y., Mat, J.: Chem **21**, 13290 (2011)
100. Xu, Z., Gao, C.: Macromolecules **43**, 6716 (2010)
101. Cai, D., Jin, J., Yusoh, K., Rafiq, R., Song, M.: Comput. Sci. Technol. **72**, 702 (2012)
102. Zhou, T., Chen, F., Tang, C., Bai, H., Zhang, Q., Deng, H., Fu, Q.: Comput. Sci. Technol. **71**, 1266 (2011)
103. Kang, H., Zuo, K., Wang, Z., Zhang, L., Liu, L., Guo, B.: Comput. Sci. Technol. **92**, 1 (2014)
104. Zhu, J., Lim, J., Lee, C.-H., Joh, H.-I., Kim, H.C., Park, B., You, N.-H., Lee, S.: J. Appl. Pol. Sci **131** (2014)
105. Huang, H.-D., Ren, P.-G., Chen, J., Zhang, W.-Q., Ji, X., Li, Z.-M.: J. Mem. Sci. **409–410**, 156 (2012)
106. Ren, P.-G., Wang, H., Huang, H.-D., Yan, D.-X., Li, Z.-M.: J. Appl. Pol. Sci. **131** (2014)
107. Bortz, D.R., Heras, E.G., Martin-Gullon, I.: Macromolecules **45**, 238 (2012)
108. Wajid, A.S., Ahmed, H.S.T., Das, S., Irin, F., Jankowski, A.F., Green, M.J.: Macromol. Mat. Eng. **298**, 339 (2013)
109. Galpaya, D., Wang, M., George, G., Motta, N., Waclawik, E., Yan, C.: J. Appl. Phys. **116**, 053518 (2014)
110. Shen, X.-J., Pei, X.-Q., Fu, S.-Y., Friedrich, K.: Polymer **54**, 1234 (2013)
111. Cooper, D.R., D'Anjou, B., Ghattamaneni, N., Harack, B., Hilke, M., Horth, A., Majlis, N., Massicotte, M., Vandsburger, L., Whiteway, E., Yu, V.: Inter. Scho. Res. Not. (2012)
112. Wei, J., Vo, T., Inam, F.: RSC Adv. **5**, 73510 (2015)
113. Sandler, J.K.W., Kirk, J.E., Kinloch, I.A., Shaffer, M.S.P., Windle, A.H.: Polymer **44**, 5893 (2003)
114. Zhao, S., Chang, H., Chen, S., Cui, J., Yan, Y.: Euro. Poly. J **84**, 300 (2016)
115. Atif, R., Shyha, I., Inam, F.: Polymers **8**, 281 (2016)
116. Kim, C.-J., Khan, W., Kim, D.-H., Cho, K.-S., Park, S.-Y.: Carbohydr. Polym. **86**, 903 (2011)
117. Luong, N.D., Pahimanolis, N., Hippi, U., Korhonen, J.T., Ruokolainen, J., Johansson, L.-S., Nam, J.-D., Seppälä, J.: J. Mater. Chem **21**, 13991 (2011)
118. Weng, Z., Su, Y., Wang, D.-W., Li, F., Du, J., Cheng, H.-M.: Adv. Ener. Mater. **1**, 917 (2011)
119. Xu, M., Huang, Q., Wang, X., Sun, R.: Indu. Cro. Prod. **70**, 56 (2015)
120. Zhao, X., Zhang, Q., Chen, D., Lu, P.: Macromolecules **43**, 2357 (2010)
121. Lee, Y.R., Raghu, A.V., Jeong, H.M., Kim, B.K.: Macromol. Chem. Phys. **210**, 1247 (2009)
122. Liang, J., Xu, Y., Huang, Y., Zhang, L., Wang, Y., Ma, Y., Li, F., Guo, T., Chen, Y.: J. Phys. Chem. C **113**, 9921 (2009)
123. Stankovich, S., Dikin, D.A., Dommett, G.H.B., Kohlhaas, K.M., Zimney, E.J., Stach, E.A., Piner, R.D., Nguyen, S.T., Ruoff, R.S.: Nature **442**, 282 (2006)
124. Liu, N., Luo, F., Wu, H., Liu, Y., Zhang, C., Chen, J.: Adv. Func. Mater **18**, 1518 (2008)

Facile Room Temperature Synthesis of Reduced Graphene Oxide as Efficient Metal-Free Electrocatalyst for Oxygen Reduction Reaction

Arpan Kumar Nayak and Akshaya Kumar Swain

Abstract A continuous global demand for energy resources poses serious threats to the human race in forms of pollution that stimulates many natural hazards. To overcome such problems, fuel cell technology seems to be a viable solution. However, it remains a challenge to develop highly efficient metal-free electrocatalysts for oxygen reduction reaction (ORR) to achieve optimal performance for the fuel cells. Herein, we demonstrate a facile room temperature synthesis of reduced graphene oxide (RGO) via chemical reduction of graphene oxide (GO) using sodium iodide (NaI) and hydrochloric acid (HCl). As-synthesized GO and RGO were employed as an efficient electrocatalyst for the ORR in 0.1 M KOH. The RGO shows higher ORR activity compared to GO due to its higher surface area and low charge transfer resistance. Thus as-synthesized RGO is found to be a viable metal-free electrocatalyst with higher current density, larger half-wave potential, and long-term operation stability for ORR via a four-electron pathway in alkaline media. The high performance of cost-effective RGO-based ORR electrodes is suitable to function as an alternative to platinum-based materials for energy conversion device applications.

1 Introduction

Carbon is the fourth abundant element on earth that forms the building block of all forms of life. Owing to its electronic configuration ($1s^2 2s^2 2p^2$), it can easily form triple, double, or single bonds when interacted with another carbon atom. Carbon also forms a variety of molecules that could lead to either simple or complex chains. There are mainly two types of nanostructures based on the bonding [1]. The first type of carbon nanostructures mainly exhibits sp^2 hybridization. The second type of carbon nanostructures contains a mixture of both sp^2 and sp^3 carbon atoms. Graphene, a two-dimensional monolayer of graphite, forms the basis of all other nanostructures of carbon exhibits a honeycomb lattice. The atoms in graphene

A. K. Nayak (✉) · A. K. Swain
Department of Physics, School of Advanced Sciences, Vellore Institute of Technology (VIT), Vellore, Tamilnadu 632014, India
e-mail: aknayakju@gmail.com

© Springer Nature Switzerland AG 2019
S. Sahoo et al. (eds.), *Surface Engineering of Graphene*, Carbon Nanostructures,
https://doi.org/10.1007/978-3-030-30207-8_10

259

are sp^2 hybridized with a molecular bond length of 0.142 nm. The graphene sheet can be used to generate all other dimensional forms. For example, graphite (three-dimensional) can be formed by simply stacking graphene sheets on top of each other. It can be rolled to form the carbon nanotubes (two-dimensional) and wrapped to yield fullerenes (one-dimensional) [2]. Thus, based on dimensionality, one can classify the carbon nanostructures. The zero-dimensional nanostructures include carbon dots, fullerenes, quantum dots, nanodiamonds, and the like. The one-dimensional nanostructures include nanoribbons, carbon nanotubes (CNTs), nanowires, nanofibers, and the like. Finally, three-dimensional nanostructures include all other graphitic forms. Georgakilas et al., reported a detailed study and the properties of carbon allotropes [1].

The isolation of graphene from graphite using a sticky tape led to the 2010 Nobel Prize for Andre Geim and Konstantin Novoselov. It was the first time to realize a two-dimensional material which could have not existed due to Mermin's theorem. Eventually, graphene becomes a wonder material for researchers due to its superlative properties that kept surprising the science community [3]. Graphene has a surface area of ~3000 m^2 per g, which is the largest among all other materials. Also, it has higher thermal conductivity than diamond even. However, the most appealing characteristic of graphene is in its high density of massless Dirac fermions with zero rest mass that are abundant at room temperature.

2   Structure of Graphene

The p_z electron in graphene is free that lies perpendicular to the plane of the graphene sheet where the atoms are sp^2 bonded (Fig. 1). The in-plane sp^2-bonded carbon atom gives rise to σ-bonds. The nearest neighboring atoms have a separation of 0.142 nm. One can easily compute this by taking the average of C–C and C=C bond lengths in benzene. The three sp^2-bonded carbon atoms form the honeycomb lattice of graphene

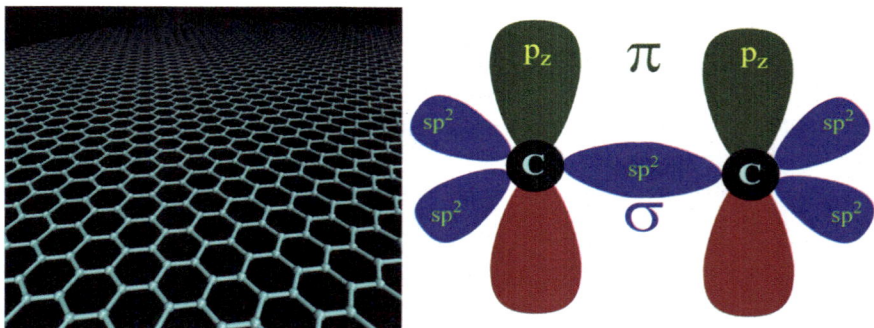

Fig. 1 Pictorial representation of graphene sheet (left) and sp^2 hybridization in graphene (right) [4]. Reprinted with permission from Ref. [4]. Copyright 2018 @ Royal Society of Chemistry

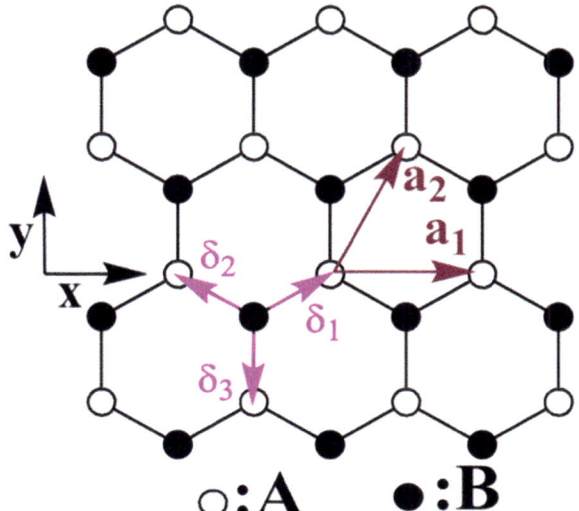

Fig. 2 Honeycomb lattice structure of graphene. The atoms in the sublattice are separated by a_0. The basis vectors of the Bravais lattice are represented by a_1 and a_2. The nearest vectors that connect the atoms from the sublattice A to sublattice B are represented by δ_1, δ_2, and δ_3 [4]. Reprinted with permission from Ref. [4]. Copyright 2018 @ Royal Society of Chemistry

[5]. However, the Bravais lattice of graphene is a superposition of two triangular Bravais lattices (sublattices A and B), thereby giving two free electrons per unit cell [6]. Thus, a semi-metallic behavior can be expected from graphene due to such basis nature of graphene. The lattice vectors of its unit cell can be written as follows,

$$\vec{a_1} = a_0\sqrt{3}(1, 0)$$

$$\vec{a_2} = a_0\sqrt{3}\left(\frac{1}{2}, \frac{\sqrt{3}}{2}\right)$$

The vectors, δ_1, δ_2, and δ_3 represent the nearest neighboring vectors connecting sublattice A to sublattice B (Fig. 2). a_0 gives the distance between the atoms from the two sublattices.

$$\delta_1 = \frac{a_0}{2}\left(\sqrt{3}, 1\right)$$

$$\delta_2 = \frac{a_0}{2}\left(-\sqrt{3}, 1\right)$$

$$\delta_3 = a_0(0, -1)$$

The superlative properties of graphene/graphene oxide form the basis of many applications. A suitable combination of materials with graphene can yield the desired properties in its composite. For example, excellent photocatalytic properties have been found with graphene–metal oxide composites [7]. Graphene–iron oxide composites prepared by Swain et al., proved to be an efficient carrier of drugs that shows an enhanced killing efficiency of cancer cells [8]. Graphene with noble metals can exhibit high efficient fuel cells, electrochemical sensors. Sun and Shi, have shown promising energy applications using graphene/polymer composites [9]. On the other hand, platinum (Pt) and its alloy have long been regarded as the best oxygen reduction reaction (ORR) catalyst in fuel cells [10, 11]. However, the high cost, limited earth abundance, poor durability of Pt catalysts hamper widely in commercial applications of fuel cells. In this prospect, a broad range of alternative catalysts was introduced to reduce/replace Pt-based catalysts [12, 13]. Recently, carbon-based nanomaterials known as metal-free ORR catalysts have been explored as alternative ORR catalysts, which is low cost and improves the fuel-cell efficiency [14, 15]. In particular, metal-free vertically aligned nitrogen-doped carbon nanotube (VA-NCNT) arrays show higher electrocatalytic activity with four-electron ORR process, long-term stability (compared to commercially available Pt/C electrode) and free from CO poisoning and fuel crossover effects in alkaline media [15]. Jiang et al., reported defective carbon nano-cages show good ORR performance, as evidenced by DFT calculations that suggest pentagon- and zigzag-edge defects are responsible for its high ORR performance [16]. On the other hand, graphene, a two-dimensional sheet like structure of sp^2-hybridized carbon plays a promising role in wide range of fields including electronics [17], supercapacitors [18], sensors [19] and catalysts [20–22]. In particular, Qu et al, shows nitrogen-doped graphene was synthesized using chemical vapor deposition of methane in the presence of ammonia for ORR via four-electron pathway [20]. In another study, Luo et al., reported pyridinic N-doped graphene by thermal chemical vapor deposition for ORR [21]. Nitrogen-doped graphene catalysts were synthesized by thermal treatment (800–1000 °C) of graphene using ammonia exhibits ORR activity followed by $4e^-$ in alkaline solution reported at Geng et al. [22]. Though there are several reports on graphene-based materials for ORR, but its low performance and high temperature for synthesis promote the researchers to study further in this direction. Recently, graphene has been synthesized using several approaches, such as chemical reduction of exfoliated graphite oxide (GO) [23, 24], chemical vapor deposition (CVD) [25], epitaxial growth [26], and electrochemical method [27]. Among the above approaches, chemical reduction of GO using sodium borohydride [28, 29], hydrazine or hydrazine hydrate [30], and hydroquinone [24], as reducing agents are widely used. However, the highly poisonous and explosive nature of hydrazine and its derivative is a major limitation for its use [31]. Fan et al., used metal ion Fe and HCl for the reduction of GO [32]. In another study, Moon et al, demonstrates the synthesis of high-quality graphene by dispersing GO in the acetic acid and hydriodic acid (HI) stored at 40 °C for 40 h [33]. Though there are several reports on the synthesis of graphene using reducing agent, still facile synthesis of graphene at room temperature in short duration remains a challenging task. Herein, we demonstrate the reduction of GO using concentrated hydrochloric acid and sodium iodide salt at room temperature. Further as-synthesized GO and RGO were employed as metal-free electrocatalyst for the oxygen reduction reaction.

3 Making of Graphene Oxide

Graphite powder, sulfuric acid (H_2SO_4, 98%), potassium permanganate ($KMnO_4$), hydrochloric acid (HCl, 35% v/v), potassium hydroxide (KOH), hydrogen peroxide (H_2O_2), and sodium iodide (NaI) were purchased from Sigma-Aldrich. All the above chemicals were analytical grade and used without further purification. GO was prepared from graphite powder using modified Hummer's method [34]. In a typical synthesis, the graphite powder (1 g) and sodium nitrate (1 g) were grind and dispersed in concentrated H_2SO_4 (46 mL) in a 250-mL beaker placed in ice-bath. Then, $KMnO_4$ (6 g) was slowly added to the above solution, maintaining the ice-bath temperature (0–5 °C) with continuous stirring for 2 h. Then, the solution mixture is taken out from ice-bath and kept at room temperature (~30 °C) with stirring for overnight. The solution mixture was moved to a 1-L beaker and 100 mL of distilled water was added and stirred for another 2 h. Then, hydrogen peroxide solution (35%) was added dropwise until the color of the solution turns into bright yellow. The excess of manganese salt was removed by adding dilute HCl solution (5% by volume). Finally, the as-prepared GO solution was repeatedly washed with distilled water to obtain the solution with neutral pH, demonstrating the complete removal of the residual salts and acids and dried at 60 °C for overnight.

RGO was synthesized at room temperature. Typically, GO (40 mg) was dispersed in 40 mL HCl (6 M) solution. The NaI (0.4 g) was added to the preceding solution and stirred for 4 h at room temperature. Finally, the graphene was collected by repeated washing and centrifuging with distilled water and dried at 60 °C for overnight.

4 Characterization of Materials

The XRD analysis of as-synthesized samples was measured using a Bruker D2 PHASER X-ray diffractometer using Cu Kα X-rays (1.54 Å) operated at a power of 40 kV × 40 mA. The Raman spectra of the samples were collected using a fiber-coupled micro-Raman spectrometer (Model TRIAX550, JY). The surface morphology of the powders was ascertained using field emission scanning electron microscopic (FESEM, JEOL JSM-7600F). The microstructure of the samples was recorded using transmission electron microscope (TEM, FEI TECNAI G^2). The multi-point Brunauer–Emmett–Teller (BET) surface area and total pore volume of the samples were studied using nitrogen adsorption–desorption isotherms at 77 K with autosorb iQ_2 volumetric physisorption analyzer (Quantachrome instruments).

The electrochemical analysis of the as-synthesized GO and RGO catalysts was carried out by cyclic voltammetry (CV), linear sweep voltammetry (LSV), chrono-amperometry (i–t), and electrochemical impedance spectroscopy (EIS) by an electrochemical analyzer (model Autolab PGSTAT; Metrohm). A three-electrode system, in which the electrocatalyst-coated glassy carbon (GC) as working electrode, graphite rod and Ag/AgCl (3.0 M KCl) as the counter, and reference electrode, respectively, were used for all the measurements. The electrochemical experiments were acquired in 0.1 M KOH solution. The LSV study was recorded with a catalyst loaded GC rotating disk electrode (RDE) at various rotation speeds from 500 to 2500 rpm. The electrode potential was calibrated against the reversible hydrogen electrode (RHE) by using the equation, $E(RHE) = E(Ag/AgCl) + 0.21 + 0.059 \times pH$. The EIS measurements were analyzed in a frequency range of 1 MHz–10 Hz at 0.3 V with a sinusoidal perturbation of 5 mV with the same electrode configuration.

The electrons transferred number in the ORR can be analyzed by well-known Koutecky–Levich (KL) equations given by

$$\frac{1}{J} = \frac{1}{J_L} + \frac{1}{J_K} = \frac{1}{B\varpi^{1/2}} + \frac{1}{J_K} \tag{1}$$

$$B = 0.62nFC_0(D_0)^{2/3}v^{1/6} \tag{2}$$

Here, J, J_L, J_K, and ω mentioned as current density, diffusion-limiting current density, kinetic-limiting current density, and RDE rotation rate, respectively. B refers the slope of the KL plots. "n" is the electron transfer number involved in ORR, "F" denotes the Faraday constant (96,485 C mol^{-1}), "C_0" shows the bulk concentration of O_2 (1.26×10^{-3} mol L^{-1}), "D_0" is the diffusion coefficient of O_2 (1.9×10^{-5} cm^2 s^{-1}) in 0.1 M KOH electrolyte and "v" is the kinematic viscosity of the electrolyte (0.01 cm^2 s^{-1}) in 0.1 M KOH electrolyte.

The structural properties, transformation of graphite powder to GO, and GO to RGO were investigated by XRD and Raman spectroscopy. Figure 3A represents the XRD pattern of GO and chemical reduction of GO to RGO using HCl and NaI at room

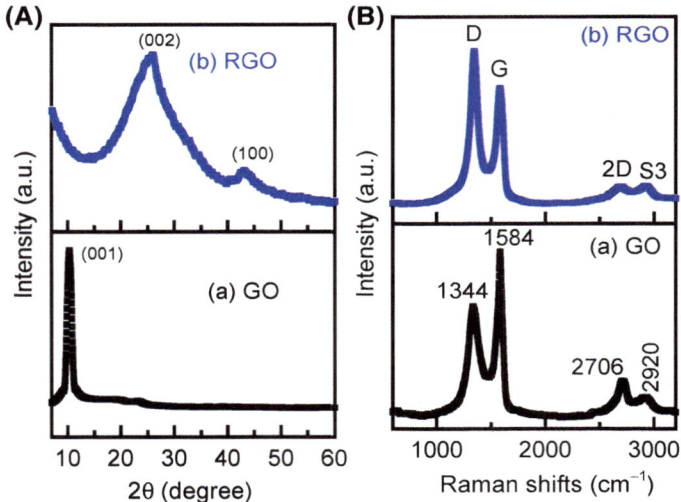

Fig. 3 **A** XRD pattern, **B** Raman spectra of (a) GO and (b) RGO, respectively

temperature. The sharp diffraction features at 10.3° in the XRD pattern (Fig. 3Aa) of GO corresponds to its intercalated (001) planes [35]. The XRD features of RGO (Fig. 3Ab) show a broad diffraction peak at 25.94° assigned for (002) plane and small hump at 43.3° attributed to (100) plane [36]. In the XRD pattern (Fig. 3Ab), it is clearly noticeable that there is no Bragg signature at the 2θ position of 10.3°, which clearly indicates the reduction of GO to RGO.

Further to understand the structural changes of as-synthesized GO to RGO, Raman spectroscopic analysis was carried out. Figure 3B illustrates the Raman spectra of GO and RGO. The Raman spectrum of GO (Fig. 3Ba) and RGO (Fig. 3Bb) shows two prominent features at 1344–1584 cm^{-1}, assigned to the D and G bands, respectively [37]. The Raman G band corresponds to the E_{2g} mode of sp^2-carbon atoms, whereas the D band relates to the structural defects/disorder [27]. The intensity ratio of D band (I_D) to the G band (I_G) was measured to be 0.65 and 1.3 for GO and RGO, respectively. The significant increase of I_D/I_G ratio after the chemical reduction of GO suggests the restoration of sp^2 carbon and is small in size of the sp^2 domain upon reduction of GO [38]. Furthermore, second order of zone boundary phonons or 2D band at 2706 cm^{-1} was found to be higher intense in GO compared to RGO, suggesting higher stacking of graphene layers in GO [39]. The S3 peak at ~2920 cm^{-1} of GO and RGO indicating the excellent graphitization, and no charge transfer occurs due to the absence of impurities [38].

The surface morphology of as-synthesized samples has been analyzed with field emission scanning electron microscopy. Figure 4a–d shows a typical FESEM image of GO and RGO synthesized as room temperature. The low magnification FESEM

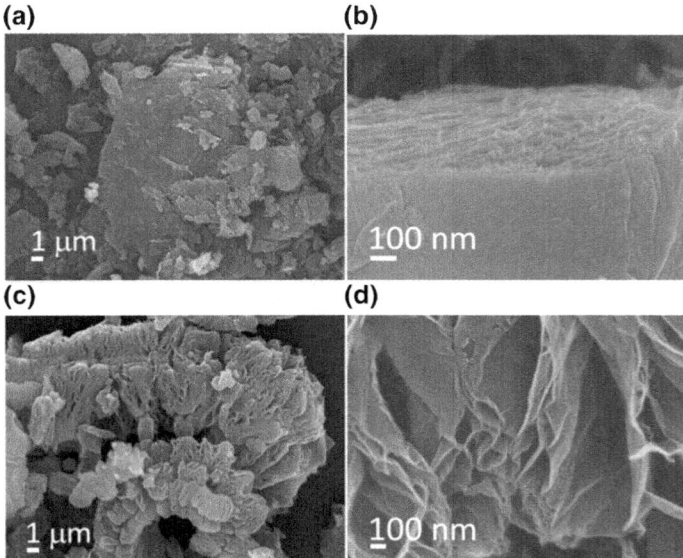

Fig. 4 FESEM image of **a, b** GO, **c, d** RGO

Fig. 5 TEM image of **a** GO
and **b** RGO

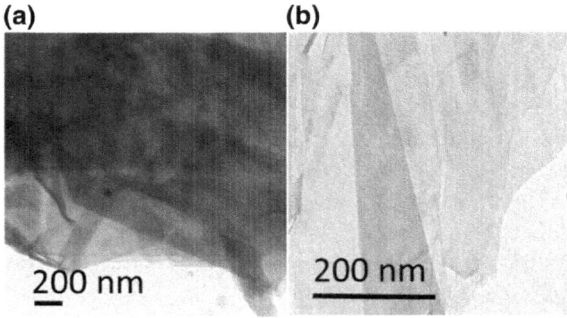

image of GO (Fig. 4a) reveals that sheets like structures with huge agglomeration. However, at higher magnification (Fig. 4b), those agglomerated sheets are found to be closely packed and stacked with several layers of sheets. After the chemical reduction of GO, the low magnification FESEM image of RGO (Fig. 4c) illustrates various sheets are self-assembled with distinct gaps between each sheets. Figure 4d shows the highly magnified image of RGO demonstrating ultra-thin sheets. It is evidenced that after the reduction, RGO sheets are not agglomerated, more gaps between each sheet, which indicate toward higher surface area of the sample. Similar results were also observed in previous report by Tiwari et al. [40].

The microstructure of as-synthesized GO and RGO was recorded using TEM as shown in Fig. 5a, b. The TEM image of GO (Fig. 5a) shows nanosheets like morphology and those nanosheets are stacked with various layers and danced. However, after the chemical reduction of GO using NaI and HCl at room temperature, the TEM image (Fig. 5b) shows ultra-thin sheets like structure with distinct layers is observed. It is noticeable that those RGO ultra-thin sheets are not stacked, indicating higher surface area of the sample. The TEM results were well corroborating with the FESEM results.

The effective surface area of the samples plays a major role to understand the improved electrocatalytic performance [41]. Figure 6 shows the N_2 adsorption–desorption isotherm plot of as-synthesized GO and RGO at 77 K. As per Brunauer classification, both the samples GO and RGO show a type IV isotherm, with a hysteresis, suggesting the porous features of the samples [42]. The adsorption curve of isotherm plot (Fig. 6) was used to measure the effective surface area of 48.5 and 147.3 m^2 g^{-1} for the samples of GO and RGO, respectively, using multi-point Brunauer–Emmett–Teller (BET) method. A larger effective surface area of RGO compared to GO indicates the chemical reduction of GO prevents their stacking and forming more gaps between the graphene layers, which was also evidenced in TEM analysis. Moreover, the total pore volume of GO and RGO was found to be 0.13 and 0.44 cm^3 g^{-1}, respectively. The larger pore volume of RGO than GO further suggests the electrolyte ions can easily interact with the electrode surface, results in higher electrocatalytic performance. Recently, Li et al. synthesized RGO by chemical reduction of GO using hydrazine hydrate observed surface area of 4.7 m^2 g^{-1}

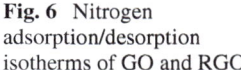

Fig. 6 Nitrogen adsorption/desorption isotherms of GO and RGO

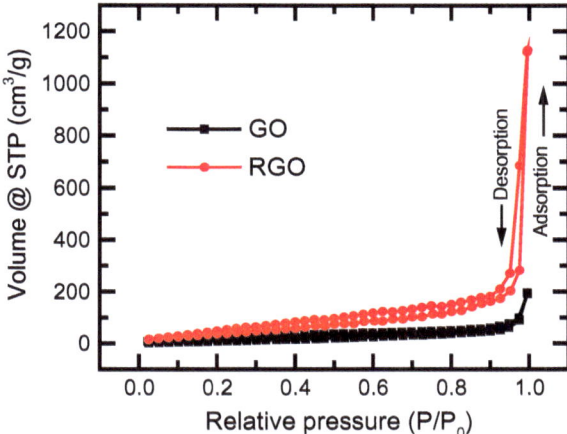

[43]. In contrast, the present synthesis RGO shows a large surface area; therefore, it can offer more active sites for electrochemical reaction, giving rise to higher ORR performance.

5 Electrocatalytic Activity of GO/RGO

The electrocatalytic activity of Go and RGO nanosheets for ORR was investigated by cyclic voltammetry (CV) in 0.1 M KOH solution saturated with nitrogen or oxygen. As shown in Fig. 7A(i), both the voltammetric profile of GO (Fig. 7Aai) and RGO (Fig. 7Abi) shows no current change within the potential range from 0.3 to 1.0 V (vs. RHE) was found in the nitrogen-saturated 0.1 M KOH solution. In contrast, a well-defined reduction peak for GO (Fig. 7Aaii) and RGO (Fig. 7Abii) was observed at 0.7 V (vs. RHE) in the same electrolyte solution saturated with oxygen. Thus, both the samples suggest a pronounced electrocatalytic activity for oxygen reduction reaction. The ORR peak current was found to be higher in RGO as compared to GO, signifying the higher ORR activity of RGO. Further, the ORR onset potential was measured to be more positive for RGO (Fig. 7A), indicating it to be a superior metal-free electrocatalyst for ORR. Figure 7B shows the linear voltammetric scans of GO, RGO, and stand Pt/C for comparison in an O_2-saturated 0.1 M KOH electrolyte with a rotation speed of 1600 rpm. The RGO shows a more positive half-wave potential (peak potential) of 0.811 V (0.759 V) for the ORR than the GO of 0.781 V (0.711 V), thus indicating that chemical reduction of GO leads to a remarkable enhancement in RGO catalysis for the ORR. Moreover, the onset potential and current density were also measured to be higher in RGO than GO (Table 1). Here the current density, onset potential, half wave potential and peak potential of RGO (Table 1) was found to be comparable with standard Pt/C, suggesting the importance of the metal free catalyst

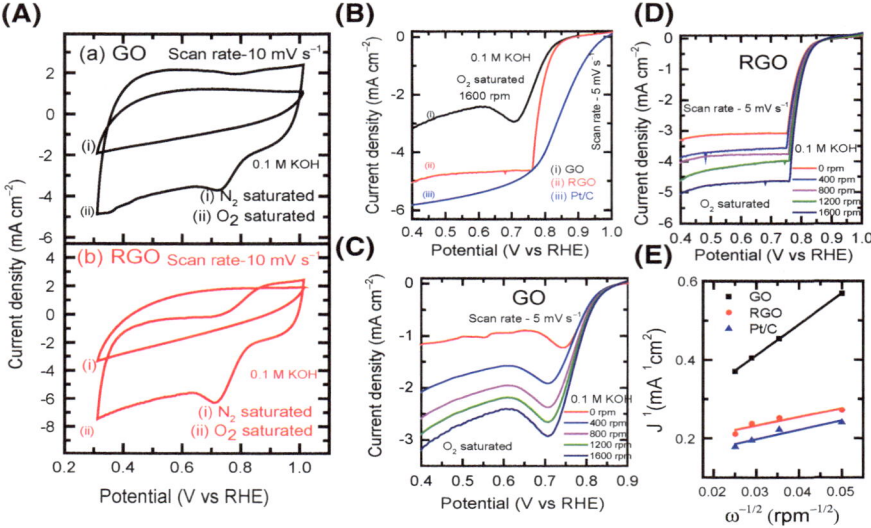

Fig. 7 **A** CV profile of in O_2- and N_2-saturated 0.1 M KOH electrolyte at a scan rate of 10 mV s^{-1} **a** GO, **b** RGO; **B** Linear sweep voltammograms at 1600 rpm using GO, RGO and Pt/C; Linear sweep voltammograms at different rotating rates, **C** GO, **D** RGO; **E** Koutecky–Levich plots of GO, RGO, and Pt/C at 0.5 V; All the measurements were studied in O_2-saturated 0.1 M KOH electrolyte

Table 1 ORR properties of GO, RGO, and standard Pt/C in 0.1 M KOH

Sample name	ORR peak potential (V vs. RHE)	Half-wave potential (V vs. RHE)	Onset potential (V vs. RHE)	ORR current density (mA cm^{-2}) @ 1600 rpm	Electron transfer no. (n)
GO	0.711	0.781	0.851	−3.16	3.55
RGO	0.759	0.811	0.863	−5.02	3.93
Pt/C	0.763	0.873	0.983	−5.83	4.0

RGO for ORR. The current density of GO (Fig. 7C) and RGO (Fig. 7D) was measured as a function of RDE rotation speed from 400 to 1600 rpm. The current density was found to be increased with the rotation speed; because at higher speed, more oxygen diffusion occurs at the electrode surface, resulting higher current density was observed. The LSV profiles illustrate three different potential regions: (i) $V > 0.9$ V is the kinetically controlled high-potential region, (ii) V within 0.7–0.9 V is the mixed diffusion kinetic-limited region, and (iii) $V < 0.7$ V is the low-potential diffusion-limited region. The electron transferred number per O_2 molecule was calculated from various RDE rotation speeds using Koutecky–Levich (KL) equation (details in experimental section). Figure 7E shows the KL plot of GO, RGO and standard Pt/C with linear relationship of J^{-1} as a function of $\omega^{-1/2}$ at a fixed potential of 0.5 V (vs. RHE). The electrons transferred number per O_2 molecule were calculated (considering $n = 4$ for Pt/C) to be 3.55 and 3.93 for GO and RGO, respectively, suggesting ORR follow four-electron path. A schematic representation of the ORR process is shown in Scheme 1. The electron transfer number of as-synthesized RGO is similar to the previous reports using graphene-based materials. Zheng et al, reported $n = 3.97$ for boron, nitrogen-doped graphene and $n = 3.5$ by Sheng et al, for boron-doped graphene [44, 45].

Further, long-term cyclic stability of RGO (Fig. 8a) was tested at a fixed potential of 0.4 V (vs. RHE) for 10 h in 0.1 M KOH using chrono-amperometric measurement. As shown in Fig. 8a, a negligible decrease of current density after 10-h measurement suggesting the high stability of the RGO for ORR. Moreover, the charge transport behavior and diffusion characteristics of GO and RGO catalysts were measured from electrochemical impedance spectroscopy (EIS) technique. Figure 8b illustrates

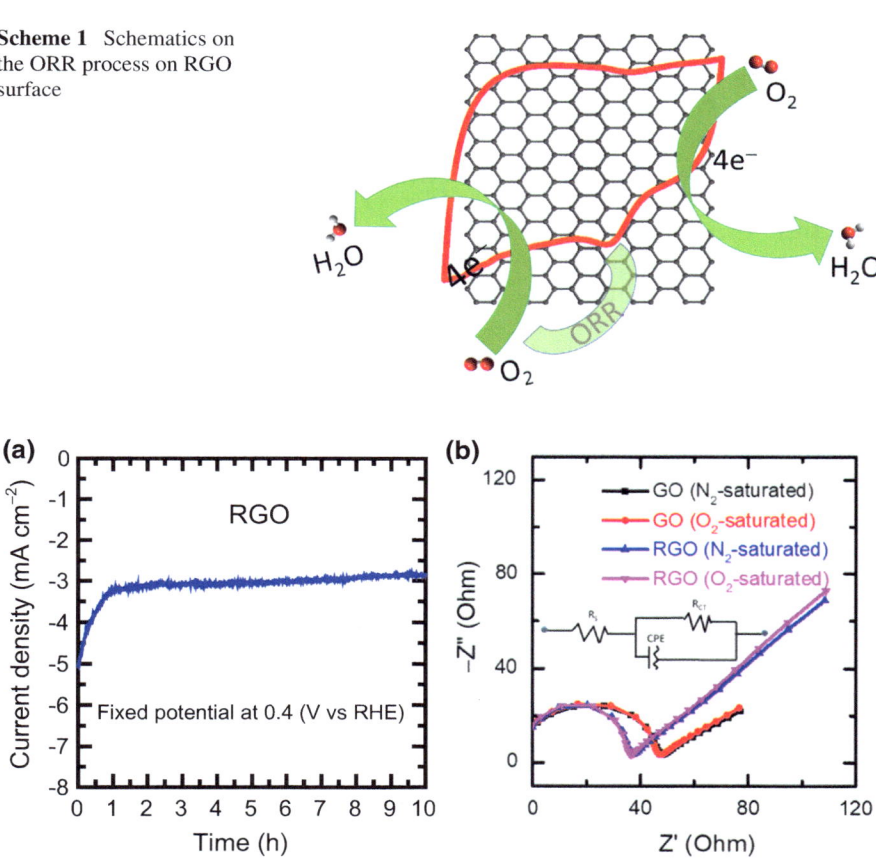

Scheme 1 Schematics on the ORR process on RGO surface

Fig. 8 **a** Chrono-amperometric response of RGO for 10 h in O_2-saturated solution at 0.4 V in 0.1 M KOH electrolyte; Nyquist plots of GO and RGO in N_2- and O_2-saturated 0.1 M KOH electrolytes, inset shows the equivalent circuit diagram

impedance Nyquist plot of GO and RGO in N_2- and O_2-saturated 0.1 M KOH electrolytes. In the Nyquist plot, the diameter of the semicircle arc demonstrates the charge transfer resistance (R_{CT}) of the electrode and was fitted with an equivalent circuit (Inset of Fig. 8b) with solution resistance (R_s), R_{CT}, and constant-phase element (C_{PE}). The measured R_{CT} values of GO and RGO were found to be in the order of GO (N_2-saturated) (50.3 Ω) > GO (O_2-saturated) (49.11 Ω) > RGO (N_2-saturated) (37.76 Ω) > RGO (O_2-saturated) (36.2 Ω). It is to be noted that the smaller semicircle with smaller R_{CT} was observed in O_2-saturated electrolyte than N_2-saturated electrolyte for all the catalysts, suggesting faster change transfer reaction in O_2-saturated electrolyte. The smaller R_{CT} value of RGO than GO further confirms the faster charge transport and higher ORR kinetics.

6 Conclusion

In summary, we report a facile cost-effective synthesis of reduced graphene oxide at room temperature using NaI and HCl. The as-synthesized GO and RGO was used as efficient metal-free electrocatalyst for oxygen reduction reaction. The RGO exhibits higher ORR activity than GO, due to its larger surface area and faster charge transfer resistance. The measured current density, peak potential, and half-wave potential of as-synthesized RGO are found to be comparable with standard Pt/C. Further, the excellent long-term stability toward ORR of RGO in alkaline electrolyte suggesting its potential for practical application. Therefore, the as-synthesized metal-free electrocatalyst RGO as an alternate to Pt/C for ORR in energy conversion device applications.

References

1. Georgakilas, V., Perman, J.A., Tucek, J., Zboril, R.: Chem. Rev. **115**, 4744 (2015)
2. Geim, A.K.: Angew. Chem. Int. Ed. **50**, 6966 (2011)
3. Geim, A.K., Kim, P.: Sci. Am. **298**, 90 (2008)
4. Tuček, J., Błoński, P., Ugolotti, J., Swain, A.K., Enoki, T., Zbořil, R.: Chem. Soc. Rev. **47**, 3899 (2018)
5. Neto, A.C., Guinea, F., Peres, N.M., Novoselov, K.S., Geim, A.K.: Rev. Mod. Phys. **81**, 109 (2009)
6. Sarma, S.D., Adam, S., Hwang, E.H., Rossi, E.: Rev. Mod. Phys. **83**, 407 (2011)
7. Xiang, Q., Yu, J., Jaroniec, M.: Chem. Soc. Rev. **41**, 782 (2012)
8. Swain, A.K., Pradhan, L., Bahadur, D., Appl, A.C.S.: Mater. Interfaces **7**, 8013 (2015)
9. Sun, Y., Shi, G.: J. Polym. Sci. B **51**, 231 (2013)
10. Yu, X., Ye, S.: J. Power Sources **172**, 145 (2007)
11. Liang, Y., Li, Y., Wang, H., Zhou, J., Wang, J., Regier, T., Dai, H.: Nat. Mater. **10**, 780 (2011)
12. Jaouen, F., Proietti, E., Lefevre, M., Chenitz, R., Dodelet, J.P., Wu, G., Chung, H.T., Johnston, C.M., Zelenay, P.: Energy Environ. Sci. **4**, 114 (2011)
13. Liang, H.W., Wei, W., Wu, Z.S., Feng, X., Müllen, K.: J. Am. Chem. Soc. **135**, 16002 (2013)
14. Liu, X., Dai, L.: Nat. Rev. Mater. **1**, 16064 (2016)

15. Gong, K., Du, F., Xia, Z., Durstock, M., Dai, L.: Science **323**, 760 (2009)
16. Jiang, Y., Yang, L., Sun, T., Zhao, J., Lyu, Z., Zhuo, O., Wang, X., Wu, Q., Ma, J., Hu, Z.: ACS Catal. **5**, 6707 (2015)
17. Li, X.L., Zhang, G.Y., Bai, X.D., Sun, X.M., Wang, X.R., Wang, E., Dai, H.J.: Nat. Nanotechnol. **3**, 538 (2008)
18. Nayak, A.K., Das, A.K., Pradhan, D.: ACS Sustain. Chem. Eng. **5**, 10128 (2017)
19. Wang, T., Huang, D., Yang, Z., Xu, S., He, G., Li, X., Hu, N., Yin, G., He, D., Zhang, L.: Nano-Micro Lett. **8**, 95 (2016)
20. Qu, L.T., Liu, Y., Baek, J.B., Dai, L.: ACS Nano **4**, 1321 (2010)
21. Luo, Z., Lim, S., Tian, Z., Shang, J., Lai, L., MacDonald, B., Fu, C., Shen, Z., Yu, T., Lin, J.: J. Mater. Chem. **21**, 8038 (2011)
22. Geng, D., Chen, Y., Chen, Y., Li, Y., Li, R., Sun, X., Ye, S., Knights, S.: Energy Environ. Sci. **4**, 760 (2011)
23. Stankovich, S., Dikin, D.A., Piner, R.D., Kohlhaas, K.A., Kleinhammes, A., Jia, Y.Y., Wu, Y., Nguyen, S.T., Ruoff, R.S.: Carbon **45**, 1558 (2007)
24. Wang, G.X., Yang, J., Park, J., Gou, X.L., Wang, B., Liu, H., Yao, J.: J. Phys. Chem. C **112**, 8192 (2008)
25. Chen, Z., Ren, W., Gao, L., Liu, B., Pei, S., Cheng, H.M.: Nat. Mater. **10**, 424 (2011)
26. Yang, W., Chen, G., Shi, Z., Liu, C.C., Zhang, L., Xie, G., Cheng, M., Wang, D., Yang, R., Shi, D., Watanabe, K.: Nat. Mater. **12**, 792 (2013)
27. Guo, H.L., Wang, X.F., Qian, Q.Y., Wang, F.B., Xia, X.H.: ACS Nano **3**, 2653 (2009)
28. Murphy, S., Huang, L., Kamat, P.V.: J. Phys. Chem. C **117**, 4740 (2013)
29. Chua, C.K., Pumera, M.: Chem. Soc. Rev. **43**, 291 (2014)
30. Park, S., An, J., Potts, J.R., Velamakanni, A., Murali, S., Ruoff, R.S.: Carbon **49**, 3019 (2011)
31. Furst, A., Berlo, R.C., Hooton, S.: Chem. Rev. **65**, 51 (1965)
32. Fan, Z.J., Kai, W., Yan, J., Wei, T., Zhi, L.J., Feng, J., Ren, Y.M., Song, L.P., Wei, F.: ACS Nano **5**, 191 (2010)
33. Moon, I.K., Lee, J., Ruoff, R.S., Lee, H.: Nat. Commun. **1**, 73 (2010)
34. Hummers, W.S., Jr, Offeman, R.E.: J. Am. Chem. Soc. **80**, 1339 (1958)
35. Mhamane, D., Ramadan, W., Fawzy, M., Rana, A., Dubey, M., Rode, C., Lefez, B., Hannoyer, B., Ogale, S.: Green Chem. **13**, 1990 (2011)
36. Lin, X., Shen, X., Zheng, Q., Yousefi, N., Ye, L., Mai, Y.W., Kim, J.K.: ACS Nano **6**, 10708 (2012)
37. Pham, V.H., Dang, T.T., Singh, K., Hur, S.H., Shin, E.W., Kim, J.S., Lee, M.A., Baeck, S.H., Chung, J.S.: J. Mater. Chem. A **1**, 1070 (2013)
38. Cui, P., Lee, J., Hwang, E., Lee, H.: Chem. Commun. **47**, 12370 (2011)
39. How, G.T.S., Pandikumar, A., Ming, H.N., Ngee, L.H.: Sci. Rep. **4**, 5044 (2014)
40. Tiwari, S.K., Huczko, A., Oraon, R., De Adhikari, A., Nayak, G.C.: J. Mater. Sci. **51**, 6156 (2010)
41. Zhang, L., Zhang, F., Yang, X., Long, G., Wu, Y., Zhang, T., Leng, K., Huang, Y., Ma, Y., Yu, A., Chen, Y.: Sci. Rep. **3**, 1408 (2013)
42. Vinayan, B.P., Nagar, R., Raman, V., Rajalakshmi, N., Dhathathreyan, K.S., Ramaprabhu, S.: J. Mater. Chem. **22**, 9949 (2012)
43. Li, G., Jing, M., Chen, Z., He, B., Zhou, M., Hou, Z.: RSC Adv. **7**, 10376 (2017)
44. Zheng, Y., Jiao, Y., Ge, L., Jaroniec, M., Qiao, S.Z.: Angew. Chem. **125**, 3192 (2013)
45. Sheng, Z.H., Gao, H.L., Bao, W.J., Wang, F.B., Xia, X.H.: J. Mater. Chem. **22**, 390 (2012)

Printed by Printforce, the Netherlands